Practical Environmental Analysis

Practical Environmental Analysis

Miroslav Radojević
Department of Chemistry, University of Brunei Darussalam

Vladimir N. Bashkin
Department of Atmospheric Sciences, Seoul National University

ROYAL SOCIETY OF CHEMISTRY

ISBN 0-85404-594-5

A catalogue record for this book is available from the British Library.

Published by The Royal Society of Chemistry,
Thomas Graham House, Science Park, Milton Road, Cambridge CB4 0WF, UK

For further information see our web site at www.rsc.org

Typeset by Paston PrePress Ltd, Beccles, Suffolk, NR34 9QG
Printed by MPG Books Ltd, Bodmin, Cornwall, UK

Preface

Environmental chemistry is becoming an increasingly popular subject in tertiary education. Courses in chemistry, environmental science, civil engineering, public health and environmental engineering all have to include environmental chemistry in their syllabuses to a greater or lesser extent. Many textbooks have appeared in recent years aiming to fulfill this requirement; however, most of these deal mainly with theoretical aspects of the subject. This book aims to supplement the existing textbooks by providing detailed, step by step instructions for experiments in environmental analytical chemistry. These could be used to teach the practical components of undergraduate and postgraduate (Diploma and Masters) degree courses. The book may also be useful to students on HNC and HND courses, and to those on training courses for technicians working in environmental or other (*e.g.* public health, sewage, water, industrial) laboratories. Relatively easy experiments, requiring only basic laboratory equipment and instrumentation, have been selected. Some of the simpler experiments may also be used by secondary school teachers of chemistry to illustrate applications of chemistry to the environment, a topic of growing concern among today's school students. Many of the experiments can serve as a basis for more extensive surveys of the environment in school science projects or undergraduate research projects.

Treatment of general and analytical chemistry was considered to be outside the scope of this book. It is assumed that the student would be familiar with basic chemical theory, and laboratory procedures and practices. Anyway, many good textbooks dealing with these topics are available and the student is referred to these books in the text where appropriate. Nevertheless, some basic practices of analytical chemistry are dealt with in the introduction and in Appendix II, especially those aspects which are relevant to environmental analysis. Also, worked examples of problems relating to analytical and environmental chemistry are included where appropriate.

The experiments aim to provide practical experience in the analysis of real environmental samples, and to illustrate the application of classical and instrumental techniques to environmental analysis. A brief introduction explaining why a particular substance is important and describing its behaviour in the environment is given before each experiment. Easy to follow experimental procedures are then outlined. Suggestions for further study, questions and exercises, and recommended further reading, are given after each experiment. Most undergraduate laboratories would be equipped with the instruments required for carrying out these experiments. If any instruments or materials are not available, instructors can select experiments that do not require them.

This book is not a reference manual for professionals working in environmental laboratories; many comprehensive texts are available for this purpose. Nevertheless, it can serve as an introductory text to those entering into employment in environmental laboratories, especially for those from a non-environmental chemistry background. There is a strong bias in the book towards inorganic analysis. This is primarily because equipment for carrying out inorganic analysis is more widely available in teaching laboratories, and it is not meant to reflect the relative importance of inorganic/organic analysis. A large number of organic compounds are present at trace levels in the environment and their determination requires the use of instruments that tend to be fairly expensive and may not be readily available in many laboratories (*e.g.* gas chromatographs equipped with mass spectrometer or electron capture detectors). Furthermore, organic analysis requires the purchase of specialised standards which also tend to be quite expensive. Experimental procedures for these compounds are best left to a future volume dealing with "advanced environmental analysis". Also, microbiological analyses, such as *coliform* and *E. coli* tests, although extremely important from a public health point of view, are not included in the present volume which is mainly restricted to purely chemical analysis. Inclusion of these tests would have entailed adding considerable background material on environmental microbiology, general laboratory procedures for microbiological analysis, *etc.*, not central to the theme of the present book. This shortcoming is regrettable, but unavoidable. These tests are described in several books dealing with microbiological analysis.

Many of the experimental procedures are based loosely on standard methods (APHA, US EPA, British Standards Institution, *etc.*). The main aim of the book is to serve as an educational tool in preparing environmental chemists for the more demanding regimen of a real environmental laboratory. If the book opens the student's eyes to the problems and demands of environmental analysis, it would have surely served its purpose.

Miroslav Radojević
Vladimir Bashkin
September 1998

Acknowledgements

We would like to thank all those whose contribution to environmental sciences over the years has given us inspiration and encouragement, notably S. E. Allen, J. P. Lodge Jr., R. M. Harrison and P. Brimble-combe. We would also like to thank F. L. Wimmer, K. R. Fernando, J. Davies, S.-U. Park and other colleagues for useful discussions and comments. Thanks are also due to the Russian Fund of Basic Research (grant no. 96-05-64368) for providing assistance to one of the authors.

Contents

CHAPTER 1

Introduction

1.1 THE ENVIRONMENT

The *environment* is the sum total of human surroundings consisting of
the atmosphere, the hydrosphere, the lithosphere and the biota. Human
beings are totally dependent on the environment for life itself. The
atmosphere provides us with the air we breathe, the hydrosphere
provides the water we drink and the soil of the lithosphere provides us
with the vegetables that we eat. In addition, the environment provides us
with the raw materials to fulfill our other needs: the construction of
housing, the production of the numerous consumer goods, *etc*. In view of
these important functions it is imperative that we maintain the environment in as pristine a state as is possible. Fouling of the environment by
the products of our industrial society (*i.e. pollution*) can have many
harmful consequences, damage to human health being of greatest
concern.

In addition to the outdoor environment, increasing concern is being
expressed about the exposure of individuals to harmful pollutants within
the *indoor environment*, both at home and at work. Levels of harmful
pollutants can often be higher indoors than outdoors, and this is
especially true of the workplace where workers can be exposed to fairly
high levels of toxic substances. *Occupational health*, *occupational medi-
cine*, and *industrial hygiene* are subjects that deal with exposure at the
workplace.

Pollution is mainly, although not exclusively, chemical in nature. The
job of the environmental analyst is therefore of great importance to
society. Ultimately, it is the environmental analyst who keeps us
informed about the quality of our environment and alerts us to any
major pollution incidents which may warrant our concern and
response.

1

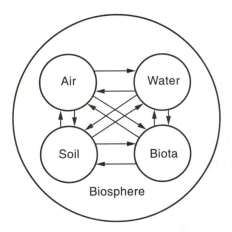

Figure 1.1 *Interactions between component parts of the biosphere*

1.1.1 Biogeochemical Cycles

The different components of the *biosphere* and their interactions are illustrated in Figure 1.1. The biosphere is that part of the environment where life exists. It consists of the hydrosphere (oceans, rivers, lakes), the lower part of the atmosphere, the upper layer of the lithosphere (soil) and all life forms. The concept of the biosphere was first introduced by the Russian scientist Vladimir Vernadsky (1863–1945) as the "sphere of living organisms distribution". Vernadsky was among the first to recognise the important role played by living organisms in various interactions within the biosphere, and he established the first-ever biogeochemical laboratory specifically dedicated to the study of these interactions. He expounded his theories in an aptly entitled book, "Biosphere", published in 1926.

The various *spheres* act as reservoirs of environmental constituents and they are closely linked through various physical, chemical and biological processes; there is constant exchange of material between them. Chemical substances can move through the biosphere from one reservoir to another, and this transport of constituents is described in terms of a *biogeochemical cycle*. Biogeochemical cycles of many elements are closely linked to the hydrological cycle. The hydrological cycle acts as a vehicle for moving water soluble nutrients and pollutants through the environment. If all the components of the cycle are identified and the amounts and rates of material transfer quantified, the term *budget* is used. Both beneficial nutrients and harmful pollutants are transported through biogeochemical cycles with far-reaching consequences. The more commonly discussed biogeochemical cycles are those of important

Figure 1.2 *The box model*

macronutrients such as carbon, sulfur, nitrogen and phosphorus, but, in principle, a biogeochemical cycle could be drawn up for any substance. The cycle is usually illustrated as a series of compartments (reservoirs) and pathways between them. Each reservoir can be viewed in terms of a box model shown in Figure 1.2.

If the input into a reservoir equals the output, the system is said to be in a steady state. The *residence time*, τ, is defined as:

$$\tau = \frac{\text{Amount of substance in the reservoir (mass)}}{\text{Flux (mass/time)}}$$

Flux is the rate of transfer through the reservoir (*i.e.* the rate of input or output). If the input exceeds the output, there will be an increase in the amount of substance in the reservoir. There are many examples of the build-up of pollution in environmental systems since pollutants are often added at rates greater than the rates of natural processes that act to remove them from the system. On the other hand, if the output is greater than the input, the amount of substance in a reservoir will decrease. An example of this is the depletion of natural resources.

It is debatable whether, in the absence of human activities, natural systems would tend towards some sort of steady-state or equilibrium. Natural systems are dynamic, and both natural and human-induced disturbances lead to change, albeit over different time scales. Natural changes to biogeochemical cycles generally take place over geological time scales, and for millennia these cycles have maintained the delicate balance of nature conducive to life. However, since the industrial revolution, and especially over the last 40 years, human activities have caused significant perturbations to these cycles. The effects of these disruptions are already becoming apparent, and are likely to become even more severe in the coming millennium. Serious environmental problems that have been caused by disruptions of biogeochemical cycles include: global warming, acid rain, depletion of the ozone layer, bioaccumulation of toxic wastes and decline in freshwater resources. Modelling of biogeochemical cycles is becoming increasingly important in understanding, and predicting, human impacts on the environment, and the possibility of using biogeochemical cycles to solve environmental

Table 1.1 *Human impacts on biogeochemical cycles*

Cycle	Human interference	Environmental consequence
Carbon	Fossil fuel combustion, clearing of forests	Global warming
Sulfur	Fossil fuel combustion	Acid rain
Nitrogen	Fossil fuel combustion, fertilisers	Acid rain, eutrophication
Phosphorus	Detergents and fertilisers	Eutrophication

Table 1.2 *Relative contribution of anthropogenic and natural sources (approximate)*

	Emissions to the atmosphere (% of total)	
Pollutant	Natural	Anthropogenic
Sulfur dioxide	50	50
Nitrogen oxides	50	50
Carbon dioxide	95	5
Hydrocarbons	84	16

problems, so-called *biogeochemical engineering*, has recently been recognised.

Some of the major human impacts on biogeochemical cycles are given in Table 1.1.

The extent of human impacts on biogeochemical cycles can be illustrated by comparing the contribution of anthropogenic emissions to the atmosphere with natural emissions (Table 1.2). For some toxic substances the contribution of industrial emissions is even more striking: the ratio of anthropogenic to natural emissions to the environment is 3:1 for arsenic, 5:1 for cadmium, 10:1 for mercury and 28:1 for lead.

1.1.2 Environmental Pollution

Pollution is commonly defined as the addition of a substance by human activity to the environment which can cause injury to human health or damage to natural ecosystems. This definition excludes "natural pollution", although natural processes can also release harmful substances into the environment. There are different categories of pollution: chemical, physical, radioactive, biological and aesthetic. This book is concerned primarily with chemical pollutants and their determination in environmental matrices.

Most substances that are considered as pollutants are actually natural constituents of the environment, albeit at concentrations which are generally harmless. It is the increase in the concentration of these natural constituents, usually by industrial activity, to levels at which they may have harmful effects that is of concern. There are, however, a few pollutants which are entirely synthetic and would not be present in the environment if it were not for human activity (*e.g.* chlorofluoro-carbons).

Pollution can be classified according to its geographical scale as *local*, *regional* or *global*. Local pollution may affect only a single field, small stream or a city (*e.g.* photochemical smog). Regional pollution may affect a part of a country, a whole country or even an entire continent. Global warming due to the greenhouse effect of CO_2 is an example of a pollution problem on a global scale. However, the distinction between these different categories is not always clear-cut. For example, many contemporary megacities extend over enormous areas and many urban conurbations may consist of several cities (Metro Manila, Los Angeles area, north-eastern seaboard of the USA, *etc.*). In such areas, photochemical smog is a regional problem. Acid rain was, until lately, considered a regional problem as it affects almost all of Europe and North America. However, acid rain has recently been identified at locations throughout the world, from tropical rainforests in Asia, Africa and South America to the polar ice caps in the Arctic. Therefore, acid rain may now be viewed as a global problem.

The number of pollution sources is constantly rising throughout the world as a consequence of growing industrial development. The driving force behind the increase in pollution is the rapidly growing population of the world (Figure 1.3). The world's population has more than doubled over the last 40 years, and over the next 30 years it is expected to increase by another 2–4 billion. The consequent demand for energy and resources required to feed, clothe and house the increasing population will be accompanied by a parallel increase in waste production. Not only is the population rising, but the standard of living is also rising, placing additional stress on the environment. In the future we may expect environmental problems to become more widespread and more severe unless measures are taken to control pollution.

Problems of environmental pollution have been widely recognised only in the latter half of the 20th century, but they have been known since antiquity. In fact, air pollution has been around since the first humans started using fire for heating, lighting and cooking. The air quality in inhabited prehistoric caves must, almost certainly, have been poor, and early humans must have been exposed to elevated levels of combustion

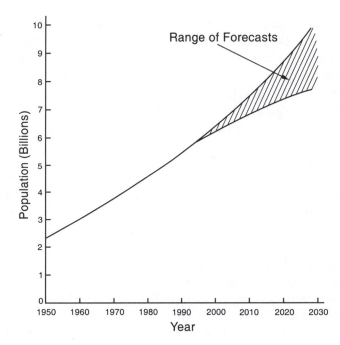

Figure 1.3 *The world's population growth*

products. In antiquity, and in the Middle Ages, the air of towns was polluted by the products of burning wood and coal, and the smelting of ores of iron and other metals. Also, in the absence of a sewage system, human and other wastes were dumped onto the streets, contributing to the rise of many epidemics. Furthermore, human beings have been throwing their wastes into surface waters since time immemorial. However, most of these early problems were local and had limited impacts, and natural processes within the environment were capable of rapidly diluting and eliminating the pollution.

It was the Industrial Revolution that greatly accelerated the release of pollutants into the environment. The Industrial Revolution originated in the north of England in the late 18th/early 19th century and quickly spread to other regions of Europe and North America. Rapid industrialisation is still going on throughout the world, especially in the developing countries of Asia, Africa and South America. The variety of pollutants and the extent of pollution is now greater than ever in history, and this trend looks set to continue well into the next millennium. Natural processes can no longer cleanse the environment of the enormous quantities of pollutants that are generated daily, and the pollution is steadily accumulating in the air, the oceans and the soil.

While natural ecosystems may accommodate a certain amount of pollution, this capacity is now being overloaded. When pollution levels reach a critical limit, harmful consequences follow. The legacy of present industrial development may have a dramatic impact on future generations.

Technological progress has been a two-edged weapon. Technology has given us enormous power over nature, to use for better or for worse. Medical and technological advances have eradicated many diseases, improved health care, provided protection against many natural disasters, increased the standard of living, eliminated many dangerous jobs, improved safety at work, *etc*. On the other hand, we now have the ability not only to destroy isolated ecosystems but all life on the earth, including human life, and not just by means of weapons of mass destruction but also by our polluting influence on the environment. The "globalisation" of what were previously minor, local environmental problems (*e.g.* acid rain), as well as the emergence of new global threats (*e.g.* destruction of stratospheric ozone, global warming), seems to indicate that we are well on our way to accomplishing this. It would be a sad indictment on the human race if it were to undo, in a small fraction of the geological time scale, what took nature millions of years to achieve: life in its many forms. Clearly, the real-world experiment which we are conducting needs to be carefully controlled if we are to slow down, or reverse, this trend.

However, it is not all gloom and doom; there have been many environmental success stories over the years. Unfortunately, these successes have so far been mainly limited to the developed nations. For example, air quality of most cities in Western Europe and North America has significantly improved compared to half a century ago. Concentrations of SO_2 and smoke have decreased steadily since the 1950s and catastrophic smog episodes are no longer a menace to urban populations. More recently, the introduction of catalytic converters could reduce the emissions of automotive air pollutants and improve urban air quality even further. "Car-free" zones have been introduced in some cities and improvements in public transport have been implemented. DDT, organotin compounds, phosphate-containing detergents and many other harmful chemicals have been banned in most developed nations. Water quality has improved in some countries compared to what it was during the Industrial Revolution. Many countries have phased out the use of lead in petrol. Strict controls have been imposed on the transport and disposal of toxic wastes. Increasing emphasis is being placed on so-called "clean technologies". Considerable research has gone into developing alternative, non-polluting energy sources (solar, wind, *etc*.). The greater general awareness of environmental problems has resulted in the public raising environmental issues and demanding greater environmental accountability from industries and governments. Unfortunately, similar improvements are not evident in developing countries, where development has been accompanied by increasing environmental devastation. Over the past 20 years, the relatively unspoiled environment of these countries has regressed to a state on a par with that of the developed countries during the Industrial Revolution. However, there is cause for optimism. As these countries increasingly adopt pollution control technologies, much as they have adopted other technologies pioneered by the developed nations, the quality of the environment could yet improve.

1.1.3 Environmental Standards

Whether a specific concentration of a particular chemical substance is harmful or not depends on many factors and is the subject of extensive research among various branches of science. The maximum level of a substance which can be allowed in the environment without any foreseeable harmful effects is called a *standard*, and the establishment of such standards is a complex and difficult process. Many standards change quite frequently as new research sheds more light on the effects of pollution or as better control technologies become available. The tendency is for standard values to decrease (*i.e.* become more stringent) with time. Most countries specify standards for many air pollutants, water pollutants, *etc*. These standards are legally enforceable and offenders may be prosecuted for infringement. There are two types of standards:

- *Quality standards*. These refer to the concentration of a pollutant in the environment. For example, air quality standards specify the concentrations of pollutants in the general atmosphere that are not to be exceeded. Such standards are used to maintain the quality of our environment in a generally unpolluted, if not pristine, state.
- *Emission standards*. These refer to the maximum levels of a pollutant that may be emitted from a particular pollution source. For example, waste waters from a particular industry must have concentrations of specific pollutants below the levels required by the emission standard.

Furthermore, there are also *guidelines*. These are not legally binding but they are recommended levels of pollutants which, if exceeded, may result in some harmful effect. A well known guideline is the World

Table 1.3 *WHO guide values for air pollutants (daily mean)*

Gas	Guide value (ppbv)
O_3	33
SO_2	48
NO_2	80

Health Organisation (WHO) recommendations for drinking water quality which cover a large number of inorganic and organic species (see Appendix III). WHO guidelines for three important air pollutants are given in Table 1.3.

The job of the environmental analyst is to test for compliance with the various standards. If it is established that permissible levels are being exceeded, technological measures may be required in order to reduce the emissions of pollutants. For example, such measures may include the installation of a water treatment plant to control the discharge of waste waters, or the operation of a flue gas desulfurisation plant to reduce atmospheric emissions from a power station. The selection and design of the most appropriate pollution control technology is made by environmental engineers. Environmental analysts may be required to assess the efficiency of the control technology, and once installed, to confirm that the problem has been eliminated and that legislative standards are being adhered to. International and national guidelines and standards for air, water, soil, sludge, crops and foods are listed in Appendix III.

1.2 ENVIRONMENTAL ANALYSIS

1.2.1 Aims of Analysis

The purpose of environmental analysis is two-fold:

- To determine the background, natural, concentrations of chemical constituents in the environment (*background monitoring*).
- To determine the concentration of harmful pollutants in the environment (*pollution monitoring*).

Background monitoring is useful in studies of general environmental processes, and for establishing concentrations against which any pollution effects could be assessed. It is, however, a fact that pollution has now affected even the most remote areas of the globe, and true background

levels of many substances are becoming increasingly difficult to determine.

The objectives of pollution monitoring are:

- To identify potential threats to human health and natural ecosystems.
- To determine compliance with national and international standards.
- To inform the public about the quality of the environment and raise
  public awareness about environmental issues.
- To develop and validate computer models which simulate environmental processes and are extensively used as environmental management tools.
- To provide inputs to policy making decisions (land-use planning,
  traffic, *etc.*).
- To assess the efficacy of pollution control measures.
- To investigate trends in pollution and identify future problems.

Environmental analysis is often used in *environmental impact assessment* (EIA) studies. These studies are carried out before any major
industrial development is given a go-ahead by the authorities and their
aim is to assess any potential impacts of the development on environmental quality. As part of EIA it is often necessary to establish the
baseline concentrations of various substances at the proposed site so that
potential impacts may be assessed.

1.2.2 Types of Analysis

The chemical substance being determined in a sample is called an *analyte*
(*i.e.* element, ion, compound). Samples are "analysed" whereas analytes
are "determined". We can broadly define two categories of chemical
analysis:

- *Qualitative analysis* — concerned with the *identification* (*i.e.* determining the nature) of a chemical substance.
- *Quantitative analysis* — concerned with the *quantification* (*i.e.*
  determining the amount) of a chemical substance.

The former answers the question: "Which substance is present?" while
the latter answers the question: "How much is present?". Results of
quantitative analysis are generally expressed in terms of *concentration*.
Concentration is the quantity of analyte (in grams or moles) per unit

amount of sample (grams or litres). Obviously, quantitative analysis involves identification as well as quantification since a numerical value must be ascribed to a particular substance. For example, qualitative analysis may simply tell us whether mercury is present in a sample of waste effluent, whereas quantitative analysis will tell us exactly how much mercury is present in the effluent. Often, so-called "spot tests" based on distinctive colour forming reactions, or complex schemes of analysis, can be used for purposes of identification. There is, however, a hidden quantitative aspect in qualitative analysis. If a result of a qualitative analysis is negative (*i.e.* the substance in question was not identified), this does not mean that the substance is absent; it merely implies that the substance is present at a level below that at which the spot test, or analysis scheme, responds. For example, mercury is present in seawater at a level of 3×10^{-5} parts per million (ppm). It is unlikely that a routine spot test would be able to identify this. The development of ever more sensitive instruments capable of quantifying even the minutest traces of substances has exposed the shortcomings of many of the cruder and insensitive schemes of qualitative analysis. Nevertheless, qualitative spot tests still remain useful in many situations where substance levels are high, and they are particularly useful where routine analyses are required, such as in industry.

Chemical analysis may also be categorised with respect to the type of substance being analysed. *Inorganic analysis* is concerned with the determination of elements and inorganic compounds, whereas *organic analysis* involves the determination of organic compounds.

Quantitative analysis may be classified according to the following categories:

- *Complete analysis* — each and every constituent of the sample is determined.
- *Ultimate analysis* — each and every element in the sample is determined without regard to the compounds present.
- *Partial analysis* — the amount of one or several, but not all, constituents in a sample is determined.

Another way of categorising types of analyses is according to the level of the substance in the sample. *Macro analysis* involves the determination of major constituents present at high concentration (%) whereas *micro analysis*, or *trace analysis*, involves the determination of constituents present in very small quantities (0.1 ppb–100 ppm). *Ultra-trace analysis* involves the determination of constituents present at levels lower than in trace analysis (<0.1 ppb).

Furthermore, there are other categories still. *Destructive analysis* involves the use of a method or technique which destroys the sample in question during analysis (*e.g.* dissolution of solid sample into acid). *Non-destructive analysis*, as the name implies, does not destroy the sample during analysis (*e.g.* X-ray fluorescence) and the sample may be re-used for other analyses. *Speciation* involves the determination of all the different forms of a class of compounds in a sample. For example, the speciation of lead in the environment would involve the analysis of all the different inorganic and organic compounds of lead.

1.2.3 Stages of Analysis

Both industry and government operate laboratories dedicated to environmental analysis. Furthermore, environmental analysis is performed by commercial analytical laboratories, serving smaller industries that find it more cost effective to sub-contract out analytical work rather than to invest in their own laboratories, and by universities and institutes that carry out research into environmental chemistry and pollution. Hence, there is a wide range of employment options available for trained environmental chemists.

An environmental analyst should be proficient at carrying out all the different stages of an analysis given below:

Sampling

⇓

Sample treatment

(extraction, concentration, *etc.*)

⇓

Analysis

⇓

Calculation and interpretation of results

1.3 SAMPLING AND STORAGE

1.3.1 Sampling

A *sample* is that portion of the physical environment which is withdrawn for chemical analysis. A sample can be aqueous (*e.g.* river water), gaseous (*e.g.* air) or solid (*e.g.* soil). The chemical substance being analysed in the sample, whether an element, ion or compound, is referred to as the *analyte* (*e.g.* Pb in dust). *Sampling* is the process by which a sample is obtained and this can be done in one of two ways:

- *Batch sampling* involves taking a sample from the environment and performing an analysis either on site or later on in the laboratory. For example, batch sampling of a wastewater effluent for pH analysis would imply that a volume (*e.g.* 100 mL) of the effluent is collected and then analysed for pH. These samples are collected at a specific time and place and are also called *grab* samples.
- *Continuous sampling* involves continuously monitoring the environmental parameter of interest. In the above example, continuous analysis of the effluent pH would involve placing a pH electrode directly into the effluent stream and recording the pH on a chart recorder or a data logger. In this way a continuous record of the effluent pH is obtained. This kind of sampling could detect important changes in the effluent that would be missed by batch sampling.

Batch sampling is the easiest and most common method of obtaining a sample and it is widely used in environmental surveys. Continuous sampling is generally combined with an instrumental method of analysis and the term given to this combination is *continuous monitoring*. This method of analysis is being adopted more extensively in many applications (*e.g.* effluent monitoring, air monitoring). Usually, the method is combined with some kind of alarm system to alert the operator when standards are being exceeded. For example, if the level of pollutant in an effluent stream is found to exceed the emission standard, an alarm may be activated and the plant personnel could switch off the process and attempt to resolve the problem. In many cities, continuous monitoring of air quality is carried out by the municipal authorities. This allows for alerts to be broadcast to the public *via* the media when air quality standards are observed to have been breached. The public is then asked to stay indoors, refrain from outdoor exercise, and so on, until the air quality improves. Such a fast response would not be possible in the case of batch sampling, which involves laboratory analysis at some later time.

Another type of sample is a *composite* sample, prepared by mixing several batch samples, usually collected at the same place but at different times. These are used to evaluate the average concentration in a medium in which the concentration may vary with time. For example, batch samples are collected every two hours over a 24 hour period and pooled into one container. The concentration in the mixture is supposed to reflect a 24 hour average.

Although sampling appears to be a relatively straightforward matter, it is generally one of the most problematic stages of an environmental analysis. The main difficulty lies is obtaining a *representative* sample. The sample represents only a small portion of the system under investigation and it is important that the sample is representative of the whole system as much as possible. In environmental analysis this is not always possible to achieve. Usually, it is easier to obtain representative samples from homogeneous than from heterogeneous systems. An environmental analyst faces unique problems of obtaining representative samples from water, air, effluent gases, dust and soil. Some of the questions that need to be addressed are:

- When and where should the sample be taken?
- How many samples should be taken?
- How much sample is required?

Some quite sophisticated statistical sampling procedures have been developed that can help the analyst answer some of these questions; nevertheless most sampling is carried out without reference to these statistical considerations. The analyst will usually decide on the best location, time and number of samples to be taken (so-called *random sampling*). Considerations of site accessibility, time and expense are often more influential factors than purely scientific considerations. Anyway, several samples are collected at each site in order to obtain some indication of variability in analyte concentration at the site, and in case some of the samples are lost, spoilt or incorrectly analysed. Obtaining as representative a sample as possible is imperative since the analyst cannot obtain the same sample again. The environment is a dynamic system which is continuously changing, so returning to the same site at a later date may give completely different results.

1.3.2 Storage

Once the sample has been collected it is transported to the laboratory for analysis. Sometimes it is possible to carry out the analysis at the site using portable test kits, or inside on-site laboratories (see Section 1.5.4), but most often the sample has to be transported some distance. It is desirable to perform the analysis as soon as possible after sample collection. On many occasions this is not possible and the sample has to be stored until the analysis can be performed. During transportation and storage it is important to preserve the integrity of the sample. Once the sample has been collected inside the sampling vessel, the following processes may threaten the integrity of the sample:

- Chemical reactions.
- Biological reactions.
- Interaction with sampling bottle material.

The analyte under investigation may be destroyed or created by chemical or biological reactions, it may be adsorbed onto the walls of the sampling bottle, or interfering substances may be leached from the walls of the bottle. The analyst has to be aware of the chemical properties of the analyte he is investigating so as to ensure storage conditions which will minimise all these possible effects. He may have to adopt different sampling and storage procedures for different analytes even if he is dealing with the same sample; for example, in water analysis, organic and inorganic analytes require different types of sampling bottle materials. Contamination of the sample during sampling, transport and storage is a real possibility and all measures must be taken to avoid this. Many of the analytes are present at trace and ultra-trace levels in environmental samples, and it is quite easy to contaminate the sample. Even placing the sample in a well-sealed plastic bottle may in some circumstances not be satisfactory: contaminant gases have been known to permeate through the walls of plastic vessels! However, if the analyst is aware of all the potential problems, he/she can generally take preventive measures. Many analytes are exceedingly reactive and require addition of a preserving agent on site. For example, when analysing for dissolved oxygen in a water sample it is necessary to "fix" the oxygen at the time of sampling otherwise its concentration will be reduced by oxidation reactions and aerobic biological processes in the sample. Samples taken for metal ion analysis are acidified to prevent adsorption onto the walls of the bottles prior to storage.

Once collected, the samples are usually stored in a refrigerator at 4 °C until analysis can be performed. The length of time for which samples can be stored varies, depending on the analyte. Recommendations for bottle material, preservative and maximum storage times for some analytes are given in Table 1.4. It should be remembered that the table gives "maximum storage times" and the samples should be analysed at the earliest convenient time.

Many sample bottle materials are available: borosilicate glass, Pyrex glass, polypropylene, polyethylene, Teflon, *etc*. Teflon (PTFE) is the most unreactive material but also the most expensive and its use is therefore precluded in surveys requiring large numbers of samples. The general recommendation is that Pyrex glass bottles be used for organic analysis and plastic bottles (polyethylene or polypropylene) for inorganic analysis. Soda-glass bottles are unsuitable since they may leach

Table 1.4 *Recommended storage conditions for some analytes in water samples. All samples should be stored in a refrigerator at 4 °C*

Analyte	Bottle material[a]	Preservative	Maximum storage time
Alkalinity	P	None	2 weeks
Ammonia	P	H_2SO_4 to pH < 2	4 weeks
BOD	P, G	None	2 days
Calcium	P	None	4 weeks
COD	P, G	H_2SO_4 to pH < 2	4 weeks
Chloride	P	None	4 weeks
Conductivity	P	None	1 week
Dissolved oxygen	G	$MnSO_4$	Analyse as soon as possible
Fluoride	P	None	4 weeks
Hardness	P	None	4 weeks
Magnesium	P	None	4 weeks
Nitrate	P	H_2SO_4 to pH < 2	4 weeks
Nitrite	P	None	Analyse as soon as possible
Pesticides	G	pH 5–9	1 week to extraction, 6 weeks after extraction
pH	P	None	Analyse as soon as possible
Phenols	G	NaOH to pH 12	1 week to extraction, 6 weeks after extraction
Phosphate	P	None	2 days
Potassium	P	None	4 weeks
Sodium	P	None	4 weeks
Sulfate	P	None	4 weeks
Suspended solids	P, G	None	1 week
Total solids	P, G	None	1 week
Trace metals (*e.g.* Pb, Fe)	P	HNO_3 to pH < 2	6 months
Volatile solids	P, G	None	1 week

[a] P = polyethylene; G = Pyrex glass.

sodium, calcium and silicate. All bottles should be cleaned before sampling according to the methods described in Appendix II. The bottles should be thoroughly rinsed with laboratory water to remove any traces of cleaning agent and filled with laboratory water when not in use.

1.4 SAMPLE TREATMENT

Although some samples may be analysed directly, most often the sample has to be prepared for analysis. A variety of sample treatment methods are used, depending on the type of sample, the analyte to be determined and the kind of analytical method to be used. The purposes of sample treatment are three-fold:

- To convert the sample and analyte into a form suitable for analysis by the chosen method.
- To eliminate interfering substances.
- To concentrate the sample.

Typical sample treatment methods include:

- *Dissolution/digestion*. A solid sample has to be dissolved in a solution before it can be analysed by most analytical methods. Various methods for decomposing and dissolving solid samples are available: acid digestion on a hot plate, refluxing, ultrasonic digestion and microwave digestion.
- *Filtration*. Aqueous samples are usually filtered. For example, when determining soluble components it is customary to filter out the suspended particles from solution as these may interfere in the analysis.
- *Solvent extraction*. Organic analytes are usually extracted into an organic solvent. This can also serve to concentrate the sample.

Various other treatments can also be applied: drying, sieving, ignition, boiling, precipitation, complexation, reduction, oxidation, *etc*. Solid samples are usually dried in an oven to remove any water before carrying out any other treatment.

1.5 ANALYTICAL METHODS

1.5.1 Selection of Method

The analyst can use either a classical method (titrimetry, gravimetry) or one of many instrumental methods. The selection of the appropriate method is based on the following criteria:

- Expected concentration of analyte in the sample.
- Number of samples to be analysed.
- Time that can be devoted to the analysis.
- Cost of the analysis.

Analytical methods employed in environmental analysis are summarised in Table 1.5. More details about methods used in this book are given in Appendix II. Methods may be classified as specific, selective or universal. *Specific* methods respond to only one analyte and are therefore not prone to interference from other substances.

Table 1.5 *Analytical methods, their principles, and typical applications to environmental analysis*

Method	Principle	Sub-classes	Typical application
Titrimetry	Addition of a standard solution to sample until reaction is complete as shown by colour change in added indicator	Neutralisation Precipitation Complexation Reduction/oxidation	Alkalinity, acidity Chloride Water hardness Dissolved oxygen
Gravimetry	Precipitation of analyte out of solution and weighing the product		Sulfate, chloride
Spectroscopy	Interaction of electromagnetic radiation with sample	UV/visible Infrared (IR) Atomic absorption (AAS) or emission (AES)	Iron in water, SO_2 in air, O_3 in air Oxidants in air Trace metals (*e.g.* Pb, Cu, Cd, Zn)
Chromatography	Separation of components in a mixture as they move down a column	Gas chromatography (GC) Ion chromatography (IC)	Pesticides Hydrocarbons in air Ions in solution (*e.g.* Cl^-, SO_4^{2-}, NO_3^-)
Electrochemical	Measurement of electrical properties	Ion selective electrodes Voltammetry	pH, fluoride Trace metals (*e.g.* Pb, Cd, Cu)

Selective methods respond to certain classes of analytes and may be prone to some interference. *Universal* methods respond to all classes of analytes.

1.5.2 Classical Analysis

Titrimetry (also called volumetric analysis) is simple, inexpensive, rapid and accurate. It requires the most rudimentary of laboratory glassware (burettes, pipettes and volumetric flasks) available in all laboratories. Titrations are generally useful for determining analyte concentrations at levels >1 mg L^{-1} and are of limited use for trace component analysis.

Gravimetry is inexpensive and accurate but tedious and slow. It, too, requires the minimum of laboratory equipment and can be carried out in all laboratories.

1.5.3 Instrumental Analysis

Instrumental methods in environmental analysis generally involve *spectroscopy* and *chromatography*. Spectroscopic methods used in the analysis of environmental samples include UV/visible, atomic absorption (AAS) or emission (AES) and infrared (IR). Most laboratories would be equipped with a colorimeter (for visible spectrophotometry) and a flame photometer (for AES), but some may also have an AAS and more advanced UV/visible instruments. Other advanced spectroscopic techniques which may not be generally available in all laboratories include: inductively coupled plasma spectroscopy (ICP), nuclear magnetic resonance spectroscopy (NMR) and Fourier-transform infrared spectroscopy (FTIR). Chromatographic methods available in many laboratories include gas chromatography (GC) and high-pressure liquid chromatography (HPLC) including ion chromatography (IC). These are all very useful for analysing environmental samples.

A variety of electrochemical techniques can be used for environmental analysis. The most common are ion selective electrodes (ISE), an example of which is the widely used pH electrode found in all laboratories. Other techniques such as coulometry, polarography and voltammetry, although useful for some environmental analyses, may not be found in many laboratories.

More advanced techniques which may be used for environmental analysis but which may not be available in many laboratories are: gas chromatography/mass spectrometry (GC/MS), X-ray methods, fluorimetry, thermogravimetry (TG) and a variety of radiochemical techniques.

Experiments in this book require only an atomic absorption spectrophotometer (AAS), an ion chromatograph (IC) and a colorimeter or UV/visible spectrophotometer.

1.5.4 Test Kits and Portable Laboratories

In many instances it is necessary to analyse pollution on the spot in the field rather than taking the sample back to the laboratory. This may be necessary for the following reasons:

- to avoid any changes in the sample composition due to chemical or biological reactions during transport to the laboratory;
- to obtain results immediately as, for example, during an emergency following a spillage of hazardous chemicals, when a delay in analysis could have grave consequences.

Many test kits and portable systems are commercially available for the on-site analysis of pollutants. These can vary in sophistication from simple colorimetric methods involving visual comparisons to portable laboratories and instruments.

Kits suitable for rapid and easy determination of various compounds have been developed, and these generally involve adding pre-measured reagents supplied in powder or pill form to the sample and matching the developed colour with colour discs which can be rotated to obtain a colour match with the reacted sample. There are several variations of this methodology available for water testing. Indicator tubes based on visual colorimetry have been developed for determining on-site air pollution (see Section 3.2.7). Test kits are also available for testing acid rains by means of visual colour comparisons after addition of a reagent. Visual methods are suitable for quick spot checks and although they provide quantitative information they are not of much use in serious environmental research.

Many different portable instruments are commercially available for the determination of numerous water and air pollutants. Instruments can range from those capable of measuring only one parameter to multi-substance analysers. Portable, battery-operated laboratories, consisting of colorimeters and spectrophotometers for comprehensive on-site analysis of a suite of pollutants, can also be purchased on the market.

1.6 STANDARDISATION AND CALIBRATION

1.6.1 Standardisation

In titrimetric analysis a standard solution is added to the sample and the concentration of analyte is determined from the volume of the standard solution used up. The exact concentration of the standard solution is not known at the onset but it can be found by titrating the standard solution with another solution called a *primary standard*. The primary standard is prepared from reagents of high purity and stability. For example, in the determination of water hardness the standard EDTA solution used as a titrant is first titrated against an accurately prepared solution of $CaCO_3$. *Standardisation* is the name given to this process of accurately determining the concentration of a standard solution.

1.6.2 Calibration

Quantifying the results of an instrumental analysis generally involves the construction of a *calibration graph*. The direct response of the instrument, or the peak height obtained on a chart recorder, is plotted against the analyte concentration for a series of standard solutions containing differing concentrations of the analyte substance. A typical calibration curve for lead (Pb) in atomic absorption spectroscopic (AAS) analysis is shown in Figure 1.4. The sample is then analysed in the same way and the concentration in the sample is then determined by interpolation. Most calibration curves are straight lines; however, with some graphs, curvature can be noted.

1.6.3 Standard Addition

An alternative way to quantify the analyte concentration is by means of the *standard addition method*. In this case, a series of solutions containing both the sample and varying concentrations of the substance to be determined are prepared by adding aliquots of a standard solution to the sample. The solutions are analysed and the response of the instrument is plotted against the concentration due to the added standard. The negative intercept on the *x*-axis gives the concentration in the sample, as shown in Figure 1.5. This method is used to eliminate matrix effects, *i.e.* interferences by other components which may be present in the sample.

Graphs obtained from the calibration method and the standard additions method can be treated using the method of least squares to obtain the best fit line through the data points (see Section 1.7.8).

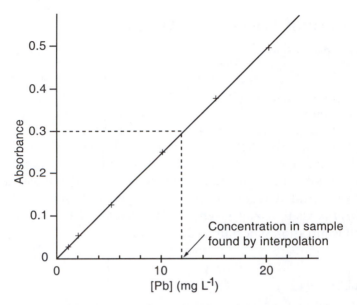

Figure 1.4 *Linear calibration graph for lead analysis by AAS. A sample producing an absorbance of 0.3 would have a Pb concentration of 12 mg L^{-1}*

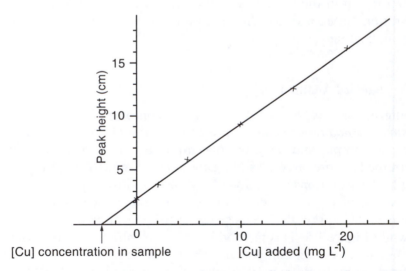

Figure 1.5 *Standard addition calibration graph. Concentration in sample is determined by extrapolation to the abscissa. In this case the concentration of copper in the sample is 3.3 mg L^{-1}*

1.6.4 Blanks

The water used for preparing various reagents and standard solutions should be of the highest purity. Doubly distilled, deionised or various waters obtained from laboratory purification systems (*e.g.* Milli-Q) can be used. With many types of analyses it is also necessary to analyse blanks. Blanks consist of pure laboratory water, while *reagent blanks* contain pure laboratory water and the various reagents used in the analysis. Sometimes *sample blanks* are also analysed. This is common practice when coloured or turbid samples are analysed by colorimetry. In this case, samples are analysed without the addition of the colour forming reagent. Blanks should be analysed frequently as they can reveal sources of contamination.

1.7 ANALYTICAL DATA

1.7.1 Concentration Units

Since analytical measurements are based on physical quantities, every analytical chemist should be familiar with the system of units. The International System of Units (SI) is widely recognised and used, although some non-SI units are still in use. Different units are related by a *conversion factor*. The most important unit, as far as the chemist is concerned, is the *mole*. The mole is defined as the amount of matter which contains the same number of elementary units as are present in 0.012 kg of pure ^{12}C.

Analytical chemists usually express the results of an analysis in concentration units, and the strengths of analytical reagents are also quoted in concentration units. Typical concentration units are defined in Table 1.6.

Normality is generally not used in modern chemistry textbooks; however, many catalogues of reagent manufacturers still employ this unit. Normality depends on the type of reaction. For acids, *equivalent* refers to a mass which has 1 g H^+. For HCl and HNO_3, normality = molarity. For a diprotic acid such as H_2SO_4, normality = 2 × molarity (*i.e.* 0.5 M H_2SO_4 = 1 N H_2SO_4). Normality and equivalents are extensively used in environmental analysis, and they are quite useful in titration calculations. Concentrations of analytes are often expressed as *milliequivalents* (meq) or *microequivalents* (μeq) per litre in environmental chemistry. Concentrations in mg L^{-1} can be converted to meq L^{-1} by dividing the ionic charge by the ionic mass (*e.g.* 1 mg L^{-1} of PO_4^{3-} = 3/94.97 = 0.03159 meq L^{-1}).

Table 1.6 *Concentration units*

Unit	Definition
(a) Solution (water)	
molarity (M)	moles per litre of solution
molality (m)	moles per kilogram of solvent
normality (N)	equivalents per litre of solution
mole fraction (x)	moles of solute per (moles of solute + moles of solvent)
wt%	(grams of solute per grams of solution) × 100%
ppm by wt	10^{-6} g solute per g solution (for aqueous solutions this is = mg L^{-1} or μg mL^{-1})
ppb by wt	10^{-9} g solute per g solution (for aqueous solutions this is = μg L^{-1} or ng mL^{-1})
(b) Gas (air)	
μg m^{-3}	10^{-6} g m^{-3}
ppmv	parts per million (10^6) by volume
ppbv	parts per billion (10^9) by volume
atm	partial pressure in atmospheres (*e.g.* 1 ppm = 10^{-6} atm at sea-level)
%	percentage by volume
(c) Solid (dust, soil, sediment, plant and animal tissue)	
μg g^{-1}	10^{-6} g g^{-1}
ppm	μg g^{-1} or mg kg^{-1}
ppb	ng g^{-1} or μg kg^{-1}

For most aqueous solutions which have a density of around 1 g mL^{-1}, 1 ppm can be assumed as being equal to 10^{-6} g mL^{-1}. Concentrations in solid samples usually refer to dry weight of sample.

1.7.2 Significant Figures

When numerical data are quoted it is customary to report all the digits known with certainty plus one uncertain digit. Thus if a burette reading of 15.38 mL is reported, this implies that numbers 1, 5 and 3 are known with certainty and that there is some doubt about number 8. This burette reading was reported to four significant figures. A burette reading of 5.64 mL would imply three significant figures. In case numerical values are obtained, say as a result of a calculation using a calculator, with more figures than are significant, then the data should be rounded off in accordance with the above rule. Further rules of significant figures are given below (N.B. somewhat different rules may be found in other books owing to a lack of unanimity on the subject):

- Zero is not a significant figure when it is the first figure in a number (*e.g.* 0.00034 has only two significant figures). A zero in any other position is significant (*e.g.* 102 has three significant figures). In order to avoid confusion it is preferable to use scientific notation when expressing results (*e.g.* 6.20×10^4 has three significant figures).

- When rounding off numbers, add one to the last figure retained if the following figure is greater than 5 (*e.g.* 0.53257 becomes 0.5326 when rounded off to four significant figures).

- Round 5 to the nearest even number (*e.g.* 0.255 becomes 0.26 when rounded off to two significant figures). If the digit just before 5 is even, it is left unchanged (*e.g.* 0.345 becomes 0.34 when rounded off to two significant figures); if it is odd, its value is increased by one (*e.g.* 0.335 becomes 0.34 when rounded off to two significant figures).

- If two or more figures are present to the right of the figure to be retained, they are considered as a group (*e.g.* 6.8[501] should be rounded off to 6.9; 7.4[499] should be rounded off to 7.4).

- In addition and subtraction the result should be reported to the same number of decimal places as there are in the number with the smallest number of decimal places (*e.g.* 143.53 + 3.078 + 0.7462 = 147.35).

- In multiplication and division the result should have an uncertainty of the same order as the number with the greatest uncertainty (*e.g.* $x = (96 \times 587)/(1067 \times 2.875) = 18.4$; the result is reported to three significant figures because the greatest uncertainty is in the number 96, the uncertainty in which is about 1%).

- In the logarithm of a number we retain the same number of digits to the right of the decimal point as there are significant figures in the original number (*e.g.* $\log_{10} 8.91 \times 10^{-6} = -5.050$).

- In the antilogarithm of a number we retain as many significant figures as there are digits to the right of the decimal point in the original number (*e.g.* $10^{3.72} = 5.2 \times 10^3$).

Integers or pure numbers (*e.g.* ionic charge) are known with absolute certainty even though the significant figures are not specified (*e.g.* 32.86/2 = 16.43; the answer is rounded off to 4 significant figures and not to one significant figure).

1.7.3 Accuracy and Precision

Accuracy is the degree of agreement of the measured value with the true value (*i.e.* it is the closeness of the measured to the true value). The true value is frequently unknown and it is therefore impossible to determine the accuracy of a method using samples. It is, however, possible to determine the accuracy of a method or technique using *standard reference materials.* Standard reference materials are samples in which the concentration of an analyte has been determined by an external (usually government) laboratory. These materials can then be analysed and the concentration compared to that determined by the external laboratory, thus providing an estimate of the accuracy. A whole range of standard reference materials are available (coal, river water, oil, biological materials, *etc.*) (see Section 1.7.10). Accuracy is inversely related to the *bias*; the greater the bias in a method, the lower is the accuracy.

Precision is the degree of agreement between different measurements carried out in the same way. This can be determined by repeatedly analysing the same sample (*replicate* analysis). Precision is related to the *scatter* in the data; the lower the scatter, the greater is the precision.

Repeatability and *reproducibility* are closely related to precision. *Repeatability* is determined by analysing replicate samples on the same day under the same conditions and this measures within-run precision. *Reproducibility* is determined by analysing replicate samples on different days when conditions may vary (re-optimising and re-calibrating instruments, preparing fresh reagents, *etc.*) and it gives between-run precision.

Preferably we would like analytical methods to be both precise and accurate but in practice this is not possible. There will always be some inaccuracy and imprecision in the method as a result of errors and the aim of the analytical chemist is to minimise the sources of these errors. Figure 1.6 illustrates the concepts of accuracy and precision.

1.7.4 Errors

Error is the difference between the measured and the true value, and as such it is a measure of the accuracy of the method. There are three types of errors.

- *Gross errors (blunders)*. These are errors due to human negligence (misreading instrument, mislabelling sample, errors in calculations, spillage, contamination, *etc.*).
- *Systematic (determinate) errors*. These errors are always of the same magnitude and they produce a bias in the method. They relate to the inaccuracy of the method and they have specific and identifiable causes (*e.g.* reagent blank, a meter with a zero error, observer bias, interference). Once these errors are identified they can be eliminated (*e.g.* removal of interfering substance) or a correction can be applied. Systematic errors may result in a method exhibiting good precision but poor accuracy.

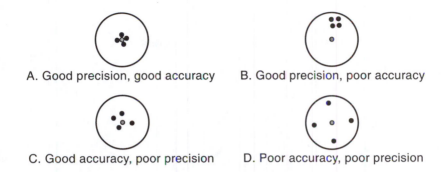

A. Good precision, good accuracy   B. Good precision, poor accuracy

C. Good accuracy, poor precision   D. Poor accuracy, poor precision

Figure 1.6 *Illustration of precision and accuracy.* ◉, *true (target) value;* ●, *measured value*

- *Random (indeterminate) errors.* These errors occur by chance, and they vary in sign and magnitude. Random errors are inherent in all measurements, even under the best conditions using the best instruments. Random errors are generally small and have an equal probability of being positive and negative. If a large number of repeated measurements are made it is found that the data exhibit a *normal (Gaussian) distribution* around the mean (see Figure 1.7). These errors cannot be eliminated; however, they can be determined by replicate measurements and reported as an uncertainty in the result.

The absolute error, $E_{absolute}$ is defined as:

$$E_{absolute} = \text{measured value} - \text{true value}$$

The relative error, $E_{relative}$ is defined as:

$$E_{relative} = (E_{absolute}/\text{true value}) \times 100\%$$

1.7.5 Reporting of Results

If possible, several replicate measurements should be made on the sample in order to test for the variability in the method and obtain some indication of the precision of the measurement. If a sufficiently large number of measurements are carried out the results are expected to exhibit a Gaussian, or so-called *normal*, distribution as shown in Figure 1.7.

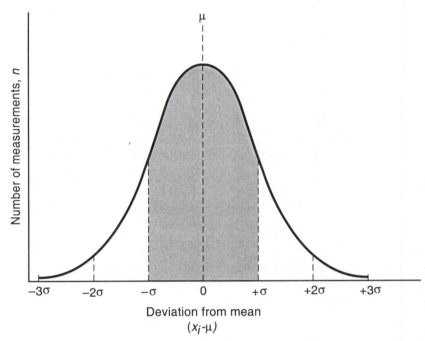

Figure 1.7 *A Gaussian or normal distribution curve. When the number of measurements (n) is infinitely large the population mean, μ, and population standard deviation, σ, are used. For real measurements, where n is finite, the sample mean, x̄, and the sample standard deviation, s, are used instead*

The *mean, x̄,* is defined as:

$$\bar{x} = \Sigma x_i / n$$

where Σx_i is the sum of the individual measurements and n is the number of measurements. The *standard deviation, s,* is defined as:

$$s = \sqrt{\frac{\Sigma(x_i - \bar{x})^2}{n-1}}$$

For a normal distribution, 68% of all data lie within one standard deviation of the mean, $\bar{x} \pm s$; 95% of the data lie within two standard deviations, $\bar{x} \pm 2s$; and 99.7% of the data lie within three standard deviations, $\bar{x} \pm 3s$. The *variance* is defined as the square of the standard deviation, *i.e.* s^2. The *coefficient of variation*, CV, is given by:

$$CV = (s/\bar{x}) \times 100\%$$

Example 1.1

Consider the following six measurements of Ca concentration in a water sample in mg L^{-1}:

16.3, 15.7, 16.8, 15.9, 16.0, 15.5

The mean is 16.03 mg L^{-1} and the standard deviation is 0.46 mg L^{-1}. As there are six data points (n) we find the Student t factor by looking up the appropriate statistical table (see Appendix V) under five degrees of freedom ($n - 1$). For $n - 1 = 5$ the Student t factor at the 95% confidence limit has a value of 2.571. We then use the following equation to calculate the 95% confidence interval for the above measurements:

$$t \times s/n^{1/2} = 2.571 \times 0.46/6^{1/2} = 0.48$$

The result of the analysis is then reported as 16.03 ± 0.48 mg L^{-1}. Note that the result (mean and confidence interval) is reported to one more significant figure than the raw data.

The mean and standard deviation of the replicate measurements should be calculated and the result should be reported as the mean $\pm$ the 95% confidence interval. Many authors choose to report the result as the mean $\pm 2s$ (*i.e.* two standard deviations); however a better method is to calculate the 95% confidence interval using the *Student t test* (see Example 1.1).

1.7.6 Rejection of Data

When several replicate determinations have been made, one result may appear to be much higher or much lower in value than the rest. It may be that the suspect value does not belong with the other values owing to inadvertent contamination of the sample or some error in the analysis. In this case the analyst has to decide whether to retain or reject the suspect value. This is achieved by applying the so-called Q *test* to the data. Q is calculated from:

$$Q = \frac{|\text{Questionable value} - \text{Nearest value}|}{\text{Highest value} - \text{Lowest value}}$$

The determined value is then compared with a value given in a table for the appropriate number of measurements. This tabulated value is

Example 1.2

Consider the following six measurements of dissolved oxygen in a water sample in mg L^{-1}:

5.6, 5.3, 5.8, 5.6, 5.7, 7.1

The result of 7.1 mg L^{-1} appears suspect and the Q-test is applied as follows:

$$Q = \frac{|\ 7.1 - 5.8\ |}{7.1 - 5.3} = 0.722$$

Critical Q for six data points has a value of 0.621 (Appendix V). Since calculated Q is greater than critical Q, the result is rejected. Had the suspect result been 6.1 instead of 7.1 the result would have been retained as in that case calculated Q = 0.375, which is less than critical Q.

called "Q-critical". Values of Q-critical for different numbers of measurements are given in Appendix V. If the calculated value is higher than the value of Q-critical, the questionable value is rejected. If, on the other hand, the calculated value of Q is lower than the value of Q-critical, the suspect value is retained.

When a number of measurements have been made, any suspect data should first be considered and the mean, standard deviation and confidence limit should be based only on the retained values (see Example 1.2). The analyst should also reject the result of any analysis in which a known error (*e.g.* gross error) has occurred.

1.7.7 Correlation Coefficient

Quite often in environmental analysis we may want to investigate whether there is a significant relationship between two variables (x and y). For example, we may want to compare the results of two different methods over a range of analyte concentrations in order to assess their suitability for a particular analysis, or we may compare the concentrations of two different analytes measured in samples at various sites in order to identify potential pollution sources or investigate some environmental process. Normally we would plot x against y and visually ascertain whether a relationship exists. However, it is not uncommon to find considerable scatter when plotting the results of environmental analysis in this way owing to the complex nature of the samples and environmental processes that influence the concentrations of substances in different samples. Such graphs are called *scatterplots*. We therefore have to rely on statistics to tell us whether a significant relationship exists between different variables.

The *correlation coefficient, r*, can be used to test whether there is a significant linear relationship between two variables (x and y). Values of r can vary between -1 and 1. Values of 1 or -1 indicate a perfect relationship between the two variables (*i.e.* all the data points would lie exactly on a straight line if plotted on a graph). Positive values of r indicate a positive relationship between x and y, *i.e.* as x increases, so does y. Negative values indicate an inverse relationship between x and y. The closer the estimated value of r is to 1 or -1, the more significant (*i.e.* stronger) is the relationship between the two variables; the closer r is to 0, the less significant is the relationship. Also, for low numbers of data pairs (n), higher values of r are needed to show statistical significance, and, conversely, for higher numbers of data pairs, lower values of r may indicate a significant relationship. Some scatterplots and correlation coefficients obtained in environmental studies are illustrated in Figure 1.8. Regression lines are drawn only for those plots where the value of the correlation coefficient indicates a significant relationship.

Pearson's correlation coefficient can be calculated manually from the following equation:

$$r = \frac{n\Sigma x_i y_i - \Sigma x_i \Sigma y_i}{([n\Sigma x_i^2 - (\Sigma x_i)^2] \times [n\Sigma y_i^2 - (\Sigma y_i)^2])^{1/2}}$$

but the process is tedious. Most scientific calculators are programmed to calculate the correlation coefficient. A correlation table (see Appendix V) is used to evaluate the level of significance of a linear relationship between x and y. The table gives critical value of r for different numbers of data pairs (n). Critical values of r are usually taken at $P = 0.05$. If the calculated value of r is greater than that of the critical value given in the table, then in all likelihood there is a significant linear relationship between x and y, *i.e.* there is less than 5% probability (1 in 20) that this could be due to random data points. If r is lower than the critical value, then there is no significant linear relationship between x and y. It should be remembered that although the correlation coefficient is a valid indicator of association between two variables, it does not imply causation.

If a large number of parameters have been determined at one sampling site, possible inter-relationships may be illustrated by a *correlation matrix*. Such a matrix is shown in Table 1.7 for various chemical parameters in rainwater sampled at a single site. The significance of each

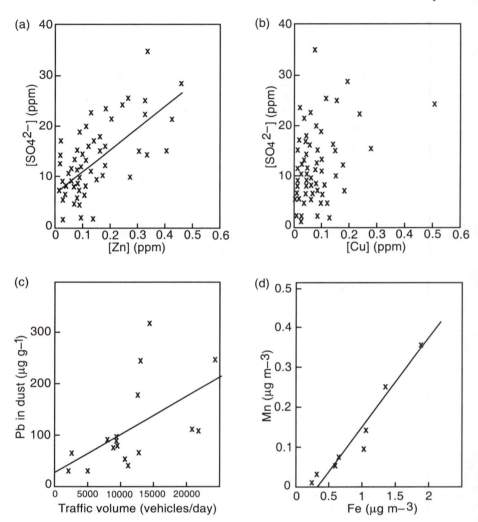

Figure 1.8 *Typical scatterplots of environmental data.* (a) *Sulfate versus zinc in rainwater* (r = 0.649); (b) *sulfate versus copper in rainwater* (r = 0.348); (c) *lead in road dust versus traffic flow* (r = 0.570); (d) *manganese versus iron in air* (r = 0.967)

correlation coefficient must be tested (see Example 1.3). It is apparent that there are many significant correlations between rainfall constituents. These may not necessarilly indicate a common source, as high correlations between two parameters may arise out of a co-dependence on a third, such as the rainfall amount in this example. On the other hand, many of the substances in Table 1.7 have common sources, *e.g.* Na, Cl, Mg, Ca and SO_4^{2-} are all present in sea-salt aerosols, while Mg, Ca and SO_4^{2-} (as $CaSO_4$) are also found in mineral dusts.

Table 1.7 *Correlation matrix[a] for rainwater samples giving values of Pearson's correlation coefficient, r, for pairs of variables*

	mm	cond.	H^+	Cl^-	NO_3^-	SO_4^{2-}	Na^+	K^+	Mg^{2+}	Ca^{2+}
Ca^{2+}	−0.585	0.671	−0.067	0.329	0.597	0.648	0.435	0.703	0.860	1
Mg^{2+}	−0.477	0.734	−0.006	0.506	0.362	0.555	0.698	0.732	1	
K^+	−0.508	0.696	0.030	0.359	0.413	0.486	0.458			
Na^+	−0.349	0.530	−0.022	0.692	0.075	0.406	1			
SO_4^{2-}	−0.337	0.619	0.369	0.582	0.831	1				
NO_3^-	−0.342	0.584	0.399	0.347	1					
Cl^-	−0.275	0.504	0.344							
H^+	0.129	0.247	1							
cond.	−0.478	1								
mm	1									

[a] Significant correlations ($P < 0.05$) are underlined; mm, rainfall amount; cond., conductivity. Taken from Radojević and Lim (1995).

Example 1.3

Concentrations of Pb and Mn in atmospheric particulate matter were determined in 10 air samples. The following results were obtained in $\mu g\ m^{-3}$:

No.	1	2	3	4	5	6	7	8	9	10
Pb	1.82	0.96	0.37	0.61	0.68	0.38	0.24	0.77	1.52	0.58
Mn	0.36	0.14	0.05	0.08	0.25	0.03	0.01	0.09	0.29	0.07

(a) Is there a statistically significant linear relationship between Mn and Pb concentrations?
(b) If there is a significant relationship, derive an equation relating Mn to Pb for these samples.

(a) The correlation coefficient, r, was calculated using the linear regression programme on a scientific calculator and found to be 0.908. The critical value of r ($P = 0.05$) for 10 data pairs (*i.e.* 8 degrees of freedom) is given as 0.632 in statistical tables (Appendix V). Since calculated r > critical r it can be concluded that there is a significant linear relationship between Pb and Mn in air.
(b) Using the same programme on the scientific calculator it was possible to determine the linear regression, taking Mn as y and Pb as x. The slope, b, was calculated to be 0.214, and the intercept, a, was found to be −0.033. The equation is:

$$Mn = -0.033 + 0.214\ Pb$$

1.7.8 Linear Regression

Once a significant relationship between two variables x and y has been confirmed, either visually or by means of the correlation coefficient, it is often necessary to draw a best-fit line through the data points, or derive an equation relating the two variables. This is typically required when preparing a calibration graph in which some measured parameter, y (*e.g.* absorbance), is plotted against concentration of standard, x.

The equation for a straight line is:

$$y = a + bx$$

In order to be able to draw the best-fit line we need to know the slope of the line, b, and the intercept on the y-axis, a. These can be calculated from the following equations:

$$b = \frac{n\Sigma x_i y_i - \Sigma x_i \Sigma y_i}{n\Sigma x_i^2 - (\Sigma x_i)^2}$$

and

$$a = \bar{y} - a\bar{x}$$

where $\bar{x}$ and $\bar{y}$ are the mean values of x_i and y_i. Most scientific calculators contain programmes for linear regression, as do computers supplied with modern instruments. With most modern instruments the calibration graph and the final result can be calculated automatically by the on-line computer.

1.7.9 Detection Limits

The detection limit (d.l.) is the smallest value that can be distinguished from a blank and there are various ways to calculate this. One method is to use the following expression:

d.l. = 3 × standard deviation of the baseline noise/sensitivity

The sensitivity is the instrument response (*e.g.* peak height on a chart recorder output) per unit amount or concentration of substance. If a result of a sample analysis is found to be below this value it is reported as being <d.l. and no numerical value is specified. Different analytical methods have different detection limits and the analyst should be aware of these when selecting the appropriate method for a specific analysis.

The practice of not giving numerical values to measurements that are

<d.l. has been repeatedly criticised. Problems arise with these kind of data when statistical analysis of results is carried out, and various assumptions have to be made. For example, values <d.l. can be replaced with zeros, the value of d.l. or some fraction of the d.l. value, but none of these are satisfactory. It has been suggested that results of all measurements should be reported, whether above or below the d.l., in order to facilitate subsequent statistical analysis.

1.7.10 Reference Materials

While precision of a method can be tested by means of replicate analysis, the accuracy of the method cannot be determined since the "true" value is unknown. One way to assess the accuracy of a method is by analysing *reference materials*. Reference materials are actual samples (*e.g.* river-water, sediment, soil) which have been carefully analysed by a government laboratory. Concentrations of various analytes in the samples are reported and the material is said to have been *certified* when accompanied by a certificate. These are called *certified reference materials* (CRM) and they can be purchased from government laboratories, *e.g.* Laboratory of the Government Chemist (LGC) in the UK, Community Bureau of Reference (BCR) of the European Community, National Bureau of Standards (NBS) in the USA. The reference material can then be analysed and the determined concentration compared with that quoted on the certificate accompanying the material to get some indication of the analytical error.

1.7.11 Quality Control

There are two types of quality control: *intralaboratory* quality control, which is carried out in one laboratory, and *interlaboratory* quality control involving several laboratories. Nowadays, many laboratories participate in government-run quality control schemes in order to earn *accreditation*. The central government laboratory routinely monitors the performance of participating laboratories and those laboratories which show a poor performance may lose their accredited status. If a laboratory loses its accredited status it may not be allowed to carry out analyses on government contract, its results would not be considered valid in a court of law in case of a dispute over infringement of environmental standards, *etc.* For a commercial laboratory to lose accredited status it may be a significant blow since customers may lose confidence in it and take their samples elsewhere.

One way of monitoring the quality of laboratory analyses is by means of a *control chart*. A typical quality control chart is illustrated in Figure 1.9. A *control sample* is routinely analysed with each batch of samples and a day-to-day record of the result of the standard analysis is kept in the form of a time series chart. At the onset of the quality control scheme the control sample is repeatedly analysed (*e.g.* 20 times) and the mean and standard deviation determined. Lines are drawn on the chart corresponding to the concentration of the control sample (mean) and corresponding $\bar{x} \pm 2s$ and $\bar{x} \pm 3s$ where s is the standard deviation of replicate measurements of the control sample. The $\bar{x} \pm 2s$ lines are called the *warning limits* and the $\bar{x} \pm 3s$ lines are called the *action limits*. The performance of the laboratory is satisfactory if the measured values of the control sample fall within $\bar{x} \pm 2s$. If the measurement falls outside the $\bar{x} \pm 3s$ limits, then there is something seriously wrong; the analytical procedure has to be reviewed, and all analytical results obtained on that day have to be rejected. Once the source of the problem has been identified and corrected, the samples have to be re-analysed before the results can be released. If the result on a particular day falls ouside the warning limits but below the action limits, there may be a hint of a problem, and the analyst has to pay close attention to the analytical method to ensure that future analyses do not exceed the action limits. Control samples can be either natural environmental samples or samples spiked with a standard of the analyte. Accredited laboratories that participate in interlaboratory quality control schemes can usually obtain control samples from the central government laboratory.

1.8 COMMON PROBLEMS OF ENVIRONMENTAL ANALYSIS

The need for data of the highest quality in environmental analysis cannot be overstated as the results of an environmental analysis may have far-reaching implications beyond the immediate laboratory. Data generated by the environmental analyst may be used by government agencies to investigate compliance with environmental emission and quality standards, to assess the status and trends of the environment, and to make policy decisions that could affect not only the environment but society at large. Furthermore, the results of an environmental analysis may be presented in a court of law when there is a dispute between a polluting industry and an aggrieved party. Therefore the environmental analyst should perform his work conscientiously and diligently. Unfortunately, the fact remains that a lot of environmental data, even that published in reputable academic journals, is of dubious quality. This, however, is not in most cases due entirely to the

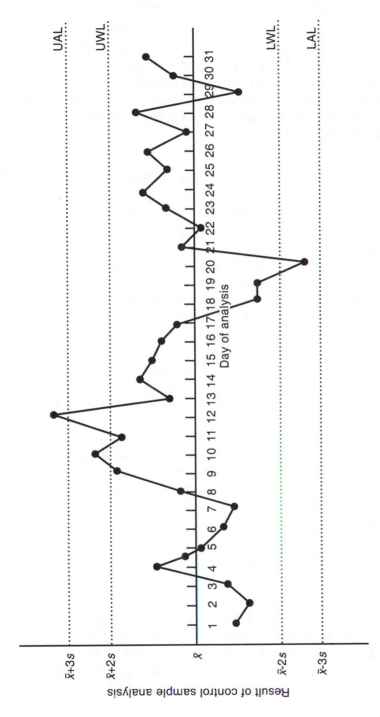

Figure 1.9 *Example of a control chart where x̄ is the mean of replicate analyses of the control sample and s is the standard deviation. Concentrations of analyte determined in control sample are plotted for each day when an analysis is carried out. UWL, upper warning limit; UAL, upper action limit; LWL, lower warning limit; LAL, lower action limit*

negligence of the analyst, but is due to the very nature of environmental analysis, which can be quite complex and problematic, to say the least. Unlike many other types of chemical analyses, such as those performed routinely in industrial laboratories, environmental analysis has additional, specific problems, of which the analyst should be aware. Some of these were mentioned earlier, and others will be dealt with throughout the course of this book, but owing to their importance they are summarised below:

- *Low concentration of analyte*, often close to, at or below the detection limit of many analytical methods.
- *Complex matrix*, with numerous other compounds (known and unknown) present in the sample, and this could lead to several other problems on this list.
- *Interferences* are more likely than in other types of analysis owing to the large numbers of compounds present.
- *Contamination* is more likely due to the low concentrations of analyte and special precautions have to be taken to avoid this during various stages of analysis.
- *Sample variety* is greater than in other types of routine analyses and samples from different sources may require modified or different procedures (*e.g.* water samples may range from highly polluted wastewaters, to very clean rainwater, to seawater with a very high electrolyte content, *etc.*).
- *No suitable method* may be available for the analyte, as the analyte may be a new pollutant not previously considered, and the analyst may have to develop a satisfactory method.
- *Speciation* may present a particular problem since a substance (*e.g.* heavy metal) may be present in different forms, and toxicity of the substance may depend on its form.
- *Reaction* of analyte, which could either increase or decrease its concentration (*e.g.* biological reactions, chemical reactions, adsorption).

Over the years, scientists have become aware of the above problems and a historical review of environmental literature reveals a steady drop in reported concentrations of many trace substances at re-visited sites not always due to a decrease in pollution but quite often to improvements in analytical methodology, elimination of interferences, *etc.* For example, much of the early data on the pH of acid rainwater has been questioned, and is probably unreliable.

1.9 A WORD OF ADVICE

Before you embark on the experimental programme you should read Appendix I, which deals with safety in and out of the laboratory. Always pay particular attention to safety issues when doing practical work. Be especially cautious when handling acids or when boiling or digesting solutions. Concentrated acids are used in many of the experiments and these are extremely dangerous. Use rubber gloves and protective glasses when handling concentrated acids. Reagents and specialised equipment required are listed for each experiment. General laboratory glassware such as balances, beakers, flasks, pipettes and burettes are not listed. It is assumed that such equipment would be available in every laboratory. The volumes required are given in the experimental procedures. Also, wherever the term "water" is used in the experimental sections it refers to laboratory grade water: distilled, deionised, doubly distilled, Milli-Q, NANOpure, *etc.* You should use water of the highest quality available.

It should be noted that modifications can readily be made to many of the experimental procedures given in the book. In some cases the required instruments may not be available but the laboratory may be equipped with some other suitable instrument. For example, an inexpensive flame photometer may be used for the determination of some alkali and alkaline earth metals instead of the more expensive atomic absorption spectrophotometer. Where an alternative method or technique is available, this is pointed out in the text. Other modifications may be necessary owing to the nature of the sample. Environmental samples are not uniform and vary considerably from location to location. For example, the concentration of some substances in rainwater may be rather different at a remote, rural location than in a polluted city with many pollution sources. In some cases, calibration standards may have to be prepared over a different range from that quoted in the experimental procedures, or samples may have to be diluted, or concentrated, as the case may be, to fit them within the range of the calibration curve. These are relatively simple modifications and should present no problem.

Some of the questions and problems may require additional reading in order to be solved. The references given under "Further Reading" at the end of each section should be sufficient for this. Suggestions for various projects are given at the end of some of the experiments. These could be carried out individually as a final year undergraduate research project, or by groups as part of science projects in schools. With a little imagination it is easy to come up with other projects based on these experiments.

1.10 EXERCISES AND INFORMATION

1.10.1 Questions and Problems

1. What is a biogeochemical cycle? Give examples of some important cycles and explain how human activities may have affected them.

2. If the mass of a substance in a reservoir is 60×10^6 kg and the flux through the reservoir is 5×10^6 kg per annum, calculate the residence time assuming a steady state.

3. Differentiate between the following pairs of concepts:
   (a) random error and systematic error
   (b) continuous analysis and batch analysis
   (c) qualitative analysis and quantitative analysis
   (d) sample and analyte
   (e) precision and accuracy
   (f) local pollution and global pollution

4. Define the following:
   (a) standard deviation
   (b) normal distribution
   (c) method of standard addition
   (d) calibration
   (e) certified reference material

5. The chloride ion concentration in a certified reference material (river water) was determined by precipitation titration with $AgNO_3$. Five replicate measurements were made and the following Cl^- concentrations were determined (in mg L^{-1}): 15.5, 14.8, 15.9, 15.2 and 15.3. The reference material was accompanied with a certificate that quoted a Cl^- concentration of 15.1 mg L^{-1}. Calculate:
   (a) the mean
   (b) the standard deviation
   (c) the variance
   (d) the coefficient of variation
   (e) the absolute error
   (f) the relative error

6. Lead (Pb) in motorway runoff water was determined by flame atomic absorption spectrometry (AAS). A stock solution labelled "S1" contained 1000 mg L^{-1} Pb. Fifty μL of this solution were diluted to 50 mL to give a standard solution labelled "S2". Successive portions (1, 2, 3, 4 and 5 mL) of solution S2 were placed in a series of 50 mL volumetric flasks and diluted to the mark with distilled water. These solutions were then analysed using AAS and the following absorbance values were recorded:

Volume of standard S2 added (mL)	1	2	3	4	5
Absorbance (A)		0.15	0.29	0.45	0.56	0.69

The sample of motorway runoff water gave a response of 0.38 absorbance units.

   (a) Plot a calibration graph with the x-axis in units of mg Pb L^{-1}

   (b) Read off the concentration of Pb in the motorway runoff sample

   (c) Convert the concentration to units of mol L^{-1}

   (d) Derive an equation relating absorbance to Pb concentration

7. A water sample was analysed for iron (Fe) by means of flame atomic absorption spectrometry (AAS) by the method of standard addition. Fifty mL aliquots of the sample were placed in 100 mL volumetric flasks and varying quantities of a standard 100 mg L^{-1} solution were added. The flasks were made up to the mark with distilled water and analysed. The output from the spectrophotometer was recorded on a chart recorder and the following results were obtained:

Volume of standard added (mL)	0	5	10	15	20
Response (mm)		16	45	77	101	125

   (a) Prepare a calibration plot for the standard addition

   (b) Determine the concentration of Fe in the sample

8. Five aliquots of a water sample were analysed for cadmium (Cd) and the following concentrations (in mg L^{-1}) determined: 38, 36, 41, 40 and 39.

   (a) Calculate the mean, standard deviation and the 95% confidence interval

   (b) Express the mean concentration in units of mol L^{-1}

   (c) Analysis of an additional aliquot of the same sample yielded a concentration of 25 ppm. Should this result be accepted or rejected?

9. The concentrations of Fe and Zn were determined in atmospheric particulate matter in eight samples and the following concentrations determined (in $\mu g\ m^{-3}$):

No.	1	2	3	4	5	6	7	8
Fe	1.82	1.02	0.57	0.62	1.29	0.30	0.23	0.99
Zn	0.27	0.02	0.13	0.02	0.71	0.13	0.15	0.14

Draw a scatterplot and calculate the correlation coefficient. Is there a significant linear relationship between Fe and Zn in these samples? If there is, plot the regression line.

10. The concentration of nitric oxide (NO) was measured in a flue gas from a coal-fired power station. Repeated measurements were made using the same sampling technique and same instrumentation. The

following six measurements were recorded (in ppmv): 501, 495, 503, 497, 488 and 531. Should the sixth measurement of 531 ppmv be retained or rejected?

1.10.2 Suggestions for Projects

1. Select one biogeochemical cycle (*e.g* nitrogen) and carry out a thorough literature search. Try to read up as much as possible on the topic and write a long essay (*ca.* 4000 words). Discuss how the cycle may have been affected by anthropogenic activities and what could be the potential consequences of this perturbation. Suggest measures that could be taken to redress the balance.
2. Select an environmental problem of your choice (*e.g.* acid rain) and carry out a thorough literature search using books, journals, Internet, *etc*. Try to read as much as possible on the topic and write a long essay (*ca.* 4000 words).
3. Initiate a quality control programme in your laboratory based on the use of quality control charts (Figure 1.9). You will need a large volume of a specific sample or an artificially prepared solution containing the analyte that you are interested in. For example, you may be conducting a survey of chloride in riverwater (but you can do the same for any analyte and any sample). At the start of the monitoring programme, take a large volume of one riverwater sample, say several litres, or prepare an artificial sample by adding NaCl to a similarly large volume of laboratory water to give a concentration in the range of that expected in real samples. This solution will be your control sample. Analyse a large number of replicates of this control sample (*e.g.* 20) and establish the warning and action limits for your quality control chart. Store this solution in a refrigerator. Each time that an analysis of riverwater samples is carried out in your laboratory, also analyse one aliquot of the stored control sample. Plot the results on the chart and monitor the quality of work in your laboratory. If the analysis is performed by different individuals on different days the quality control chart can be used to assess the performance of different analysts.

1.10.3 Reference

M. Radojević and L. H. Lim, A rain acidity study in Brunei Darussalam, *Water, Air, Soil Pollut.*, 1995, **85**, 2369–2374.

1.10.4 Further Reading

1.10.4.1 Analytical Chemistry Textbooks. The following introductory texts provide good coverage of basic analytical techniques (both classical and instrumental) and laboratory practices. Students unfamiliar with the theory and practice of analytical chemistry should read one of these textbooks:

L. G. Hargis, "Analytical Chemistry, Principles and Techniques", Prentice Hall, Englewood Cliffs, NJ, 1988.

J. S. Fritz and G. H. Schenk, "Quantitative Analytical Chemistry", 5th edn., Allyn and Bacon, Boston, 1987.

D. A. Scoog, D. M. West and F. J. Holler, "Fundamentals of Analytical Chemistry", 5th edn., Saunders, New York, 1988.

1.10.4.2 Environmental Chemistry Textbooks. The following books provide a good introduction to environmental chemistry and the problems of pollution:

P. O'Neill, "Environmental Chemistry", 2nd edn., Chapman and Hall, London, 1993.

B. J. Alloway and D. C. Ayers, "Chemical Principles of Environmental Pollution", Blackie, London, 1994.

J. E. Andrews, P. Brimblecombe, T. D. Jickells and P. S. Liss, "An Introduction to Environmental Chemistry", Blackwell, Oxford, 1996.

R. M. Harrison, S. J. de Mora, S. Rapsomanikis and W. R. Johnston, "Introductory Chemistry for the Environmental Sciences", Cambridge University Press, Cambridge, 1991.

R. M. Harrison (ed.), "Understanding our Environment: An Introduction to Environmental Chemistry and Pollution", 2nd edn., Royal Society of Chemistry, Cambridge, 1992.

R. M. Harrison (ed.), "Pollution: Causes, Effects, and Control", 2nd edn., Royal Society of Chemistry, Cambridge, 1992.

The following books are excellent introductory texts to the subject of environmental analysis and they provide supplementary material to the experiments in the present book. At least one of these should be read by students undertaking experiments in environmental analysis for the first time:

C. N. Sawyer, P. L. McCarty and G. F. Parkin, "Chemistry for Environmental Engineering", 4th edn., McGraw-Hill, New York, 1994.

R. N. Reeve, "Environmental Analysis, Analytical Chemistry by Open Learning", Wiley, Chichester, 1994.

S. E. Allen (ed.), "Chemical Analysis of Ecological Materials", 2nd edn., Blackwell, Oxford, 1989.

CHAPTER 2

Rainwater Analysis

2.1 ACID RAIN

Rainwater, being an essential component of the hydrological cycle, plays an important role in the global cycling of water-soluble chemicals. Rainwater serves as a source of many essential nutrients in terrestrial and aquatic ecosystems. It also acts as a cleanser of the atmosphere, washing out pollutants from the air and introducing them into surface waters and soils, where they can have damaging effects on natural ecosystems.

One of the major environmental concerns of the present day is the phenomenon of acid rain. Unpolluted, pure rain is slightly acidic due to the absorption of atmospheric CO_2. The pH of water in equilibrium with atmospheric CO_2 is approximately 5.6 (see Example 2.1) and hence only rain events with a pH lower than this value are classified as "acid rain". Rainwater pH varies between 3 and 9, with samples generally having values between pH 4 and 6 (Figure 2.1). High pH values may arise from the presence of alkaline dust materials in rainwater, as for example at New Delhi which experiences basic rainfall with pH values as high as 9. Values of pH below 3 have been recorded in acid fog samples.

The pH is defined as the negative logarithm of the hydrogen ion activity:

$$pH = -\log_{10} a_{H^+}$$

For dilute aqueous samples, such as rainwater, the activity is equal to the concentration of H^+. The pH is a measure of the free acidity, *i.e.* the concentration of strong acids in dissociated form, and not the total acidity present.

Rainwater acidity was first measured more than 100 years ago during the industrial revolution. The first documented report of acid rain was made by a French scientist, Ducros, who in 1845 published a paper

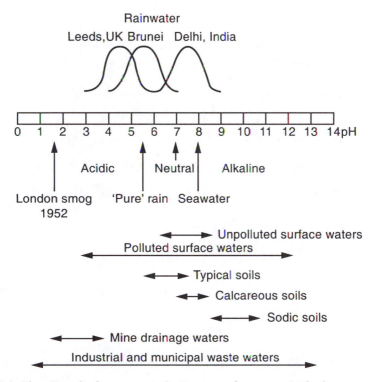

Figure 2.1 *The pH scale showing typical pH ranges of rainwater and other environmental samples*

entitled "Observation d'une pluie acide". Robert Angus Smith, the world's first air pollution inspector, reported acid rain in the city of Manchester in his book "Air and Rain" published in 1872. However, it is only in the last 30 years that widespread concern has been expressed over acid rain. In the 1960s, Scandinavian scientists began to link the mysterious disappearance of fish from lakes and streams to wind-blown pollution from the UK and central Europe. This led to a renewed interest in the phenomenon of acid rain. Nowadays, the term "acid rain" encompasses not only rainwater but the deposition of all acidic pollutants, whether as rain, fog, cloud, dew, snow, dust particles or gases.

Although first recognized as a regional problem in Europe and North America, where many of the effects have been extensively studied and documented, acid rain is now observed throughout the world, even at sites far removed from industrial sources of pollution, such as the polar ice caps and tropical rainforests. Within one generation, acid rain has grown from being a local and regional nuisance to a major global environmental problem.

Example 2.1

Calculate the pH of rainwater in equilibrium with atmospheric CO_2 (CO_2 = 0.036%) at 25 °C, given that K_H = 0.031 mol L^{-1} atm^{-1}; K_1 = 4.3 × 10^{-7} mol L^{-1}; K_2 = 5 × 10^{-11} mol L^{-1}.

The relevant equilibria to consider are the dissolution of CO_2 gas in water and the dissociation of dissolved CO_2 in aqueous solution:

$$CO_2 + H_2O \rightleftharpoons CO_2 \cdot H_2O \qquad K_H = [CO_2 \cdot H_2O]/pSO_2$$

$$CO_2 \cdot H_2O \rightleftharpoons HCO_3^- + H^+ \qquad K_1 = [HCO_3^-][H^+]/[CO_2 \cdot H_2O]$$

$$HCO_3^- \rightleftharpoons CO_3^{2-} + H^+ \qquad K_2 = [CO_3^{2-}][H^+]/[HCO_3^-]$$

We can set up the following electroneutrality balance:

$$[H^+] = [OH^-] + [HCO_3^-] + 2[CO_3^{2-}]$$

We can assume that the solution will be acidic due to the above dissociation reactions:

$$[OH^-] \ll [H^+]$$

Also, since $K_2 \ll K_1$, we can assume that the concentration of CO_3^{2-} will be insignificantly small:

$$[CO_3^{2-}] \ll [HCO_3^-]$$

Eliminating OH^- and CO_3^{2-} from the electroneutrality balance, we are left with:

$$[H^+] = [HCO_3^-]$$

Substituting into the expression for K_1 we get:

$$K_1 = \frac{[H^+]^2}{K_H \times pCO_2}$$

and

$$[H^+] = (K_H \times K_1 \times pCO_2)^{1/2}$$

Substituting the values of K_H, K_1 and pCO_2 expressed in atmospheres (since 0.036% is the partial pressure of CO_2, and the total pressure is 1 atm, 0.036% = 0.036/100 atm = 3.6 × 10^{-4} atm).

$$[H^+] = (0.031 \times 4.3 \times 10^{-7} \times 3.6 \times 10^{-4})^{1/2} = 2.19 \times 10^{-6} \text{ mol } L^{-1}$$

$$pH = -\log_{10}[H^+] = -\log_{10} 2.19 \times 10^{-6} = 5.66$$

2.1.1 Causes of Acid Rain

A simplified illustration of the acid rain phenomenon is given in Figure 2.2. Acid rain originates from the conversion of pollutant gases emitted from the chimney stacks of coal- and oil-burning power stations, smelters, refineries, chemical plants and motor vehicles. In the past, SO_2 was the major contributor to acid rain; however, the contribution of NO_x to acid rain has been increasing steadily. Nowadays the contribution of these two pollutants to rainwater acidity is roughly equal, owing to the increasing number of cars on the roads and the reduction in SO_2 emissions.

Sulfur dioxide gas is oxidised in the atmosphere to sulfuric acid, which is readily absorbed by rain. The oxidation proceeds *via* a variety of mechanisms. In the gas phase, SO_2 is oxidised predominantly by hydroxyl ($OH^\bullet$) radicals. Sulfur dioxide may also dissolve in cloud droplets according to the following equilibria to produce dissolved SO_2, bisulfite (HSO_3^-) and sulfite (SO_3^{2-}) ions:

$$SO_2 + H_2O \rightleftharpoons SO_2 \cdot H_2O \qquad K_H = [SO_2 \cdot H_2O]/pSO_2$$

$$SO_2 \cdot H_2O \rightleftharpoons HSO_3^- + H^+ \qquad K_1 = [HSO_3^-][H^+]/[SO_2 \cdot H_2O]$$

$$HSO_3^- \rightleftharpoons SO_3^{2-} + H^+ \qquad K_2 = [SO_3^{2-}][H^+]/[HSO_3^-]$$

where K_H is Henry's law constant for SO_2 and K_1 and K_2 are the first and second dissociation constants. At pH values typical of rainwater (3–6), dissolved SO_2 is present mainly as the bisulfite ion (Figure 2.3). The total of dissolved species is designated by S(IV):

$$[S(IV)] = [SO_2 \cdot H_2O] + [HSO_3^-] + [SO_3^{2-}]$$

In cloud droplets, S(IV) can be oxidised to sulfate (SO_4^{2-}) by dissolved hydrogen peroxide (H_2O_2) and ozone (O_3). In polluted air, trace metals such as iron and manganese may catalyse the oxidation of SO_2 by oxygen in cloud drops. The reaction is pH dependent: the rate of oxidation increases with increasing pH. Oxides of nitrogen, produced during combustion, are converted to nitric acid in the atmosphere, mainly by gas phase reactions. Hydrochloric acid, present in power station plumes at low concentrations, makes only a minor contribution to acid rain.

Natural processes can also give rise to rainwater acidity. Many of the precursor gases are produced during volcanic eruptions and forest fires. Also, reduced sulfur compounds such as hydrogen sulfide and dimethyl

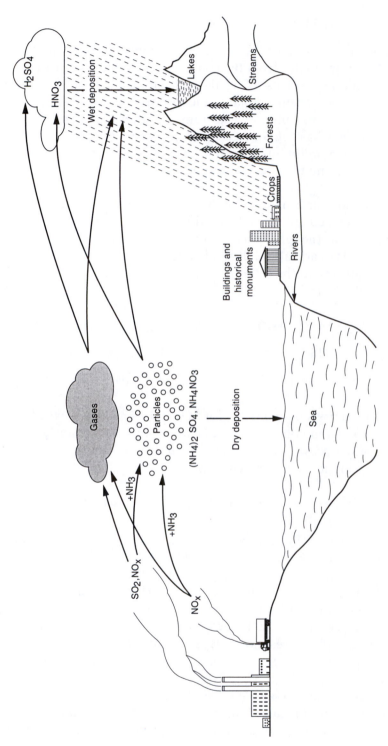

Figure 2.2 *Illustration of the pathways of acid rain*

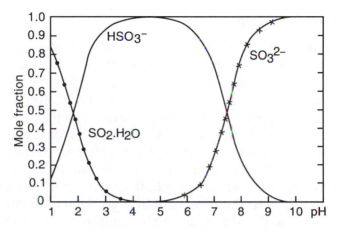

Figure 2.3 *Mole fraction of S(IV) species as a function of pH*

sulfide are emitted by plankton and by biological degradation in coastal and inland marshes and soils. These compounds are oxidised to SO_2 in the atmosphere. In addition, weak organic acids such as formic and acetic acids, originating from natural sources, can make a significant contribution to rainwater acidity, especially at unpolluted rural and remote sites. Currently, about half of the global emissions of sulfur and NO_x compounds originate from anthropogenic sources. Some 68% of anthropogenic emissions of SO_2 and NO_x originate in Europe and North America, but the contribution of developing countries, especially in Asia, is rising. Plumes of pollutants can travel hundreds or even thousands of kilometers from their source before being converted to acid rain. This so-called *long-range transport of air pollutants* (LRTAP) has become a major political issue because of the trans-boundary movement of pollution between countries.

Sulfur compounds are removed from the atmosphere by deposition processes. Gases dissolve in cloud droplets, whereas dust particles are incorporated into cloud droplets either as *cloud condensation nuclei* (CCN), around which the droplets grow, or by scavenging mechanisms. These processes are referred to as *rainout*. The incorporation of dust particles and gaseous molecules into falling raindrops is called *washout*. Both rainout and washout are termed *wet deposition*. *Dry deposition* involves sedimentation of dust particles and diffusion of gaseous molecules onto surfaces. Dry deposition processes are generally more important close to the source of pollution.

2.1.2 Effects of Acid Rain

Since the initial link between acid rain and dying fish was made some 30 years ago, a number of other harmful effects of acid rain have been uncovered. Some of these effects, given in Table 2.1, have been conclusively proven, while others still remain a source of controversy due to the poor understanding of some of the mechanisms of damage.

Acid rain can acidify lakes and rivers, leading to the mobilization of soluble aluminium which, together with low pH, has been implicated in fish deaths, although the mechanism is far from simple. The toxic effects are dependent upon the species of fish, the concentration of calcium in the water and the pH. Acid rain can also accelerate the corrosion of limestone and iron, thus damaging important historical monuments and other building structures; it may adversely affect plant growth, leading to forest decline and crop damage; and it could even lead to increasing levels of toxic metals in water supplies.

Table 2.1 *The harmful effects of acid rain*

Effect	Mechanism	Evidence
Fish deaths	Toxicity is mainly due to the release of Al from sediments. The Al causes clogging of the gills with mucus.	Salmon and brown trout have disappeared from thousands of weakly buffered lakes and rivers in Scandinavia, Canada, the USA and lochs in Scotland.
Forest decline	Poorly understood. Various hypotheses have been proposed, one of which is that acid rain releases toxic Al from soil which damages the roots and interferes with the uptake of nutrients. Other contributing stresses include drought, cold, disease and other pollutants (*e.g.* O_3).	First observed in Germany in mid-1970s. Nearly 25% of all the trees in Europe have been classified as damaged. Typically, fir trees lose needles and the crowns of Norway spruce become thinner and needles turn brown. Deciduous trees are also affected.
Building damage	SO_2 reacts with limestone ($CaCO_3$) to form gypsum ($CaSO_4 \cdot H_2O$), which is more soluble and can readily be washed off by rain.	Air pollution damage of historical monuments has been documented in Europe, North America and Asia. Notable cases are the city of Cracow in Poland, the Acropolis in Athens and the Taj Mahal in India.
Health damage	Accelerating heavy metal migration in food chains and accumulation of heavy metals in drinking water and foodstuffs.	?

Recently, the concept of *critical loads* has been developed in order to quantify and assess the potential of damage to sensitive receptor areas. A critical load is the threshold concentration of a pollutant at which harmful effects begin to be observed and it is usually determined using sophisticated models. Weakly buffered soils, such as those found in Scandinavia, exhibit particularly low critical loads for acidity. Maps of critical loads of various regions of the world are being prepared in order to predict effects and assist in the control of acidic pollution.

2.1.3 Control of Acid Rain

As a consequence of the widespread scientific and public concerns over the harmful effects of acid rain, national governments and international organisations have instituted numerous research studies of acid rain, and legal and technological measures have been taken to reduce the problem.

The United Nations Economic Commission for Europe (UNECE) Convention on Long-Range Transboundary Air Pollution, which went into force in 1983, aims to control acid rain through the reduction of sulfur emissions. The UNECE protocol signed in Helsinki in 1985 required signatories to reduce annual sulfur emissions by 30% from their 1980 values by 1993. An even stricter protocol was adopted in 1994, requiring reductions of between 13% and 70% of 1980 levels to be implemented in stages by 2000, 2005 and 2010. The European Union issued a directive in 1988 requiring SO_2 reductions from 40% to 60% by 1998 and 50% to 70% by 2003, with reference to 1980 emissions. Also, it required the reduction of NO_x emissions by 40% by 1998. Similar legislation was also passed in the USA.

Emissions of precursor gases can be reduced by taking measures before, during or after fuel combustion. These measures include:

- Fuel switching (*i.e.* changing from high-sulfur coals to low-sulfur coals or natural gas).
- Fuel desulfurisation (*e.g.* washing of coal to remove sulfur, coal gasification).
- Novel combustion technologies (*e.g.* fluidised bed combustion (FBC), low-NO_x burners).
- Flue gas desulfurisation (*e.g.* scrubbing of the flue gas with limestone slurry, seawater or some other solution that absorbs SO_2).
- Selective reduction of NO_x to N_2 (*e.g.* injection of NH_3 into the flue gases over a bed of catalyst).
- Three-way catalytic converters to reduce NO_x emissions from motor vehicles.

Introduction of these control measures has led to reduction in emissions in many of the developed nations; however, emissions in many developing countries are rising. Post-emission controls include liming of affected lakes and soils to neutralise the acidity. This does not, however, represent a long-term solution to the problem since liming has to be repeated as the water becomes acidic again with time.

2.2 SAMPLING AND ANALYSIS

The growing concern over acid rain has led to the establishment of many national and international monitoring programmes that analyse rainwater on a regular basis. One such world-wide network is operated as part of the Global Atmosphere Watch (GAW) by the World Health Organisation (WHO). The main chemical parameters which are determined in these surveys are: pH, conductivity, sulfate, nitrate, chloride, ammonium, sodium, potassium, calcium and magnesium concentrations. Increasingly, organic acids (formic and acetic) and trace metals are also being determined. Common techniques of acid rain analysis are summarised in Table 2.2. Concentration ranges of the various ions in rainwater are given in Table 2.3.

Although rainwater analysis may appear to be fairly straightforward, it is in fact quite problematic. Sources of error common with other types of environmental samples are present and these are greatly exacerbated because of the low concentrations of ions present in rainwater. Many

Table 2.2 *Common techniques of rainwater analysis*

Analytical technique	Analyte
Conductimetry	Conductivity
Ion chromatography (IC)	Cl^-, NO_3^-, SO_4^{2-}, NH_4^+, Na^+, K^+, Mg^{2+}, Ca^{2+}, formate, acetate
Flame atomic absorption spectroscopy (FAAS)	Na^+, K^+, Mg^{2+}, Ca^{2+}
Flame atomic emission spectroscopy (FAES)	Na^+, K^+, Mg^{2+}, Ca^{2+}
Inductively coupled plasma (ICP)	Na^+, Mg^{2+}, Ca^{2+}, trace metals
Graphite furnace atomic absorption spectroscopy (GFAAS)	Trace metals
Ion selective electrodes (ISE)	H^+, Cl^-, NO_3^-, Na^+, K^+, NH_4^+, some trace metals
Voltammetry	Trace metals

Table 2.3 *Concentrations of major ions in rainwater*

Ion	Concentration (mg L^{-1})
Chloride	0.02–60
Nitrate	0.1–20
Sulfate	0.1–30
Sodium	0.02–30
Potassium	0.02–2
Magnesium	0.005–2
Calcium	0.02–4
Ammonium	0.03–4

rainwater samples, especially at remote locations, are almost as pure as laboratory grade waters. Therefore, great care must be taken when handling, storing and analysing rainwater samples in order to maintain sample integrity.

Sources of error in acid rain analysis include:

- contamination by biological materials (*e.g.* insects) during sampling
- evaporation from samples
- absorption/desorption of gases during sampling or in the laboratory
- chemical or biological reactions during sampling or storage
- interaction with bottle materials
- inadvertent contamination during sample handling, treatment or analysis

Ammonia and organic acids are especially prone to reaction during storage and the use of additives such as chloroform or thymol is recommended for preserving samples for these analyses.

Rainwater samples can be of two types: *wet-only* or *bulk*. Wet-only samples are collected using specialised automatic samplers which open only during the shower and close during dry periods, therefore eliminating contamination by falling dust particles during dry periods. Samples collected with such samplers are termed *wet deposition* or *rainwater-only* samples. Originally, rainwater was collected into buckets or funnels inserted in bottles, so-called *bulk collectors*, and these samplers are still in use today, especially at remote or inaccessible sites. These collectors sample not only rainwater (wet deposition) but also dust particles and gases during dry periods between showers (dry deposition) and are said to collect *bulk deposition*. These samplers have the advantage that they do not require electric power and they can be placed at remote and inaccessible locations. The effect of dry deposition may be minimized by exposing the collectors in the field just before the shower and removing them just after the shower. However, in many instances it is desirable to sample both the wet and dry deposition in order to get an estimate of the total deposition of a substance. In this case, bulk samplers are left exposed in the field for periods from one day to one month. In many surveys it is common to leave the samplers exposed a week at a time. However, the use of bulk samplers to evaluate total deposition has been criticized, and it is recommended that the dry deposition be evaluated on the basis of airborne concentrations.

The pH of rainwater is one of the most difficult measurements to perform accurately, and doubts have been expressed about many measurements reported in the literature. This is illustrated in Table 2.4, which summarises the measurements of pH in a dilute acid solution as carried out by the National Bureau of Standards in the USA. Even in these controlled experiments performed by well-trained technicians under best conditions, the measurements vary by as much as 50% when expressed as H^+ concentrations. Problems of pH analysis are due to:

- difference in ionic strength between samples and buffers used to standardize pH electrodes
- stirring errors
- absorption of laboratory gases (*e.g.* NH_3, CO_2)
- chemical reactions (*e.g.* dissolution of alkaline dust particles in suspension)
- biological reactions (*e.g.* degradation of weak acids such as formic and acetic acids)

Table 2.4 *Response of new, research grade combination pH electrodes in 10^{-4} mol L^{-1} HCl (from a study by the National Bureau of Standards, Washington)*

| Electrode | Recorded pH value | | |
	Quiescent	Stirred	Difference
1	3.944	3.816	0.128
2	3.864	3.661	0.203
3	3.908	3.741	0.167
4	3.899	3.841	0.058
5	3.875	3.755	0.120
Range	0.080	0.180	0.145

Adapted from Koch and Marinenko (1983).

2.3 DATA ANALYSIS AND INTERPRETATION

Theoretically, the sum of cations must equal the sum of the anions when expressed in equivalents, if all the components have been analysed correctly. This is called the *electroneutrality balance* and it can be used as a check on the validity of an analysis:

$$\Sigma\text{cations} = \Sigma\text{anions}$$

If the ratio of Σcations/Σanions is less than 0.85 or greater than 1.15 the data are considered questionable, in which case the results are rejected and the sample analysed again. When performing an ion balance calculation it is important to use the H^+ concentrations based on pH measurements made at the time of the chemical analysis and not on field pH values or pH measurements made after sample collection, unless the chemical analyses were performed immediately upon collection. This is because there can be considerable changes in pH during storage as a result of reaction.

An additional check on the analysis involves calculating the conductivity of the sample on the basis of the measured concentrations:

$$\kappa_c = \Sigma\lambda_i C_i$$

where κ_c is the conductivity in μmho cm^{-1}, λ_i is the equivalent conductance of the ith ion adjusted for units (see Table 2.5) and C_i is the concentration in μeq L^{-1} of the ith ion. The calculated conductivity

Table 2.5 *Equivalent conductance at infinite dilution and 25 °C adjusted for the units specified*

Ion	Equivalent conductance (μmho cm^{-1}) per μeq L^{-1}
H^+	0.35
Cl^-	0.076
NO_3^-	0.071
SO_4^{2-}	0.08
NH_4^+	0.074
Na^+	0.050
K^+	0.074
Mg^{2+}	0.053
Ca^{2+}	0.060
HCO_3^-	0.044

Table 2.6 *US EPA criteria for re-analysis of*
samples

Measured conductance (μmho cm^{-1})	%CD
<5	> ± 50
5–30	> ± 30
>30	> ± 20

is compared to the measured value, κ_m, to give the conductivity difference, %CD:

$$\%\mathrm{CD} = 100(\kappa_c - \kappa_m)/\kappa_m$$

The US EPA criteria for rejection of results and re-analysis of samples are given in Table 2.6. See also Example 2.2.

Samples collected in rainwater surveys generally exhibit a negative relationship between ion concentration and precipitation amount. This typical dilution effect is illustrated in Figure 2.4 for one component.

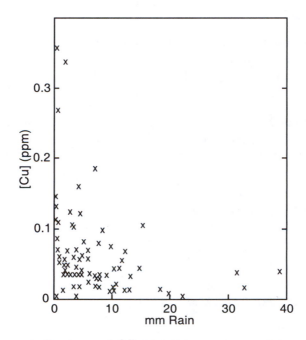

Figure 2.4 *Concentration versus rainfall amount*

Example 2.2

A rainwater sample was analysed and the following concentrations determined (in mg L^{-1}): 0.34 Na^+, 0.14 K^+, 0.08 Mg^{2+}, 0.44 Ca^{2+}, 0.78 Cl^-, 0.72 NO_3^-, 0.86 SO_4^{2-} and 0.015 NH_4^+. The pH was 5.03 and the conductivity 10.9 μmho cm^{-1}. Calculate the conductivity and compare to the measured conductivity. Does the sample need to be re-analysed?

First of all, convert the concentrations to units of μeq L^{-1} as follows:

$$(\mu eq\ L^{-1}) = 1000 \times (mg\ L^{-1})/\text{equivalent mass}$$

where the equivalent mass is the same as the ionic mass for monovalent ions, but half the ionic mass for divalent ions, as illustrated below:

	H^+	Na^+	K^+	Mg^{2+}	Ca^{2+}	Cl^-	NO_3^-	SO_4^{2-}	NH_4^+
mg L^{-1}	–	0.34	0.14	0.08	0.44	0.78	0.72	0.86	0.015
eq mass	1.01	22.99	39.10	12.15	20.04	35.45	62.01	48.04	18.05
μeq L^{-1}	9.33	14.79	3.58	6.58	21.96	22.00	11.61	17.90	0.83

Calculate the conductivity, κ_c by multiplying the concentrations in μeq L^{-1} with the appropriate factor in Table 2.5:

$$
\begin{aligned}
\kappa_c = \Sigma\lambda_i \times C_i = &\ (9.33 \times 0.35) + (14.79 \times 0.05) + (3.58 \times 0.074) \\
&+ (6.58 \times 0.053) + (21.96 \times 0.06) + (22.00 \times 0.076) \\
&+ (11.61 \times 0.071) + (17.9 \times 0.08) + (0.83 \times 0.074) \\
= &\ 9.93\ \mu\text{mho cm}^{-1}
\end{aligned}
$$

Comparing to the measured value we can calculate the conductivity difference in % as:

$$\%CD = 100(\kappa_c - \kappa_m)/\kappa_m = 100(9.93 - 10.9)/10.9 = -8.9\%$$

The US EPA criteria (see Table 2.6) stipulate that if %CD is $> \pm 30\%$ for samples having measured conductivities in the range of 5–30 μmho cm^{-1}, they should be re-analysed. Since, for the sample in this example, %CD = -8.9%, we can conclude that there is good agreement between measured and calculated conductivities and the results can be accepted as valid.

Generally, low rainfall amounts result in high concentrations, whereas heavy rains result in more dilute samples. Also, initial stages of rainfall exhibit higher concentrations of ions than later stages owing to the initial washout of the dusts and gases in the atmosphere below cloud base. A

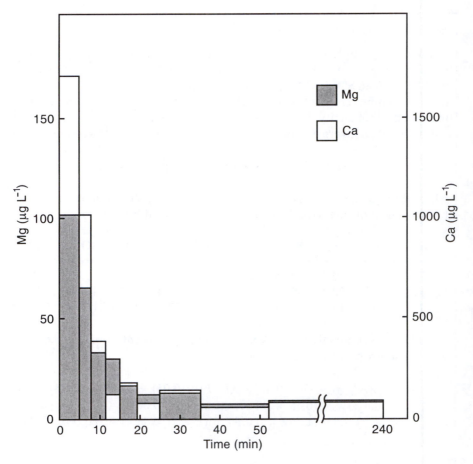

Figure 2.5 *Concentrations of some ions in rainwater during a tropical rainstorm as a function of time from the beginning of the shower*

typical profile of some ions in one particular storm is shown in Figure 2.5 illustrating this effect. The lower concentrations associated with later stages of rainstorms are generally considered to reflect cloudwater concentrations as they result primarily from rainout.

2.3.1 Questions and Problems

1. Discuss the chemical processes that lead to the formation of acid rain.
2. Discuss the effects of acid rain.
3. How can the acid rain problem be minimised?

4. Why is liming not an effective long-term solution to surface water acidification?

5. Outline the main analytical techniques and sampling methods that are used in rainwater surveys. List the major analytes and suggest appropriate analytical techniques for each analyte.

6. Calculate the concentrations of HSO_3^- and SO_3^{2-} in equilibrium with 5 ppbv SO_2 at: (a) pH 4, (b) pH 7 and (c) pH 9. Use the following equilibrium constants for SO_2: K_H = 1.24 mol L^{-1} atm^{-1}, K_1 = 1.74 × 10^{-2} mol L^{-1} and K_2 = 6.24 × 10^{-8} mol L^{-1}.

7. Calculate the pH of water in equilibrium with 10 ppbv SO_2 using the equilibrium constants given in the previous question.

8. Convert the following concentrations to μeq L^{-1}:
   - (a) 3.5 mg L^{-1} SO_4^{2-}
   - (b) 25 mg L^{-1} PO_4^{3-}
   - (c) 18 mg L^{-1} NO_3^-
   - (d) 98 μg L^{-1} Ca^{2+}

9. A rainwater sample was analysed and found to have the following composition in mg L^{-1}: 1.47 Na^+, 1.53 K^+, 0.29 Mg^{2+}, 2.56 Ca^{2+}, 0.39 Cl^-, 2.01 NO_3^-, 0.38 SO_4^{2-} and 0.12 NH_4^+. The pH of the sample was 5.01 and the conductivity was 17.2 μmho cm^{-1}. Check the correctness of the analysis by means of the electroneutrality balance and conductivity calculation. Should the results of the analysis be rejected or not?

10. Bicarbonate (HCO_3^-) is not normally included in checks of validity as it is not normally determined in rainwater. Calculate the concentration of HCO_3^- in the above example (question 9) assuming the sample is in equilibrium with atmospheric CO_2. What difference would inclusion of HCO_3^- in the electroneutrality balance and the conductivity calculation make to your conclusions (use equilibrium constants given in Example 2.1).

2.4 pH, CONDUCTIVITY AND MAJOR ANIONS

2.4.1 Methodology

The pH and conductivity are determined immediately on fresh samples using electrochemical methods. Major ions may be determined in stored samples. In the present experiment, chloride, nitrate and sulfate concentrations are determined by ion chromatography (IC) at the same time as the pH and conductivity. The samples are stored at 4 °C in a refrigerator for subsequent cation analysis by atomic absorption spectrometry (AAS).

2.4.2 Bottle Preparation

2.4.2.1 Materials
- Polypropylene bottles (any size between 150 mL and 1 L)

2.4.2.2 Experimental Procedure. Take several polypropylene bottles (150–1000 mL), label them and weigh them on an open-top balance. Record the weights of the empty bottles. Before sampling rainwater it is necessary to clean the sampling bottles. Take the weighed 150 mL polypropylene bottles and repeatedly rinse them with laboratory grade water. Then fill the cleaned bottles with laboratory grade water to the brim and cap. Some bottles may be filled with water and analysed as blanks. You should take these blank bottles into the field and back to the laboratory, and store in the same way as your samples. Analysis of these blanks may reveal just how rigorous you are in your work; if any contamination shows up, it means that you are being sloppy!

Although chemical cleaning reagents such as chromic acid or DECON 90 may be used for cleaning bottles, extreme care must be taken to avoid contamination of the bottles by the cleaning reagent itself. Owing to the great purity of many rainwater samples, even minute traces of contaminants could lead to errors.

2.4.3 Sampling

2.4.3.1 Materials
- Polypropylene funnel (15–25 cm in diameter) and bottle, or plastic bucket
- Retort stand and clamp

2.4.3.2 Experimental Procedure. The main objective is to obtain a representative sample and to avoid any contamination. Find a suitable sampling location, giving consideration to the following criteria. The sampler should be placed in an open area with no obstructions above or near the sampler which may impede the flow of falling rain, or contaminate the sample (*i.e.* do not place sampler under, or near trees, near tall buildings, next to walls, *etc.*). Good locations for placing samplers are flat roofs of building or in a clearing such as a field. The sampler should be placed at a height of 1 m above the surface if possible.

Take a 150–1000 mL sampling bottle, empty out the water and shake it to dislodge any droplets of water. Insert into the bottle the polypropylene funnel and expose the sampler in the field at your chosen sampling site. You may have to use a retort stand and clamp, or some other method, to hold the bottle and funnel firm and prevent the sampler from toppling over in the wind. If you do not have a funnel, you may use a bucket to collect the sample and transfer it to a bottle after collection. Be sure to note exactly the time and date when you expose the sampler in the field, as well as the time and date when you remove the sampler from the site or change over the sampling bottle. Ideally, you should expose the sampler just before a shower and bring it indoors immediately after the shower, but this may not be always practical. During very heavy showers the bottle may fill up very quickly. In this case, do not allow the bottle to overflow but keep changing the bottles over, recording exactly the time of exposure of each bottle. During slight drizzle you may have to expose the sampler though several rain events to collect sufficient sample volume for analysis (*i.e. ca.* 100 mL). Remember to note the location and time of sampling, as well as any other observations which may help you in interpreting your results (*e.g.* meteorological factors relating to the time of your sampling as reported in the local newspapers or on television; visual observations of nearby forest fires; proximity of any air pollution sources; presence of insects or bird droppings in your sample; presence of dust particles in your sample, *etc.*). You may use larger sampling bottles (1 or 2 L) and leave the bottles exposed for longer periods (1 day, 1 week, *etc.*) if you are carrying out long-term deposition surveys rather than studying individual rain events. If you are using an automatic wet-only collector to sample rain, then follow the appropriate operating instructions.

Ideally, you should bring the collected samples immediately to the laboratory for pH and conductivity analysis. If this is not possible, leave the sampling bottle in your home refrigerator until such time as you can take it to the laboratory (keeping samples in a refrigerator used for food is not a good idea if you are carrying out a research project into organic contaminants in rainwater). Ensure that your samples are labelled in such a way that they will not be confused with samples of other students.

Before carrying out any analysis, weigh the bottle with the rainwater sample on an open-top balance. From the difference between the weights of the bottle containing the sample and the same bottle when it was empty, calculate the volume of rainwater collected.

2.4.4 Conductivity

2.4.4.1 Materials
- Conductivity meter
- Potassium chloride

2.4.4.2 Experimental Procedure. Prepare a standard solution for calibration of the conductivity meter as follows. Weigh out accurately 5.1 g of KCl and dilute to one litre with pure water in a 1 L volumetric flask. Dilute 10 mL of this solution to 1 L with pure water in a 1 L volumetric flask. This is your working standard, which has a conductivity of 100 μmho cm^{-1}.

Read the instruction manual for the conductivity meter or ask the demonstrator to explain the operation of the instrument. Using the conductivity meter, carry out the measurements as follows. Place the working KCl standard solution in a small beaker and suspend the conductivity cell in the solution, holding it approximately 1.5 cm above the bottom of the beaker and making sure that it is not in contact with any of the beaker walls. Adjust the conductivity reading to 100 μmho cm^{-1}. Rinse the cell with pure water, and carry out the measurement on your sample in the same way. You do not need to calibrate the meter between each sample but you must rinse the cell between each reading. Also analyse water blanks. These should have conductivities no higher than your pure water. Always measure the conductivity before the pH. Leakage of electrolyte from the pH electrode can cause a significant change in the conductivity of highly pure, low conductivity, samples. When not in use store the electrode by immersing in laboratory water.

2.4.5 pH

2.4.5.1 Materials
- Buffer solutions, pH 4 and 7
- pH meter, preferably with combination pH electrode and temperature compensation probe
- Magnetic stirrer and stirring bar

2.4.5.2 Experimental Procedure. Ideally, the pH measurement should be made immediately after sample collection but this may not always be possible in practice. Use the pH meter and electrode as follows. First, calibrate the combination pH electrode and meter using a two-point calibration with buffer solutions of pH 7 and 4. A technician or demonstrator will show you how to operate the pH meter; otherwise follow the instructions in the instrument manual. After calibration, immerse the pH electrode in the rainwater sample and take a measurement. You may swirl the solution but allow it to come to rest before taking the measurement. Alternatively, you may place a magnetic stirring bar in the solution and stir for 15 s on a magnetic stirrer. After turning the stirrer off, allow the pH to stabilise (*ca.* 2 min) before taking the reading.

You may notice that in some solutions the pH value drifts. In case of drift, allow the reading to stabilise before recording the pH value. If the reading does not stabilise, and the electrode performance is satisfactory, record the pH value reached after 5 min and report this. The drift could be due to chemical changes going on in the sample and not due to a faulty electrode. You can test the performance of the electrode by measuring the pH of a dilute acid solution (*e.g.* HCl or H_2SO_4) of pH 4 and noting the pH measured and the response time. Report the pH to two decimal places. Rinse the electrode with laboratory water and swab dry between each reading. When taking readings, suspend the electrode in the test solution, making sure that it does not touch the walls of the beaker. When not in use, leave the electrode soaking in laboratory water or buffer solution. New electrodes should be cleaned with a weak acid solution (0.01 M) and subsequently stored in pH 4 buffer. If the electrode response is sluggish, rejuvenate it according to the directions given in Appendix II.

2.4.6 Major Anions (Chloride, Nitrate, Sulfate)

2.4.6.1 Materials
- Ion chromatograph (*e.g.* Shimadzu) equipped with conductivity detector, and integrator or chart recorder.
- Anion separator column (*e.g.* Shim-pack IC-A1).
- Syringe and microfilters.
- Stock solutions of Cl^-, NO_3^-, SO_4^{2-}, 1000 mg L^{-1} each. These can be purchased from suppliers; otherwise prepare your own by dissolving following amounts of salts in water and making up to 1 L: 1.6485 g NaCl, 1.3707 g $NaNO_3$ and 1.8141 g K_2SO_4. Dry the salts in an oven at 110 °C for 2 h before weighing.
- Mobile phase: 2.5 mM phthalic acid, 2.5 mM tris(hydroxymethyl)-aminomethane, pH = 4.0. Filter through a Millipore HATF (0.45 μm) filter before use.

2.4.6.2 Experimental Procedure. In the example given here, ions are separated using non-supressed ion chromatography. Prepare and filter the mobile phase for ion chromatography. A technician or demonstrator will turn on the ion chromatograph and integrator for you, and make all the necessary adjustments. Typical operating conditions with one particular type of instrument (Shimadzu) are given in Table 2.7. You may have to optimise the chromatographic and conductivity detector conditions to obtain suitable peaks. The anion separator column will vary, depending on the brand of instrument that you will use. The Shim-pack IC-A1 column is for use with a Shimadzu ion chromatograph.

Table 2.7 *Ion chromatograph and operating conditions*

Shimadzu Ion Chromatograph HIC-6A
Shimadzu SCL-6B System Controller
Shimadzu LC-6A Liquid Chromatography Pump
Shimadzu CDD-6A Conductivity Detector

Shim-pack IC-A1 column
Flow rate $= 1.5$ mL min^{-1}
Gain $= 1$ μS cm^{-1}

Dilute 1000 mg L^{-1} stock solutions to prepare mixed calibration standards, each containing the following concentrations of Cl$^-$, NO$_3^-$ and SO$_4^{2-}$: 0.1 mg L^{-1}, 0.2 mg L^{-1}, 0.5 mg L^{-1}, 0.8 mg L^{-1} and 1.0 mg L^{-1}. (*N.B.* You may have to prepare more concentrated standards if your samples contain higher levels of these ions.) First analyse a pure water blank, then analyse the calibration standards. You should filter a small portion of your rainwater sample using a microfilter and syringe and inject it into the injection port. One method of injection is to inject 1 mL of sample into a 200 μL sample loop set in the "load" position and then switch to the "inject" setting to sweep the sample into the ion chromatgraph. A typical chromatogram of a rainwater sample is shown in Figure 2.6. If your samples are too concentrated and the peaks are off-scale, dilute the samples and analyse once again. Measure the peak heights and prepare calibration graphs of peak height against concentration for each ion. Estimate the concentration in your samples by interpolation from the calibration graph. Also analyse the water blanks to check the purity of your laboratory water.

Note

If using a different chromatography system, follow the instructions in the *methods book* supplied by the manufacturer regarding the column and mobile phase you should use. If you are analysing anions using suppressed chromatography on a Dionex ion chromatograph equipped with the AS4A column, prepare a 2.8 mM NaHCO$_3$ + 2.2 mM Na$_2$CO$_3$ solution to use as the mobile phase.

2.5 MAJOR CATIONS

2.5.1 Alkali Metals and Alkaline Earth Metals (Na, K, Mg, Ca)

2.5.1.1 Methodology. Stored rainwater samples which were previously analysed for pH, conductivity and anions are analysed for alkali and

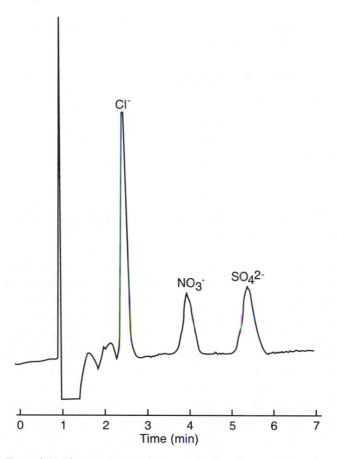

Figure 2.6 *Typical ion chromatogram of a rainwater sample*

alkaline earth metals (Na, K, Mg, Ca) by atomic absorption spectroscopy (AAS) in an air/acetylene flame.

2.5.1.2 Materials
- Atomic absorption spectrometer
- Nitric acid (HNO_3)
- Stock solutions of Na, K, Ca and Mg, 1000 mg L^{-1} each. These can
  be purchased from suppliers; otherwise you can prepare your own
  solutions from soluble salts by dissolving the following quantities of
  salts in water and diluting to 1 L: 2.542 g NaCl, 1.907 g KCl, 2.497 g
  $CaCO_3$, 4.952 g $MgSO_4$. Add about 10 mL concentrated HCl and
  1.5 mL concentrated HNO_3 to the Ca and Mg solutions, respectively, before making up to volume. Dry the salts at 110 °C for 2 h
  before weighing (especially important for NaCl)

Table 2.8 *Recommended wavelengths, detection limits and working range of AAS (air/acetylene flame)*

Element	Wavelength (nm)	Detection limit (mg L^{-1})	Working range (mg L^{-1})
Na	589.0	0.002	0.03–2
K	766.5	0.005	0.1–2
Ca	422.7	0.003	0.2–20
Mg	285.2	0.0005	0.02–2

- Lanthanum solution. Dissolve 67 g of lanthanum chloride ($LaCl_3.7H_2O$) in 1 M HNO_3 by gently warming. Cool and dilute to 500 mL with pure water

2.5.1.3 Experimental Procedure. A technician or demonstrator should turn on the flame atomic absorption spectrophotometer and show you how to use the instrument. You may also consult the instrument manual for specific details of the operating parameters. Recommended wavelengths, detection limits and working ranges are given in Table 2.8.

Prepare mixed standards containing the following concentrations (in mg L^{-1}) of ions by dilution from 1000 mg L^{-1} standards: 0.1, 0.2, 0.5, 0.8, 1.0, 1.5 and 2.0 mg L^{-1}. One mL of dilute HNO_3 should be added to the calibration standards before diluting to the mark in a 100 mL flask. Analyse the nitric acid using atomic absorption spectroscopy for each of the metal ions to ensure that it is free from impurities. Analyse the standards and rainwater samples by aspirating into the flame of the AAS instrument equipped with the appropriate hollow cathode lamp and operating at the appropriate wavelength for the metal being analysed. Record the absorbance readings or, better still, obtain peak heights on a chart recorder connected to the AAS instrument. Aspirate water between each sample and standard, allowing the baseline to stabilise. Aspirate each sample or standard long enough to obtain a steady reading. Prepare calibration graphs of peak height (or absorbance) against metal concentration and evaluate the concentrations of the ions in rainwater by reading off from the calibration graph.

Notes

1. In case any of the samples are too concentrated and their peaks register off-scale, dilute and analyse once more. On the other hand, you may prepare more concentrated standards.
2. Ca and Mg determination is prone to interference by phosphate and therefore you should repeat the determination as follows. Measure

out 50 mL portions of the working standard solutions prepared above and to each of these add 5 mL of the lanthanum solution. Also measure out 50 mL of the rainwater sample and add 5 mL of the lanthanum solution (if less than 50 mL of sample is available, decrease the amount of lanthanum solution to give the same volume ratio of 10:1). Analyse at the Ca wavelength and plot a calibration graph. If the concentration determined differs significantly from the one determined above, there was interference in the first analysis so reject the Ca data from the first run. Repeat for Mg.

3. If an atomic absorption spectrometer is not available in your laboratory, but a flame photometer is, you may use the flame photometer to determine Na, K, Mg and Ca. Flame photometry is based on atomic emission and is suitable for these analyses. Use the same standards as you would have used in atomic absorption analysis.

4. You may also use inductively coupled plasma atomic emission spectrometry (ICP-AES) if it is available in your laboratory. However, the ICP may not be sensitive enough for K determination in some rainwater samples as the detection limit for K is 100 μg L^{-1}. The instrument should be set up by a technician. Analyse standards and samples in the same way as you would if using AAS.

2.5.2 Ammonium

2.5.2.1 Methodology. Ammonium ion is determined by colorimetry using the indophenol blue method. This is based on the reaction of phenol hypochlorite and ammonium to give a blue indophenol dye described by Weatherburn (1967). The reaction is accelerated by the addition of sodium nitroprusside. The colour intensity is determined using a spectrophotometer at 625 nm.

2.5.2.2 Materials
- Spectrophotometer.
- Reagent A. Dissolve 5 g phenol and 25 mg sodium nitroprusside in water and dilute to 500 mL. Store in dark bottles in a fridge. This can be used for up to one month.
- Reagent B. Dissolve 2.5 g sodium hydroxide and 4.2 mL sodium hypochlorite (5% chlorine) in water and dilute to 500 mL. Age the solution for a few days prior to use. Store in dark bottles in a fridge. This can be used for up to one month.
- Stock ammonium solution, 1000 mg L^{-1}. This is commercially available. Otherwise you can prepare a 100 mg L^{-1} ammonium

stock solution by dissolving 0.367 g $(NH_4)_2SO_4$ in water and making up to the mark in a 1 L flask. Dry salt in oven at 110 °C for 2 h before weighing.

2.5.2.3 Experimental Procedure. Place 0.5 mL of the rainwater sample into a test tube and add 5 mL of solution A followed by 5 mL of solution B. Cover the mouth of the tube with parafilm and shake vigorously. Leave test tube in rack and place in a warm water bath at 37 °C for exactly 15 min or at room temperature for 30 min. Measure the absorbance of the resulting blue-green colour in a 1 cm cell at 625 nm on a UV/visible spectrophotometer with water in the reference cell. Repeat the same with a reagent blank (*i.e.* 0.5 mL of laboratory water + 5 mL reagent A + 5 mL of reagent B). Subtract the absorbance of the reagent blank from the absorbances of the samples and standards. Prepare a series of NH_4^+ standard solutions in volumetric flasks with concentrations in the range of 0.1–10 μg mL^{-1} by dilution from a 1000 or 100 mg L^{-1} stock solution. Place 0.5 mL of each of the calibration standards in a series of test tubes, add 5 mL of solution A followed by 5 mL of solution B and proceed as with the sample. Read off the concentration of NH_4^+ in the rainwater sample directly from the calibration graph.

Notes

1. You may increase the volume of sample taken for analysis if the concentration of ammonium is too low. Instead of 0.5 mL you may use 1, 2 or 5 mL of sample. In this case, prepare calibration standards over a lower concentration range and increase the volume of each standard analysed to correspond to the volume of sample taken for analysis.
2. You may determine ammonium using an ammonia ion selective electrode if one is available in your laboratory. Follow the specific instructions that come with the electrode. Plot the potential (mV), determined in a series of ammonium standards, against the ion concentration, as shown for the fluoride ion electrode in Section 4.13, and read off the concentration in the sample.

2.6 HEAVY METALS

You may determine heavy metals (*e.g.* Pb, Fe, Mn, Cu, Cd, Co, Cr, Zn) using the AAS technique as described for water samples in Section 4.16. You will have to acidify samples immediately upon collection. Follow exactly the given procedure.

2.7 EXERCISES AND INFORMATION

2.7.1 Questions and Problems

1. Prepare a table summarising all your measurements. Calculate the arithmetic average and volume weighted mean of the pH and the ionic concentrations (see Example 2.3).

Example 2.3

Calculate the volume-weighted mean pH of the following six rainwater samples:

pH	4.61	5.94	5.16	5.60	4.49	5.06
Rainfall amount (mm)	34.1	13.6	2.7	1.6	32.0	69.8

In rainwater surveys it is common practice to calculate the volume weighted mean (VWM) of ionic concentrations rather than the arithmetic average. The VWM takes into account the effect of dilution by the rainfall amount and it is useful in comparative studies of acid rain. The volume weighted mean is calculated from:

$$\text{VWM of X} = \Sigma([X_i] \times R_i)/\Sigma R_i$$

where $[X_i]$ is the concentration of substance X, and R_i is the rainfall amount. The sample volume may be used instead of the rainfall amount if all the samples are collected using funnels of the same diameter.

To calculate the VWM of the pH, the pH values must first be converted to H^+ concentrations which are then used in the above equation, and the calculated VWM of H^+ concentrations is then converted back to pH units.

$$\Sigma R_i = 34.1 + 13.6 + 2.7 + 1.6 + 32.0 + 69.8 = 153.8 \text{ mm}$$

VWM of H^+ =

$$\frac{\Sigma[(10^{-4.61} \times 34.1) + (10^{-5.94} \times 13.6) + (10^{-5.16} \times 2.7) + (10^{-5.60} \times 1.6) + (10^{-4.49} \times 32.0) + (10^{-5.06} \times 69.8)]}{153.8}$$

$$= 2.52 \times 10^{-3}/153.8 = 1.64 \times 10^{-5} \text{ mol L}^{-1}$$

The VWM of the pH $= -\log_{10} 1.64 \times 10^{-5} = 4.78$

The arithmetic average of the pH values is 5.14, which is quite different from the volume-weighted mean.

2. Assuming a concentration of CO_2 in air of 0.036%, calculate the concentrations of carbonate and bicarbonate in each of your samples at 25 °C. Compare these concentrations to those of the other ions and indicate whether carbonate and bicarbonate would contribute significantly to the anion concentration in your rainwater samples (use equilibrium constants given in Example 2.1).

3. Calculate the mm of rain for each of your samples using the expression:

$$\text{mm rain} = 10 \times \frac{\text{Volume of sample (cm}^3)}{\text{Funnel area (cm}^2)}$$

4. Calculate and tabulate the deposition rates of H^+, and other anions, cations and heavy metals that you have determined. The deposition rate (or flux) is defined as:

$$\text{Deposition rate} = \frac{[\text{Ion concentration}] \times \text{volume of sample}}{\text{Funnel area} \times \text{sampling period}}$$

Remember to be consistent in your choice of units. If you express the sample volume in cm^3 (*i.e.* mL) and the funnel area in cm^2, then you should express the concentration in either $mol\ cm^{-3}$ or $g\ cm^{-3}$. If your sampling period is expressed in h, then your calculated deposition rate would come out in either $mol\ cm^{-2}\ h^{-1}$ or in $g\ cm^{-2}\ h^{-1}$. The sampling period, and deposition rate, may also be expressed in days or weeks, depending on the length of the sampling period. What does the deposition rate tell you and why is it useful?

5. Is there any relationship between the mm of rain and the concentrations of the individual ions that you have determined? Draw scatterplots for each ion.

6. Consider all the data you have gathered. Is there any relationship between the different ions which you have measured? Calculate correlation coefficients between pairs of chemical parameters and prepare a correlation matrix. Discuss the likely causes of these relationships.

7. Discuss the possible sources of various anions, cations and heavy metals in rainwater.

8. In rainwater surveys it is common to express the concentration of ions in terms of a *non-sea salt* or *excess* fraction. This represents that fraction of the ion originating from sources other than the rainout and washout and sea salt particles (*e.g.* anthropogenic pollution or

other natural sources such as volcanoes or continental dust particles). The sea salt fraction, X_{SS}, of a particular ion is defined as:

$$X_{SS} = (X_{SEA}/R_{SEA}) \times R_{RAIN}$$

Where X_{SEA} is the concentration of ion X in bulk seawater, R_{SEA} is the concentration of the reference ion (usually Na^+) in bulk seawater and R_{RAIN} is the concentration of the reference species (*i.e.* Na^+) in rainwater. The excess or non-sea salt fraction, X_{NS}, is then defined as:

$$X_{NS} = X_T - X_{SS}$$

where X_T is the concentration of ion X measured in the rainwater (*i.e.* total concentration = sea salt + non-sea salt). When calculating sea salt and non-sea salt contributions, be sure to employ concentrations expressed in units of equivalents per litre and not moles per litre. Although there is no difference between the two units for monovalent ions, there is a difference for divalent ions.

Use your measurements to calculate the sea salt and non-sea salt fractions of sulfate, chloride, potassium, magnesium and calcium using Na^+ as the reference and the following ratios for seawater (estimated on the basis of concentrations in seawater expressed in eq L^{-1}):

X_{SEA}/R_{SEA}	= Ratio value
SO_4^{2-}/Na^+	= 0.12
Cl^-/Na^+	= 1.16
K^+/Na^+	= 0.021
Ca^{2+}/Na^+	= 0.044
Mg^{2+}/Na^+	= 0.227

Discuss your results in terms of the likely origin of the various ions.
9. Calculate the conductivity for your samples on the basis of the measured ionic concentrations. Compare these values to the measured conductivity values and assess the validity of your analysis. Analyse again any samples that do not meet the criteria of acceptance.
10. Calculate the electroneutrality balance for your samples and assess the validity of your analysis. Analyse again any samples that do not meet the acceptance criteria.

11. Draw a typical ion chromatograph and describe the function of each component.
12. Draw a typical atomic absorption spectrometer and describe the function of each component.
13. Outline the principles of atomic absorption spectrometry.
14. Outline the principles of ion chromatography.

2.7.2 Suggestions for Projects

1. Investigate the geographical distribution in rainfall composition by collecting samples at various sites simultaneously. Illustrate the sampling sites and results on a map of the area.
2. Compare rainwater composition at an urban site with an adjacent rural site.
3. Investigate the effect of a forest canopy on the composition of rain. In this case, place one sampler under the canopy of a tree and the other sampler in an open field. The rain falling through the canopy is called *throughfall*. Analyse both rainwater and throughfall samples and compare the results. What conclusions can you draw about the interaction of rainwater with the tree canopy? Investigate throughfall under different tree species. Is the canopy acting as a source or sink of ions in rainwater reaching the ground?
4. If wind direction data are available, attempt to correlate the composition of various components with the wind direction and attempt to identify sources of the chemical components in rain.
5. Samples may be collected on a continuous basis throughout the year. In this case a computerised data base should be set up for students to enter the results of their measurements. Seasonal effects and trends in rainwater chemistry may be investigated by inspecting the data base.
6. If you live in the tropics, or if you experience heavy rainfall in your location, study the change in chemical composition of rain during individual rainstorm events. Keep changing the sampling bottle at regular intervals during the rain event, or change the bottle as soon as you have sufficient sample for analysis. After analysing the samples, plot the concentrations against time from the commencement of the rain event to obtain a profile.
7. Determine the composition of snow using the same analytical techniques. Collect fresh snow and analyse in the laboratory after allowing it to thaw.

2.7.3 References

W. F. Koch and G. Marinenko, Atmospheric deposition reference materials: measurement of pH and acidity; paper presented at the 76th Annual Meeting of the Air Pollution Control Association, Atlanta, GA, June 19–24, 1983.

M. W. Weatherburn, Phenol–hypochlorite reaction for determination of ammonia, *Anal. Chem.*, 1967, **39**, 971–974.

2.7.4 Further Reading

M. Radojević and R. M. Harrison (eds.), "Atmospheric Acidity: Sources, Consequences and Abatement", Elsevier, Amsterdam, 1992.

G. Howells, "Acid Rain and Acid Waters", Ellis Horwood, New York, 1990.

F. B. Smith, An overview of the acid rain problem, *Meteorol. Mag.*, 1991, **120**, May, 77–91.

J. R. Kramer, Analysis of precipitation, in "Handbook of Air Pollution Analysis", 2nd edn., eds. R. M. Harrison and R. Perry, Chapman and Hall, London, 1986, pp. 535–561.

F. Pearce, Acid rain, in "The New Scientist Inside Science", Penguin, London, 1992, pp. 149–159.

M. Radojević, Taking a rain check, *Anal. Eur.*, 1995, August, 35–38.

V. Mohnen, The challenge of acid rain, *Sci. Am.*, 1988, **259**, August, 14–22.

CHAPTER 3

Air Analysis

3.1 INTRODUCTION

3.1.1 The Atmosphere

Air is the most important of the vital substances required for life. Human beings can survive for up to a month without food, up to one week without water, but deprived of air they can survive no longer than a couple of minutes. We breathe between 10 and 25 m^3 of air daily and any toxins present are also inhaled. Clean air is therefore a necessity for human life.

The air required for sustaining life is contained in the lowest section of the atmosphere and it is this air at ground level that we will be mainly concerned with here. The troposphere, located below about 10 km, contains the air that we breathe and it is the location of weather processes. The lowest 1 km of the atmosphere, the so-called *boundary layer*, is where most of the air pollution takes place. The composition of air at sea level is given in Table 3.1.

In fact, most chemicals can be found in the atmosphere; even a radioactive element such as radon is found in the atmosphere, albeit at a minuscule concentration of 6×10^{-18}%. In addition to gases, the atmosphere also contains varying concentrations of suspended particles originating from sea spray, wind-blown desert dusts, forest fires, volcanic eruptions, *etc.* Another section of the atmosphere of interest to envir-

Table 3.1 *Major components of air at sea level*

Gas	Concentration (% by volume)
Nitrogen	78.09
Oxygen	20.95
Argon	0.93
Carbon dioxide	0.036
Water	0–3

onmentalists is the stratosphere. The so-called ozone layer, which prevents harmful UV rays from reaching the earth's surface, is located in the stratosphere.

3.1.2 Air Pollution

Throughout this century the quality of the air has been steadily deteriorating, giving rise to numerous cases of illness and even death. In fact, human beings have been exposed to air pollution ever since the discovery of fire, and air quality in towns throughout antiquity and the Middle Ages was probably poor. However, during this century, air pollution has evolved to the point where it not only affects the health of the population in the cities, it has also become a major global problem with the potential to change dramatically the climate, and geography, of the entire planet. Major air pollution concerns of the present day are given in Table 3.2.

The distinction between some of these problems is not always clear-cut. For example, urban smog experienced by many cities may be a combination of sulfurous and photochemical smog, and pollutants associated with both photochemical and sulfurous smog are also the cause of acid rain. Chlorofluorocarbons, which destroy the ozone layer, are also strong absorbers of infrared radiation and hence contribute to global warming, as does SO_2 and many other air pollutants.

Air pollutants are released into the atmosphere from *stationary*

Table 3.2 *Major air pollution concerns*

Problem	Cause	Effects
Sulfurous smog	SO_2 and smoke from industrial and domestic sources	Respiratory disease, reduced visibility, damage to materials and vegetation
Photochemical smog	Exhaust gases from motor vehicles	Damage to health, materials, vegetation
Global warming	CO_2 from power stations and CH_4 from paddy fields	Rising temperatures, climate change, flooding of low-lying areas
Depletion of O_3 layer	Chlorofluorocarbons from aerosol cans, refrigerators, *etc.*	Increasing incidence of skin cancer due to increasing penetration of UV radiation
Acid rain	SO_2 and NO_x from power stations and motor vehicles	Damage to aquatic and terrestrial ecosystems, and materials
Haze	Pollutants from forest fires	Reduced visibility, health effects

Table 3.3 *Major anthropogenic sources of air pollution*

Source	Pollutants
Power stations	CO_2, SO_2, NO_x, particulates
Other industry	CO_2, SO_2, hydrocarbons, particulates
Motor vehicles	CO, NO_x, Pb, hydrocarbons, particulates
Domestic	Particulates

sources, such as power stations and other industries, and from *mobile sources*, such as motor vehicles, airplanes and ships. Major sources of pollution are listed in Table 3.3. Many of the air pollutants also have natural sources. Typical examples are volcanoes, which emit large quantities of dust particles, SO_2, HCl and CO_2; bogs and marshes, which emit H_2S; and lightning, which generates NO_x. For many pollutants the natural sources emit much more than anthropogenic sources when averaged over the globe. However, anthropogenic emissions of SO_2 and NO_x are almost equal to those from natural sources.

Combustion of fossil fuels (coal and oil) for electricity generation is a major source of air pollution. The combustion process may be described in terms of the following simplified reactions:

$$3HC \text{ (fuel)} + 2O_2 \rightarrow CO_2 + H_2O + CO + HC$$

$$S \text{ (fuel)} + O_2 \rightarrow SO_2$$

$$N_2 \text{ (air)} + O_2 \rightarrow 2NO$$

Under ideal conditions, combustion should produce only CO_2 and H_2O; however, incomplete combustion also releases CO and unburnt hydrocarbons. Because sulfur is present in coal (0.2–7%) and oil (0.3–4%), it is oxidised to SO_2 during combustion. Nitrogen is also present in fuels; however, most NO results from atmospheric N_2 dissociation and oxidation at high temperature during combustion (the Zeldovich mechanism). Typical concentrations found in flue gases of coal-fired power stations are: 12% CO_2, 4.5% H_2O, 1500 ppmv SO_2, 100 ppmv CO, 500 ppmv NO, 40 ppmv N_2O, 20 ppmv NO_2, 150 ppmv HCl and 20 ppmv HF. Many other pollutants are also produced during combustion: heavy metals, hydrocarbons and smoke. Natural gas is a considerably "cleaner" fuel but it still generates CO_2 and NO_x.

Generally, what goes up must come down, so that eventually most of the pollution is removed at the earth's surface. How long a particular pollutant remains in the air depends on its *atmospheric residence time*. Residence times for some pollutants are given in Table 3.4. These are

Table 3.4 *Residence times of some atmo-spheric constituents*

Species	Residence time
O_2	10^4 years
N_2	10^6 years
CO_2	5–10 years
CO	0.1 years
SO_2	2–7 days
H_2O	10 days
NO_2	4 days
CH_4	5 years

determined mainly by the physicochemical properties of the pollutants, their reactivity, solubility, diffusion coefficient, *etc.*

Pollutants which are directly emitted from a chimney stack or an exhaust pipe are called *primary* pollutants. Typical examples are SO_2 from chimney stacks, and CO, NO and hydrocarbons emitted by motor vehicles. Primary pollutants can react in the atmosphere to produce *secondary* pollutants. For example, SO_2 can be converted to sulfuric acid, while emissions from motor vehicles can react in sunlight to produce ozone and other oxidants. Quite often, secondary pollutants can be more harmful than the primary pollutants.

Concentrations of air pollutants vary considerably, depending on the proximity of the pollution source, the source strength, meteorological conditions and the reactivity of the atmosphere. The most important air pollutants are SO_2, NO_x (NO + NO_2), CO, O_3 and atmospheric aerosol. These are called *criteria pollutants* because of their potential harmful effects on the health of the public. In the USA these pollutants are regulated by National Ambient Air Quality Standards (NAAQS). NAAQS and standards in other countries are listed in Appendix III. These standards are set in order to protect the health of the public. Typical ranges of pollutant concentrations at urban, rural and remote sites are given in Table 3.5.

3.1.3 Urban Smog

Urban centres have suffered from *smog* episodes ever since the beginning of the Industrial Revolution. The early incidences of urban smog in nineteenth century England were mainly due to hydrogen chloride, a highly corrosive gas emitted by the alkali industries (see Section 3.7). The Alkali Acts of the 1860s and 1870s quickly eliminated this problem.

In the first half of the twentieth century, urban smog was caused

Table 3.5 *Concentrations of air pollutants*[a]

	Concentration		
Pollutant	*Urban*	*Rural*	*Background*
SO_2	10–500	5–50	*ca.* 1
NO_x (NO + NO_2)	10–400	5–50	1–3
CO	2000–50 000	20–200	20–200
O_3	50–500	40–100	10–20
Particulates	20–300	10–60	0–10

[a] All in ppbv except particulates in $\mu g\,m^{-3}$.

mainly by SO_2 and smoke in combination with fog. In fact, the term smog is a contraction of the words *smoke* and *fog*. Smoke and SO_2 were emitted by various industries and from household chimneys due to the domestic burning of coal for heating. This type of smog has variously been called *sulfurous*, *wintertime* or *classical* smog, and it was quite common in London where it was known as the *pea-souper*. This smog greatly reduced visibility and damaged public health. Most affected were the young, the elderly and those suffering from respiratory diseases. Smog incidents often resulted in increasing admissions to hospitals and even deaths. Some of the more severe incidents are listed in Table 3.6. Excess deaths were assigned to bronchitis, pneumonia and cardiac disease and were recorded mainly among the elderly. The infamous London smog of December 1952 killed some 4000 people. During the episode there was a high degree of correlation between SO_2 and smoke, and both were correlated with excess deaths. Since the 1960s the occurrence of sulfurous smog has been virtually eliminated from cities in Western Europe and North America. This has largely been due to the banning of domestic coal use, building of taller chimney stacks, relocation of industries to the outskirts, introduction of control measures such as low sulfur coals and natural gas, *etc.* All this has resulted in reduction of levels of soot and SO_2 in urban centres. However, some urban centres

Table 3.6 *Selected air pollution disasters*

Date	*Place*	*Attributed mortality*
December 1930	Meuse Valley, Belgium	63
October 1948	Donora, Pennsylvania, USA	20
December 1952	London, England	4000
November 1953	New York City, USA	200
December 1962	London, England	700

in Eastern Europe and the developing world still experience sulfurous smog.

Many urban areas today suffer from another type of air pollution problem, the so-called *photochemical* or *summertime* smog. This smog is due mainly to the emissions from motor vehicles. The number of cars in the world is increasing at a phenomenal rate; it has grown tenfold since 1950 and it is expected to double in the next 20–30 years. Exhaust gases from motor vehicles contain CO, NO and hydrocarbons which react in the sunlit atmosphere to produce NO_2, O_3, peroxyacetyl nitrate (PAN), atmospheric aerosols and other compounds. The conversion of primary pollutants emitted by vehicles to secondary pollutants characteristic of smog episodes is accomplished by photochemical reactions initiated by sunlight. Many of these reactions involve free radicals such as the hydroxyl ($OH^•$) radical, which readily reacts with many compounds in the atmosphere. Photochemical smog was first observed in Los Angeles in the 1940s and is nowadays a daily occurrence in many cities throughout the world. In more temperate climates it is experienced mainly during the summer, whereas in tropical and subtropical climates it can be observed all year round. Athens, Bangkok, Mexico City and many other cities are covered with an almost permanent blanket of photochemical haze characterised by the reddish-brown hue of NO_2. An idealised daily cycle of pollutants characteristic of photochemical smog is shown in Figure 3.1. The effects of smog are most prominent at mid-

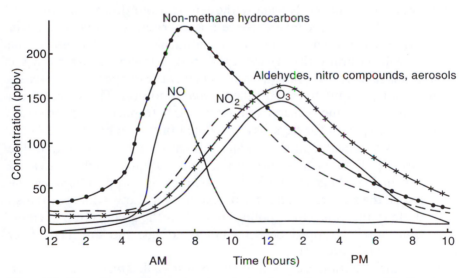

Figure 3.1 *Diurnal variation in primary and secondary pollutants responsible for photochemical smog*

day, when solar intensity is highest, and the smog tends to recede at night. Photochemical smog can reduce visibility, although not to the same extent as sulfurous smog. It can also damage materials, cause eye irritation and, most importantly, it can seriously affect the health of urban residents, especially the young, the elderly and those suffering from respiratory disease. It is widely believed that photochemical smog can exacerbate or provoke asthma attacks. Although measures are being taken to combat photochemical smog, through the introduction of catalytic converters, control of traffic and the greening of cities by building parks and planting trees, we are still a long way from solving the problem.

Lead pollution was widely present in urban environments in the past owing to its use as a petrol additive. However, as a result of the introduction of unleaded petrol, lead concentrations have decreased in many cities, especially in the developed countries. Lead pollution still remains a problem in some cities in developing countries. Even in cities where leaded petrol has been phased out several years ago, lead pollution has not been completely eliminated. Studies have shown that lead emitted by motor vehicles over many decades in the past remains in the urban environment in the form of settled dust which can readily be re-suspended into the atmosphere.

The occurrence of air pollution episodes also depends on the prevailing weather conditions and the topography. Many of the severe episodes have taken place during a low-lying *inversion*, which restricts the dispersion of pollution. Normally the temperature decreases with altitude in the lower troposphere. Since hot air rises, and since air lower down is warmer than the air above, pollution released close to the ground will naturally rise. However, during a low-lying inversion the situation is reversed. Within an inversion layer the temperature increases with altitude, leading to the so-called *stable* condition. The air below an inversion layer, being cooler, cannot rise and is trapped. Sometimes the condition can persist for days, as during the London smog of 1952. Also, many cities are located in valleys and this can further contribute to the build-up of pollutants, as shown in Figure 3.2. Figure 3.3 illustrates the situation in Los Angeles, where the pollution is trapped between the mountains, the sea-breeze and the inversion layer. Low wind speeds also contribute to the build-up of pollution, as does the architecture of cities. Pollution can be trapped between tall buildings, the so-called *street canyon* effect (Figure 3.4).

Urban air pollution is especially severe in so-called *megacities*, defined as cities with current or projected populations of 10 million or more by the year 2000. There are 24 such megacities; these are the generally

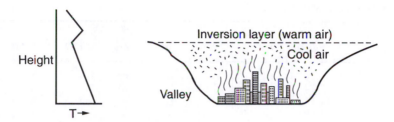

Figure 3.2 *Effect of topography on the build-up of pollutants in valley*

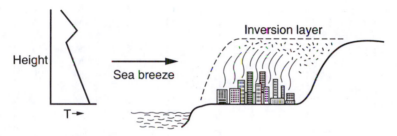

Figure 3.3 *Effect of topography on the build-up of pollutants in Los Angeles*

Figure 3.4 *Effect of the street canyon on the build-up of pollutants in a city*

Table 3.7 *Overview of air quality in 20 megacities*[a]

Megacity	Population in 1990 (million)	SO_2	NO_2	O_3	CO	SPM	Pb
Bangkok	7.16	L	L	L	L	S	M
Beijing	9.74	S	L	M	–	S	L
Bombay	11.13	L	L	–	L	S	L
Buenos Aires	11.58	–	–	–	–	M	L
Cairo	9.08	–	–	–	M	S	S
Calcutta	11.83	L	L	–	–	S	L
Delhi	8.62	L	L	–	L	S	L
Jakarta	19.42	L	L	M	M	S	M
Karachi	7.67	L	–	–	–	S	S
London	10.57	L	L	L	M	L	L
Los Angeles	10.47	L	M	S	M	M	L
Manila	8.40	L	–	–	–	S	M
Mexico City	19.37	S	M	S	S	S	M
Moscow	9.39	–	M	–	M	M	L
New York	15.65	L	L	M	M	L	L
Rio de Janeiro	11.12	M	–	–	L	M	L
Sao Paulo	18.42	L	M	S	M	M	L
Seoul	11.33	S	L	L	L	S	L
Shanghai	13.3	M	–	–	–	S	–
Tokyo	20.52	L	L	S	L	L	–

[a] Taken from UNEP and WHO (1994).
Key: S = serious problem; WHO guideline exceeded by more than a factor of two. M = moderate to heavy pollution; WHO guideline exceeded by up to a factor of two (short-term guidelines exceeded on a regular basis at certain locations). L = low pollution; WHO guidelines are normally met (short-term guidelines may be exceeded occasionally). – = No data available or insufficient data for assessment.

recognised large metropolitan areas (Los Angeles, New York, London, Shanghai, *etc.*). A recent World Health Organisation (WHO) survey of air quality in 20 of these megacities concluded that "urban air pollution is a major environmental health problem deserving high priority for action". The survey compared the concentrations of major air pollutants (SO_2, NO_2, O_3, CO, Pb and SPM) in the megacities with WHO guidelines (see Appendix III). Results of this WHO survey, conducted in 1992, are summarised in Table 3.7. While SO_2 and lead levels have been decreasing in many of the megacities, several megacities still continue to have serious SO_2 and lead pollution problems. High suspended particulate matter (SPM) was identified as a major problem followed by ozone (see Example 3.1). Many megacities do not have adequate monitoring facilities, as indicated by the gaps in the table. The WHO concluded that only 6 of the 20 megacities surveyed have satisfactory monitoring networks.

Example 3.1

Assuming steady state, derive an equation relating the ozone concentration to the concentrations of NO and NO_2 in the atmosphere.

NO is oxidised to NO_2 by reaction with O_3:

$$NO + O_3 \rightarrow NO_2 + O_2$$

and the rate of this reaction is given by:

$$d[NO_2]/dt = -d[NO]/dt = k\,[NO]\,[O_3]$$

where k is the second-order reaction rate constant and $d[NO_2]/dt$ is the rate of NO_2 formation.

NO_2 is photodissociated in sunlight:

$$NO_2 \rightarrow NO + O$$

The rate of this reaction is:

$$-d[NO_2]/dt = J\,[NO_2]$$

where J is the first-order photodissociation constant.

Atomic oxygen reacts with O_2 to form ozone:

$$O + O_2 \rightarrow O_3$$

At steady-state:

$$\text{Formation of } NO_2 = \text{Destruction of } NO_2$$

$$d[NO_2]/dt = -d\,[NO_2]/dt$$

$$k[NO]\,[O_3] = J\,[NO_2]$$

and:

$$[O_3] = J\,[NO_2]/k[NO]$$

In order for high O_3 concentrations to form, both high values of J and high NO_2/NO ratios are required. Values of J are highest around noon when sunlight is most intense.

3.1.4 Haze

In recent years there has been a steadily increasing incidence of regional haze episodes due to smoke from forest fires. *Haze* is defined as the presence of very small particles (less than a few micrometres in diameter) in the air, normally associated with dry, sunny weather, resulting in reduced visibility and *haziness*. Although urban photochemical smog can also give rise to haziness, the cause of regional haze and its chemical characteristics are different from urban smog. The increasing frequency of haze in tropical and subtropical regions is due to a combination of socio-economic (*slash and burn* farming practices, clearing of rainforests for development, *etc.*) and climatological (dry weather due to the El Nino phenomenon) factors. The haze can spread over several countries at a time owing to the long-range transport of smoke particles from source areas. The pollutants emitted by forest fires can sometimes travel several thousand kilometres, and the haze can persist for several months at a time. Visibility decreases dramatically during haze episodes, to less than a few km and at times even down to less than 100 m. Recent haze episodes have occurred in South East Asia, Central America, southern USA and Brazil. The haze could have potentially serious health effects on the large populations exposed to unusually high pollutant levels. During extremely severe haze episodes the concentration of particles less than $10 \, \mu$m in diameter can climb to $1 \, \text{mg m}^{-3}$ or more, much higher than levels associated with urban photochemical smog. In addition to soot particles, which can contain many adsorbed organic compounds such as polynuclear aromatic hydrocarbons (PAHs), numerous gases are also emitted by forest fires.

3.1.5 Control of Air Pollution

Reduction of air pollution centres around controlling emissions from power plants and other industries, and from motor vehicles. As the primary pollutants responsible for urban smog (SO_2 and NO_x) are the same as those responsible for acid rain, essentially the same technologies are employed.

The quality of the air we breathe is regulated by means of ambient air quality standards. WHO guidelines and typical standards for several countries are given in Appendix III. Emissions from industries and motor vehicles are regulated by means of emission standards. Strategies and technologies can be adopted to reduce the emissions of pollutants. In any case, it is necessary to measure the pollution in order to assess compliance with the standards.

Common technologies of air pollution control include:

- *Flue gas desulfurisation.* Many methods are available but the most common method involves absorption of SO_2 in the flue gas into a limestone slurry to produce gypsum, a usable product.
- *De-NO_x.* Ammonia is injected into the flue gas to reduce NO_x species to N_2 with, or without, the presence of a catalyst.
- *Catalytic converters.* Motor vehicle exhaust gases are passed over a three-way catalyst to reduce emissions of CO, NO_x and hydrocarbons.
- *Dust control technologies.* Many different methods are available to control dust emissions, such as fabric filters, cyclones, electrostatic precipitators, *etc.*

3.1.6 Concentration Units

Various units are used to express the concentration of air pollutants, including %, atm, ppmv, ppbv, $\mu g\,m^{-3}$, and molecules cm^{-3}. Aerosol concentrations are generally expressed in mass per volume units (*e.g.* $\mu g\,m^{-3}$). Gas concentrations are expressed in units of mass per volume or as volume mixing ratios. For example, a mixing ratio of 1 part per million (ppmv) of CO implies that, in one million volumes of air, one volume consists of CO, if the CO were separated from the rest of the air. That is, in 1 m^3 of air, 1 mL would be occupied by CO (1 mL per 10^6 mL). Similarly, a mixing ratio of 1 part per billion (ppbv) implies that, in 1 billion (10^9) volumes of air, 1 volume is occupied by the pollutant gas, or one $\mu L\,m^{-3}$ (*i.e.* 10^{-3} mL per 10^6 mL). At one atmosphere total pressure, 1 ppm = 10^{-6} atm and 1 ppb = 10^{-9} atm. It is usual to indicate that the mixing ratio is on a volume by volume basis, *i.e.* ppmv, ppmV or ppm (v/v) so as to distinguish it from the ppm unit in solution, which is on a mass by mass basis.

At *standard temperature and pressure* (STP), defined as 1 atm (760 mmHg) pressure and 273 K (0 °C) temperature, the mixing ratio can be related to the mass by volume units by means of the following equation:

$$\mu g\,m^{-3} = 1000 \times ppmv \times molecular\ mass/22.41$$

where 22.41 is the volume, in litres, occupied by one mole of gas at STP. However, measurements are seldom made at 273 K (0 °C) and it is necessary to calculate the volume occupied by one mole of gas at the temperature and pressure appropriate to the measurement (see Example 3.2). At 25 °C the volume occupied by 1 mole of gas is 24.5 L.

Example 3.2

The concentration of NO was measured in the flue gas from a coal-fired power station and found to be 490 ppmv at 120 °C and 765 mmHg. Convert this to mass by volume units (molecular mass of NO = 30.01).

$$1 \text{ mol of ideal gas at STP (760 mmHg, 273 K)} = 22.41 \text{ L}$$

Use the relationship $P_1 V_1 / T_1 = P_2 V_2 / T_2$ to calculate the volume occupied by 1 mole of gas at 120 °C and 765 atm.

$$V_2 = P_1 V_1 T_2 / T_1 P_2 = 760 \times 22.41 \times 393/273 \times 765 = 32.05 \text{ L}$$
$$490 \text{ ppmv} = 490 \times 30.01/32.05 = 459 \text{ mg m}^{-3}$$

3.1.7 Measurement of Air Pollution

Air pollution monitoring can be of two types: *ambient monitoring* or *source monitoring*. Ambient monitoring is the determination of pollutant concentrations in the ambient air and it is carried out in order to assess the quality of the air that we breathe. Many cities and towns are equipped with at least one, and often with several, air quality monitoring stations. *Source testing* is the determination of concentrations in industrial flue gases or motor vehicle exhausts and it is used to assess compliance with emission standards. Results of these measurements are often reported as emission rates or emission factors.

Most air quality monitoring stations (AQMS) are nowadays equipped with instruments capable of measuring real-time concentrations of pollutants. Instruments most commonly used for continuously measuring criteria gases at AQMSs are based on the following techniques:

- CO: non-dispersive infrared (NDIR) spectroscopy
- O_3: UV absorption spectroscopy
- SO_2: UV fluorescence spectroscopy
- NO_x: chemiluminescence

These instruments can record concentrations as low as 1 ppbv with very fast response times (seconds to minutes). Modern AQMSs are highly automated with computer control and data storage, requiring minimum technical supervision and maintenance. A number of stations covering an entire city or region can form part of a monitoring network in which individual stations can be linked to a central station. Many stations are nowadays linked to the Internet, providing regular updates on pollution levels in their respective areas.

3.1.8 Reporting Air Quality

The air pollutant concentrations measured at air quality monitoring stations are compared to the air quality standards in force, and if the air quality is not satisfactory then alerts, warnings or emergencies may be declared. The public is kept informed about the ambient air quality through bulletins in the media (television, radio, newspapers), the Internet and via digital displays along major roads. In cases of severe pollution episodes the public may be advised to stay indoors, and the authorities may take measures to ameliorate the situation such as temporarily closing down certain offending industries or limiting traffic.

In the US, the Pollutant Standards Index (PSI) is widely used to provide information to the public about daily pollution levels. The PSI was developed by the US EPA on the basis of extensive studies of the health effects of urban air pollution carried out for more than 25 years. Some other countries have also adopted the US PSI, while some have developed their own air quality indices. In the US, each of the criteria pollutants is measured, and its concentration is converted to a PSI sub-index value using a conversion chart. The highest sub-index value is then reported as the PSI. Different averaging times are used for different pollutants when computing the PSI: 24 h running average for PM_{10}, 8 h running average for CO, 1 h average for O_3, 1 h average for NO_2 and 24 h running average for SO_2. The relationship between the PSI and pollutant concentration is a segmented linear function with breakpoints, *i.e.* straight lines can be drawn between the breakpoints indicated in Table 3.8. There is no PSI for NO_2 values below 200 because in the US there is no short-term National Ambient Air Quality Standard (NAAQS) for NO_2. The PSI extends on a scale from 0 to 500. As there is no prior experience in the US with PSI values > 500, the PSI is not extended beyond this value. On the basis of the PSI the public can determine whether air pollution levels are good (0–50), moderate (50–100), unhealthy (100–200), very unhealthy (200–300) or hazardous (> 300). Also, the PSI enables the authorities to decide what measures to take in order to protect public health and reduce the level of pollution. The relationship between PSI values, pollutant concentrations, health effects and preventive measures is summarised in Table 3.8.

The Department of the Environment, Transport and the Regions (DETR) in the UK categorises air pollution according to the concentrations of SO_2, O_3 and NO_2 as shown in Table 3.9.

Table 3.8 *The Pollutant Standards Index (PSI)*

PSI	Notification stage	Pollutant breakpoint concentration ($\mu g\,m^{-3}$) (except CO in $mg\,m^{-3}$)						Health effects	Health advisory	Preventive action
		SO_2	NO_2	O_3	CO	PM_{10}	TSP^a			
50	–	80	–	120	5	50	75	Few or none	None	None
100	–	365	–	235	10	150	260	Mild aggravation of symptoms in susceptible persons. Irritation symptoms in the healthy population.	Persons with existing heart or respiratory ailments should reduce physical exertion and outdoor activity.	Potential restrictions on industrial activities.
200	Alert	800	1130	400	17	350	375	Significant aggravation of symptoms and decreased exercise tolerance in persons with heart or lung disease. Widespread symptoms in the healthy population.	Elderly and persons with existing heart and lung disease should stay indoors and reduce physical activity.	Restrict incinerator use. Restrict open burning of leaves and refuse.
300	Warning	1600	2260	800	34	420	625	Premature onset of some disease in addition to significant aggravation of symptoms and decreased exercise tolerance in healthy persons.	Elderly and persons with diseases should stay indoors and avoid physical exertion. General population should avoid outdoor activity.	Prohibit incinerator use. Severely curtail power plant operations. Cut back operations in manufacturing industries. Require public to limit driving by using car pools and public transportation.

Table 3.8 (*continued*)

PSI	Notification stage	Pollutant breakpoint concentration ($\mu g\,m^{-3}$) (except CO in $mg\,m^{-3}$)						Health effects	Health advisory	Preventive action
		SO_2	NO_2	O_3	CO	PM_{10}	TSP^a			
400	Emergency	2100	3000	1000	46	500	875	Death could occur in some sick and elderly people, and even healthy people would likely experience symptoms that would necessitate restrictions on normal activity.	All persons should remain indoors, keeping windows and doors closed. All persons should minimise physical exertion.	Prohibit most industrial and commercial activity. Prohibit almost all private use of motor vehicles.
500	Significant harm	2620	3750	1200	57.5	600	1000	–	–	–

ᵃ TSP = total suspended matter. TSP is no longer used in evaluating the PSI; PM_{10} is used instead.

Table 3.9 *UK DETR guide values for air pollution*[a]

Pollutant	Level of pollution			
	Low	Moderate	High	Very high
O_3	< 50	50–89	90–179	≥ 180
SO_2	< 100	100–199	200–399	≥ 400
NO_2	< 150	150–299	300–399	≥ 400
PM_{10}	< 50	50–74	75–99	≥ 100

[a]In ppbv except PM_{10} in $\mu g\ m^{-3}$.

3.1.9 Questions and Problems

1. Differentiate between the following pairs of concepts:
   (a) primary pollutant and secondary pollutant
   (b) classical smog and photochemical smog
   (c) point source and area source
2. Would you expect the concentration of a substance in the atmosphere to depend on its residence time? Elaborate.
3. List the criteria pollutants and their major sources.
4. Why are pollutants such as carbon dioxide and chlorofluorocarbons not considered as criteria pollutants?
5. Which control strategies would you suggest to reduce the frequency and intensity of photochemical smog in urban areas?
6. An instrument used to determine the concentration of SO_2 in ambient urban air gave a result of 35 ppbv. The measurement was made at sea-level (1 atm) and the air temperature was 22 °C. Convert this reading to mass by volume units.
7. Convert the following mass by volume concentrations to volume mixing ratios at STP:
   (a) $32\ \mu g\ m^{-3}\ NO_2$
   (b) $150\ \mu g\ m^{-3}\ O_3$
   (c) $25\ \mu g\ m^{-3}\ C_6H_6$
8. What is the Pollutant Standards Index (PSI) and how is it determined?
9. Using the data in Table 3.8 plot graphs of PSI against pollutant concentration for each of the criteria pollutants.

3.1.10 Suggestions for Projects

1. Carry out a study of air pollution in your area (town/city) to include:
   - source inventory
   - review of existing pollution control technologies
   - review of pertinent national and local legislation

First of all, compile a list of all the sources of air pollution and then try to identify and quantify the individual pollutants. Contact various industrial plants in your area and request information relevant to compiling the inventory (fuel consumption, fuel composition, composition of flue gases, average emission rates, *etc.*). Also request information from each industry about any air pollution control technologies that are in use. You may consider preparing a questionnaire and circulating it among industries in your area. A good place to start is the Public Relations Office of each company. For vehicular sources you will need information on the number and type of motor vehicles, typical fuel consumption and composition, road traffic densities, *etc.* You could calculate emission factors from the data that you gather (*e.g.* fuel composition and fuel consumption), or you may use already available emission factors. You may also be able to obtain useful information from other sources, including: the pollution control department of your local government authority (*e.g.* town council), the department responsible for road traffic, car manufacturers, *etc.* You may have to make some assumptions or justified guesses where data are lacking. Prepare a map of your area, identifying the location of each source and compile the data you have gathered into visually appealing and comprehensible formats (tables, graphs, pie-charts, bar-charts, *etc.*). Prepare a report on your findings and suggest means by which air quality could be improved further in your locality. Consider doing this project as a joint exercise with several other students, dividing the work load between you and preparing a joint report. You could discuss the results of your work, problems you encountered, uncertainties in your estimates, recommendations for further improvement, *etc.*, with your fellow colleagues during a seminar.

2. Measure SO_2, NO_2, O_3 and PM_{10} in your locality using the methods described in this chapter and compare these with concentrations measured by the local air quality monitoring station (if one is available) on the same date. Some municipal air quality monitoring stations are displaying current, and past, measurements on the Internet. Convert the measured concentrations into PSI values using the charts prepared in question no. 9 above. If you do not have the necessary equipment to sample PM_{10}, then determine TSP (total suspended particles) and convert to a PSI value using the TSP data in Table 3.8 (you will have to draw a chart of PSI against TSP). How would you categorise the ambient air quality at the time of sampling: good, moderate, *etc.*? Compare your results with Table 3.8. Do you expect any adverse health effects at the pollutant levels that you have measured? If so, what advice should be given to the public and what

preventive measures should be taken by the local government authorities? Also classify the ambient air pollution according to the UK DETR's air pollution guidelines (Table 3.9) on the basis of your measurements, and compare your results with international, national and local air quality standards in force in your area.

3.1.11 Reference

UNEP and WHO, Air pollution in the world's megacities, *Environment*, 1994, **36**, March, 4–22.

3.1.12 Further Reading

A. C. Stern, R. W. Boubel, D. B. Turner and D. L. Fox, "Fundamentals of Air Pollution", 2nd edn., Academic Press, Orlando, 1984.

H. Bridgman, "Global Air Pollution: Problems for the 1990s", Wiley, Chichester, 1994.

R. M. Harrison, A fresh look at air, *Chem. Br.*, 1994, **30**, 987–992.

R. M. Harrison, Important air pollutants and their chemical analysis, in "Pollution: Causes and Effects", ed. R. M. Harrison, Royal Society of Chemistry, Cambridge, 1990, pp. 127–155.

A. G. Clarke, The atmosphere, in "Understanding our Environment", ed. R. M. Harrison, Royal Society of Chemistry, Cambridge, 1992, pp. 5–51.

T. E. Graedel and P. J. Crutzen, The changing atmosphere, *Sci. Am.*, 1989, **261**, September, 28–36.

M. Radojevic, Burning issues, *Chem. Br.*, 1998, December, 38–42.

3.2 SAMPLING

3.2.1 Introduction

Instruments are now available for all the criteria pollutants as well as for many volatile organic compounds. However, owing to their high cost, these instruments may not be available in many teaching laboratories. Reliable measurements of air pollutants can be made using relatively inexpensive methods involving the sampling of air pollution followed by chemical analysis in the laboratory. Before the advent of instrumental techniques these methods were widely used by municipal monitoring stations and some are still recognised by the US EPA as *reference methods*. Reference methods are those methods against which all other methods, including in some cases instrumental methods, are tested. Methods which are found to give results in agreement with the reference methods are designated *equivalent* methods. Samplers are classified as *active* or *passive*. Active samplers require the use of a sampling pump.

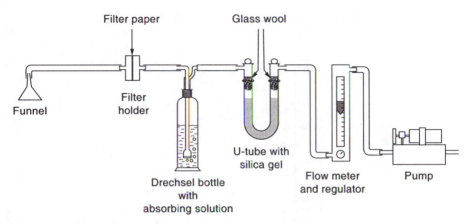

Figure 3.5 *The absorption train for sampling gases*

3.2.2 Absorption Train

This method involves the absorption of the pollutant gas into a reagent solution. Typical apparatus is illustrated in Figure 3.5. Air is drawn through a filter to remove any aerosol particles which may interfere with the analysis, and then bubbled through a reagent solution held inside a Drechsel flask. The air then passes through a trap that protects the pump from any moisture. Flaked calcium chloride, silica gel or glass wool can be used in the trap. The flow rate is controlled by a flow regulator and measured by means of a flow meter. The volume of air sampled over a sampling period can then be calculated. Alternatively, the volume of air may be determined by a gas meter. Instead of a Drechsel bottle, a midget impinger may be used (Figure 3.6). Impingers require lower amounts of reagents and lower sampling volumes, and hence shorter averaging periods. Specific reagents have been developed for SO_2, NO_2, NH_3, O_3, Cl_2, HCl, H_2S and HCHO, among others. After sampling, the solution is brought to the laboratory and generally analysed by titrimetry or spectrophotometry.

The reagent has to be specific to the pollutant gas and the absorption should be quantitative, or at least the absorption efficiency should be known. The absorption efficiency can be assessed by placing several absorbers in series and seeing if there is any carryover into the down-stream absorbers. The main disadvantages of the absorption train are the rather long sampling times (up to 24 hours) required to obtain sufficient sample for analysis and interferences by other pollutants. Many of the reagents are prone to interference from non-target gases and usually some method of eliminating the interference, such as a pre-scrubber, has to be adopted. This method, unlike continuous instru-

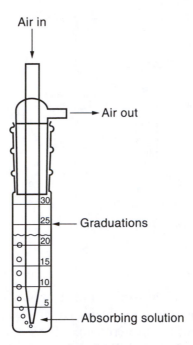

Figure 3.6 *Midget impinger. Non-graduated impingers are also available*

mental methods, will not reveal short-term peak concentrations which could be hazardous to the health.

You will be using the absorption train to sample for gases in the experiments that follow. Good locations for placing the sampling train are balconies, offices or laboratories. The inverted funnel could be placed outside an open window and the remainder of the sampling train placed on a table close to the window. You will need an electrical wall socket for the pump. Alternatively, a hole could be drilled through the window frame. Urban air is usually sampled at 10 m height; however, a height of 2 m above the pavement is more representative of human exposure. Sampling from the second floor of a building facing the street is ideal in this respect.

3.2.3 Impregnated Filters

In this method a filter is impregnated with a reagent solution specific to the gas of interest. Air is first sucked through a filter to remove aerosol particles and then through the impregnated filter. After sampling, the impregnated filter is taken to the laboratory, extracted and analysed. Impregnated filters have been used for SO_2, NO_2, NH_3, H_2S, HCl and HNO_3. It is possible to place several filters in series in a *filter pack* and

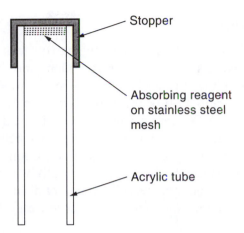

Figure 3.7 *Diffusion tube*

sample several gases simultaneously. One of the advantages of this method is its greater robustness and ease of transport compared to absorption trains. The main disadvantage is the variable sampling efficiency, which varies with the ambient gas concentration and the relative humidity.

3.2.4 Diffusion Tubes

A typical diffusion tube is illustrated in Figure 3.7. A reagent is adsorbed on a stainless steel mesh placed at one end of a small tube. The other end of the tube is left open. The tube is inverted to protect it from rain and dust, and exposed for a period of two weeks to one month at the sampling site. After exposure the tube is brought to the laboratory where the absorbed gas is determined by spectrophotometry or some other analytical technique. This sampling method is based on the diffusion of the gas up the tube; hence the name *diffusion tube*. As the rate of diffusion is proportional to the concentration, the ambient concentration can be calculated from the quantity of gas absorbed using a formula that includes a factor based on the diffusion coefficient of the gas. Diffusion tubes have been developed for SO_2, NO_2, NH_3, H_2S, O_3, benzene, xylene and toluene. This is a very low cost method of air monitoring and it is particularly useful in surveys of the spatial distribution of pollution. As such surveys require sampling at many sites, the use of instrumental methods is precluded by their high cost. The main disadvantage is the low precision of diffusion tubes: variation of 10% or more between replicate analyses is not uncommon.

3.2.5 Denuder Tubes

In this method, air is sucked through a long, narrow, cylindrical tube, the walls of which have been coated with a reagent specific to the gas of interest. After sampling, the walls of the tube are washed off and the solution analysed. Denuder tube methods have been developed for SO_2, NO_2, NH_3, HCl, HNO_3 and organic gases. The advantage of this method is that it avoids interferences by aerosol particles common with some of the other sampling methods. Under conditions of laminar flow in the denuder tubes, gases diffuse to the walls whereas the aerosol particles pass through the tube. Disadvantages of diffusion tubes include the laborious wall coating and extraction procedures.

3.2.6 Solid Adsorbents

This method is widely used for sampling of organic gases. A stainless steel or glass tube is filled with a packing material and air is then sucked through the tube (Figure 3.8). Gases are adsorbed onto the packing material during sampling. Common materials used as packing include: activated carbon, molecular sieve, silica gel, Poropak, Chromosorb and Tenax. After sampling, the tube is brought to the laboratory and analysed by means of gas chromatography. The adsorbed gas is either extracted into an organic solvent or thermally desorbed into the gas chromatograph. Volatile organic hydrocarbons (VOCs), including polynuclear aromatic hydrocarbons (PAHs), can be determined with this method.

3.2.7 Indicator Tubes

Commercially available glass *indicating tubes*, also called *gas detector tubes*, can be used for rapid on-site testing of gaseous pollutants (Figure 3.9). Tubes are available for numerous gases and the concentration of the

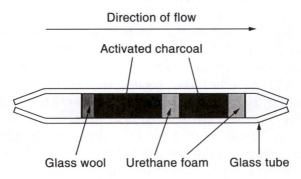

Figure 3.8 *Solid adsorbent tube*

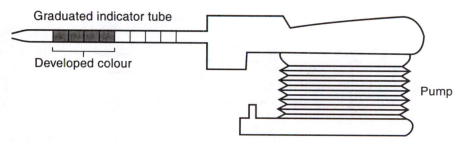

Figure 3.9 *Gas indicator tube and pump*

gas is determined by visual colour comparison. These tubes are purchased sealed and one end of the tube is broken open in the field. The glass tube is placed in a manufacturer's holder and the recommended volume of air pumped through the tube. The colour of the solid packing changes on reaction with the pollutant gas and the concentration is read off directly from the calibration scale printed on the tube. The length of the packing which changes colour corresponds to the concentration. In earlier tubes the colour was matched with printed colour charts. These tubes are generally designed for higher pollutant concentrations encountered in workplace atmospheres; however, some are capable of measuring concentrations in ambient air. There are several manufacturers of these tubes (*e.g.* Draeger). Specific tubes have been developed for more than 300 different gases.

3.2.8 Cryogenic Trapping

In this method the gas is condensed into a glass trap placed inside a Dewer flask containing a cold bath solution. Air is drawn at a low flow rate through the tube. The cold bath solution must have a temperature below the boiling point of the gas being sampled. This method has generally been used for sampling organic vapours followed by gas chromatographic analysis.

3.2.9 Grab Sampling

Grab samples can be obtained by means of evacuated sampling flasks, sampling bags and gas syringes (Figure 3.10). These are generally used in conjunction with gas chromatographic analysis. Flasks and bags are fitted with a septum. The needle of a gas syringe can be inserted through the septum and an aliquot withdrawn and injected into the gas chromatograph.

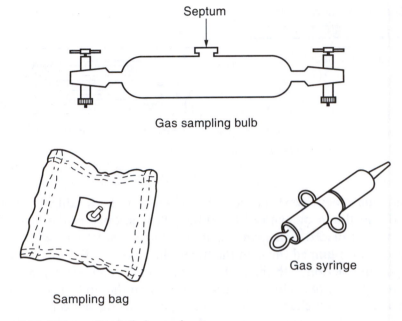

Figure 3.10 *Gas sampling bulb, bag and syringe*

3.2.10 Aerosol Sampling

This is described in Section 3.8.

3.2.11 Questions and Problems

1. What is the difference between active and passive sampling?
2. How would you test the efficiency of an absorption train?
3. Which sampling methods would be more appropriate for volatile organic compounds and why?
4. Discuss the different errors that could arise from the use of a pre-filter when sampling gases.

3.2.12 Further Reading

R. M. Harrison and R. Perry, "Handbook of Air Pollution Analysis", 2nd edn., Chapman and Hall, London, 1986.

J. P. Lodge, Jr., "Methods of Air Sampling and Analysis", 3rd edn., Lewis, Boca Raton, 1988.

3.3 SULFUR DIOXIDE (SO₂)

3.3.1 Introduction

Sulfur dioxide was one of the the first air pollutants to cause widespread concern and one of the first to be measured. In the past, SO_2, in combination with smoke, was responsible for many of the air pollution disasters, including the infamous London smog of 1952. Although concentrations of SO_2 have decreased over the last 40 years in the major urban centres in developed countries, SO_2 still remains a major air pollutant, especially in many developing countries. It can cause health problems on its own, in combination with smoke or in the form of sulfate aerosol. Furthermore, SO_2 is a precursor of sulfuric acid which is a major contributor to acid rain, a problem of global proportions (see Chapter 2). It can also corrode stone, metal and other building materials, as well as having a direct effect on vegetation. Effects of SO_2 at different concentrations are summarised in Table 3.10.

Nowadays, SO_2 concentrations in European and North American cities are relatively low and typical values range between 10 and 20 ppbv. In the past, concentrations of several hundred ppbv were common; however, introduction of control measures in the late 1950s and early 1960s has resulted in significant reduction of ambient SO_2 concentrations. Nevertheless, urban centres of Eastern Europe, China and many developing countries still experience high SO_2 levels. For example, daily maximum concentrations of SO_2 in Mexico City vary between 100 and 200 ppbv.

Major sources of SO_2 are coal- and oil-fired power stations, smelters,

Table 3.10 *Effects of SO₂*

Concentration (ppbv)	Effect
20–40	Possible effects on plants[a]
100–150	Decline in lung function of children[b]
200	Threshold of taste recognition
200	Aggravated bronchitis[b]
300	Threshold for odour recognition
200–400	Increased mortality[b]
1000	Immediate clinical symptoms with asthmatics
1600	Reversible bronchial constriction in unimpaired individuals
8000	Throat irritation
10 000	Eye irritation
20 000	Immediate coughing

[a] Possible synergism with O_3, NO_x, acid rain and heavy metals.
[b] In combination with black smoke.

sulfuric acid manufacturing plants and diesel powered vehicles. In recent years, increasing attention has focused on shipping as a source of SO_2. The S content of fuels (by weight) is: coal 0.2–7%, oil 0.3–4%, petrol 0.1%. Natural gas has negligible S content as a result of treatment. Untreated natural gas can have anywhere between 0 and 90% hydrogen sulfide (H_2S), and only gas deposits with H_2S < 35% are exploited. This is reduced further to < 4 ppmv by desulfurisation. The sulfur content of wood is also very small. In the atmosphere, SO_2 is oxidised to H_2SO_4 by various chemical reactions. The more important ones include gas phase reactions and oxidation of SO_2 dissolved in cloud droplets (see Example 3.3).

Monitoring of ambient SO_2 is routinely carried out by government agencies at many urban sites. Routine SO_2 source emission monitoring is also required for electric power plants, petroleum refineries, smelters (Pb, Cu, Zn), sulfuric acid manufacturing plants and some other sources. Absorption of SO_2 into a reagent solution followed by subsequent chemical analysis is a standard method of measurement that has been used for decades. Two of these methods have been widely deployed. In one of these, SO_2 is absorbed into a dilute aqueous solution of hydrogen peroxide, after passage through a filter to remove the dust particles:

$$SO_2 + H_2O_2 \rightarrow H_2SO_4$$

The acidity is determined by titrating with standard alkaline solution. However, this method is not specific and is subject to various interferences. Any atmospheric component which dissolves to give a strong acid will produce a positive response. Also, alkaline substances such as NH_3 will neutralise the acidity and give a negative response. Despite its shortcomings, this method was widely used in many earlier surveys including the UK National Survey of Air Pollution in the 1960s and 1970s. An alternative would be to absorb the SO_2 in H_2O_2 solution as above and then analyse the solution for sulfate by ion chromatography. This is not subject to interference as only SO_2 contributes to the sulfate. Another method involves the absorption of SO_2 into a potassium tetrachloromercurate (TCM) solution followed by spectrophotometric analysis (the West–Gaeke method).

None of these methods can give real-time, continuous measurements of SO_2, but instead produce long-term averages over periods of one day or several hours. Nowadays, classical methods have largely been replaced with instruments capable of giving sensitive, and specific, direct readings of SO_2. The most common instruments are based on UV

Example 3.3

Calculate the percentage of SO_2 oxidised in one hour in an atmosphere containing 1.02×10^7 molecules cm^{-3} of $OH^{\bullet}$ radicals, given that the second-order reaction rate constant for the oxidation of SO_2 by $OH^{\bullet}$ is 6×10^{-13} cm^3 molecule^{-1} s^{-1}.

The oxidation of SO_2 by $OH^{\bullet}$ can be represented by the following reaction:

$$SO_2 + OH^{\bullet} \rightarrow HSO_3^{\bullet}$$

This is a second-order reaction and the rate of oxidation is given by:

$$Rate = -d[SO_2]/dt = k\,[SO_2]\,[OH^{\bullet}]$$

where k is the second-order reaction rate constant. Converting to the incremental form and re-arranging the equation we get:

$$-\Delta SO_2/SO_2 = k\,[OH^{\bullet}]\,\Delta t$$

where $-\Delta SO_2/SO_2$ is the fractional conversion of SO_2 within the incremental time Δt. Taking $\Delta t = 3600$ s (*i.e.* 1 hour) and multiplying by 100 we get the percentage conversion:

$$-\Delta SO_2/SO_2 = 100 \times k\,[OH^{\bullet}]\,\Delta t = 100 \times (6 \times 10^{-13})$$
$$\times (1.02 \times 10^7) \times 3600 = 2.2\%$$

Thus 2.2% of SO_2 is oxidised in 1 hour.

---

fluorescence. Nevertheless, the West–Gaeke method involving collection in TCM remains, to this date, a reference method in the US.

3.3.2 Methodology

Sulfur dioxide is collected in a solution of tetrachloromercurate (TCM) followed by spectrophotometric analysis originally developed by West and Gaeke (1956). Air is bubbled through a solution of potassium tetrachloromercurate (TCM) to form a stable sulphito-mercurate complex:

$$[HgCl_4]^{2-} + 2SO_2 + 2H_2O \rightarrow [Hg(SO_3)_2]^{2-} + 4Cl^- + 4H^+$$

Formaldehyde and acid-bleached pararosaniline are added to form purple-red pararosaniline methyl sulfonic acid, the absorbance of which is measured after 30 min at 560 nm using a spectrophotometer. This method is suitable for SO_2 concentrations between 10 ppbv and 5 ppmv. The sampling efficiency of the method is $\geq 98\%$ at flow rates up to $15\,L\,min^{-1}$.

3.3.3 Materials

3.3.3.1 For Absorption Train
- Fritted bubbler (Drechsel bottle), 250 mL, or impinger, 30 mL
- Air pump
- Rotameter
- Polypropylene tubing
- U-tube
- Funnel (glass or polypropylene)
- Silica gel (regenerate by drying in oven until blue)
- Glass wool

3.3.3.2 For Analysis
- Spectrophotometer
- Potassium tetrachloromercurate absorbing reagent (TCM), K_2HgCl_4, 0.04 M, prepared by dissolving 10.9 g of mercury(II) chloride and 5.96 g of potassium chloride and bringing to volume in a 1 L volumetric flask.
- Pararosaniline stock solution, 1%. Take 1.00 g of pararosaniline and dissolve in water. Bring to volume in a 100 mL volumetric flask.
- Pararosaniline reagent. Take 10 mL of the 1% pararosaniline solution and mix with 15 mL of concentrated HCl. Dilute to 250 mL in a volumetric flask.
- Formaldehyde solution, HCHO, 0.2%, prepared by diluting 5 mL of 40% formaldehyde solution to 1000 mL with water.
- Stock sulfite solution, prepared by dissolving 2.0 g of Na_2SO_3 in water and making up to the mark in a 1 L volumetric flask. The water for making up this solution should first be boiled and cooled. This solution is unstable and should be freshly prepared before use and immediately standardised. This solution contains $2\,mg\,mL^{-1}$ of Na_2SO_3 or *ca.* $1\,mg\,mL^{-1}$ of SO_2. The actual concentration is determined by standardising with sodium thiosulfate (see below).
- Working sodium sulfite solution. Dilute 1 mL of the standard solution to 100 mL with water. This solution contains *ca.* $10\,\mu g\,mL^{-1}$ of SO_2. Calculate the exact concentration based on the standardisation of the standard sodium sulfite solution.

- Iodine solution, 0.1 N
- Sodium thiosulphate solution, 0.1 N
- Sodium bicarbonate
- Starch indicator

3.3.4 Experimental Procedure

3.3.4.1 Sampling. Construct the absorption train as follows: inverted funnel, Drechsel bottle, U-tube with silica gel (glass wool over the silica gel at both ends of the tube), rotameter and pump, using the polypropylene tubing to connect the various components. The absorption train should be the same as that in Figure 3.5, with the exception of the filter holder and filter paper, which should be omitted. Atmospheric sulfate particles do not interfere with the West and Gaeke method, and the quantity of sulfites in atmospheric particles is too low to cause significant error. Place 100 mL of the TCM absorbing reagent into the Drechsel bottle and connect all the components of the sampling train using polypropylene tubing. Adjust the flow rate to 1 L min^{-1} and sample ambient air. You could sample for 24 h or for shorter periods. Longer sampling periods could be used if you expect very low ambient SO$_2$ concentrations. Wrap the Drechsel absorber with aluminium foil in order to shield from light. The sample collected in TCM is stable for long periods; however, it is recommended that analysis should be performed within a few days. Prior to sampling, mark the position of the top of the TCM solution in the absorber. After sampling, add pure water to the mark to compensate for any evaporation of the solution during sampling. Measure the flow rate at the beginning and end of the sampling period and calculate the average flow rate. Use this in the calculation.

3.3.4.2 Analysis. Disconnect the Drechsel bottle from the sampling train. Withdraw a 20 mL aliquot of the solution and place in a 50 mL volumetric flask. Add 5 mL of 0.2% formaldehyde and 5 mL of the pararosaniline reagent. Make up to the mark and allow the colour to develop for 30 min. Measure the absorbance at 560 nm in a 1 cm cell with pure water in the reference cell.

Prepare a series of calibration standards by pipetting aliquots of the sodium sulfite working solution containing between 1 and 30 µg SO$_2$ (*i.e.* 0.1 to 3 mL) into 50 mL volumetric flasks containing 20 mL of tetrachloromercurate solution. Also prepare a blank containing 20 mL tetrachloromercurate solution without any additions of sodium sulfite. To each of these add 5 mL of formaldehyde and 5 mL of pararosaniline reagent and make up to the mark with water. Allow to stand for 30 min and measure the absorbance at 560 nm.

Prepare a calibration curve of absorbance *versus* µg of SO_2 (you will need to express the amount of Na_2SO_3 in your standards as SO_2). Read off the absorbance of the sample and calculate the concentration of SO_2 in the air in terms of $\mu g\, m^{-3}$ and volume mixing ratios (ppmv or ppbv):

$$SO_2(\mu g\ m^{-3}) = \frac{1000 \times M \times V_1}{V_2 \times V}$$

where M is the mass of SO_2 in µg read off from the calibration graph, V_1 is the volume of absorbing solution used for sampling (*i.e.* 100 mL in above case), V_2 is the volume of absorbing solution analysed (*i.e.* 20 mL in above case) and V is the volume of air sampled in litres:

$$V = \text{sampling duration (min)} \times \text{flow rate (L min}^{-1})$$

3.3.4.3 Standardisation of the Sodium Sulfite Stock Solution. Take 25 mL of the standard sodium sulfite solution and add to 50 mL of 0.1 N iodine solution inside a conical flask. Add approximately 2 g of sodium bicarbonate. The iodine oxidises the sulfite to sulfate as follows:

$$Na_2SO_3 + I_2 + H_2O \rightarrow Na_2SO_4 + 2HI$$

The flask is shaken and the excess iodine titrated with 0.1 N sodium thiosulfate solution. The titration reaction is:

$$I_2 + 2S_2O_3^{2-} \rightarrow 2I^- + S_4O_6^{2-}$$

When the original brown colour changes to a pale yellow, add a few drops of starch solution to give a deep blue colour. Continue the titration until the blue colour disappears and record the volume of sodium thiosulfate consumed. Repeat the titration two more times and take the average volume. The weight of anhydrous Na_2SO_3 present in the 25 mL sample of the stock solution is given by:

$$\text{Weight } Na_2SO_3 = 126\ (50 - V)/20\,000 \text{ g}$$

where V is the volume (mL) of the 0.1 N $Na_2S_2O_3$ solution needed to titrate the excess iodine.

Notes

1. If samples are freshly collected, delay analysis for 20 min in order to

allow any ozone to decompose. SO_2 absorbed in TCM solution is stable for up to 1 month if stored in a refrigerator at 4 °C. Therefore, you could conduct an extensive sampling survey first and analyse stored samples in one go at a later date.

2. You can use sample flow rates of up to $15\,L\,min^{-1}$ without significant loss of efficiency. You may have to experiment to determine the optimum sampling duration and flow rate to obtain aqueous solutions that fall within the range of the calibration curve.

3. Concentrations of NO_2 exceeding 2 ppmv can cause a fading of the colour. In ambient sampling, such high concentrations are not expected and the method given above may be used. In case you expect your air sample to contain high levels of NO_2 (as in the sampling of industrial gases) you can modify the TCM reagent. Weigh 0.6 g of sulfamic acid and dilute to 1 L with the TCM reagent. Use this to collect the sample, to prepare the calibration standards and reagent blank. The sample should be analysed within 48 h of collection.

4. If your sample is too concentrated and gives a response outside the range of the calibration curve, you may analyse less than 20 mL of the sample solution.

5. You may use a midget impinger to sample the air at a flow rate of $1\,L\,min^{-1}$. In this case, use 10 or 20 mL of the TCM solution in the impinger. Mark the top of the solution on the impinger and add water after sampling to compensate for loss of any reagent. Use all of the 20 mL to analyse for absorbed SO_2 by following the same procedure as given above. In this case the factor $V_1/V_2 = 1$ in the above equation.

6. An alternative method involves bubbling of ambient air into a 1 vol H_2O_2 absorbing solution, prepared by diluting 10 mL of 100 vol H_2O_2 (30%) to 1 L with water, inside a Drechsel bottle or impinger. SO_2 is converted to SO_4^{2-} which can then be analysed by ion chromatography using the method described for anions in rainwater (Section 2.4.6). It is necessary to place a prefilter upstream of the bubbler in order to remove SO_4^{2-} present in suspended aerosol. A Whatman 41 cellulose filter (47 mm) can be used. Also, you will need to heat the filter holder and sampling line upstream of the filter holder in order to prevent losses of SO_2 by condensation in the line or filter assembly. You may do this by directing hot air from a hair dryer onto the filter assembly.

7. Some recommend that the volume of air sampled be converted to *standard conditions* before calculating the gas concentration. With regard to the pressure, the *standard atmosphere* (1 atm =

760 mmHg $= 101.3$ kPa) is always used, but unfortunately there is no firm agreement as to which temperature should be used; values of 0, 20 and 25 °C have all been used. Many seem to prefer reporting results at 25 °C. To convert the sampling volume to 25 °C and 760 mmHg, use the following equation:

$$V_s = \frac{298.15 \times V \times P}{760 \times T}$$

where V_s is the volume of air at 25 °C and 760 mmHg, V is the measured volume, P is the average atmospheric pressure in mmHg and T is the average temperature of air in K.

3.3.5 Questions and Problems

1. Convert the concentrations you have determined to volume mixing ratios (ppbv).
2. Compare your measurements to the standards given in Appendix III and comment.
3. What are the major sources of SO_2 in your area and how can they be reduced?
4. List the harmful environmental consequences of SO_2 pollution.
5. Compare the West–Gaeke method with the H_2O_2 method for SO_2 determination.
6. In the H_2O_2 method, titrimetry or ion chromatography can be employed for analysis. Suggest some other analytical methods that could be employed to determine SO_2 collected in dilute H_2O_2 solution.
7. Why is it necessary to employ a prefilter with the H_2O_2 method but not with the TCM method for SO_2?
8. Express 40 ppbv of SO_2 in mass per volume units at 1 atm and 30 °C.
9. SO_2 dissolves in cloudwater to give the bisulfite ion (HSO_3^-) which may be oxidised to bisulfate by dissolved hydrogen peroxide (H_2O_2):

$$HSO_3^- + H_2O_2 \rightarrow HSO_4^- + H_2O$$

Calculate the rate of this reaction in mol L^{-1} min^{-1}, given that the second-order reaction rate constant $k = 1.05 \times 10^3$ L mol^{-1} s^{-1}. Assume that the concentration of H_2O_2 in the gas phase to be 2 ppbv, the pH of cloudwater 4.6 and the gas phase concentration of SO_2 to be 20 ppbv. Henry's law constant for H_2O_2: $K_H = 1.09 \times 10^5$ mol L^{-1} atm^{-1} and for SO_2: $K_H = 1.24$ mol L^{-1} atm^{-1} and $K_1 = 1.74 \times 10^{-2}$ mol L^{-1}.

3.3.6 Suggestions for Projects

1. Measure and compare SO_2 concentrations at various locations: industrial, residential and rural.
2. If you have access to a tall building you could determine SO_2 at different heights above the ground. Plot the vertical profile of SO_2.
3. Measure SO_2 concentrations in the same locality during dry weather and during periods of prolonged rain. What do the results show? Investigate any possible relationship between rainfall amount and SO_2 in air. Sample on many rainy days and plot the SO_2 concentration *versus* rainfall amount (mm). You may obtain rainfall amounts for the dates when you sample SO_2 from your local meteorological station or you may determine the rainfall amount yourself (see Chapter 2). Calculate the correlation coefficient and determine its level of significance.
4. Determine SO_2 concentrations at various distances from an obvious source of SO_2 (*e.g.* coal-burning power plant) and plot them on a graph. You may also determine simultaneously NO_2 and plot this on the graph, too.

3.3.7 Reference

P. W. West and G. C. Gaeke, Fixation of sulfur dioxide as disulphito-mercurate(II) and subsequent colorimetric estimation. *Anal. Chem.*, 1956, **28**, 1816–1819.

3.3.8 Further Reading

P. O'Neill, "Environmental Chemistry", Allen & Unwin, London, 1985, pp. 108–124.

American Society for Testing of Materials (ASTM), Standard test methods for sulfur dioxide content of the atmosphere (West–Gaeke method), D2914-78, in "Annual Book of ASTM Standards, Section 11, Water and Environmental Technology, Vol. 11.03, Atmospheric Analysis: Occupational, Health and Safety", ASTM, Philadelphia, 1987, pp. 107–126.

J. P. Lodge Jr. (ed.), "Methods of Air Sampling and Analysis", 3rd edn., Lewis, Boca Raton, 1988, pp. 493–498.

R. M. Harrison and R. Perry (eds.), "Handbook of Air Pollution Analysis", 2nd edn., Chapman and Hall, London, 1986, pp. 293–296.

3.4 NITROGEN DIOXIDE (NO$_2$)

3.4.1 Introduction

Although concentrations of SO_2 in air have been decreasing in the industrialised countries of Europe and North America over the last 30 years owing to legislation aimed at curbing SO_2 emissions, the use of low-sulfur coals, the adoption of flue gas desulfurisation technology and the increasing utilisation of natural gas, emissions of NO_x have continued to rise, so much so that NO_x pollution is now a major issue in many countries.

Nitrogen oxides (NO_x) include nitric oxide (NO) and nitrogen dioxide (NO_2). The major source of NO_x pollution is fossil fuel combustion in power stations, industrial boilers and motor vehicle engines. During combustion, N_2 in the air and nitrogen in the fuel are oxidised to NO. Nearly half of all anthropogenic NO_x emissions come from mobile sources, the remainder coming from power stations and industrial boilers. Incinerators and chemical plants are minor contributors to NO_x pollution. Natural sources of NO_x include volcanoes, bacterial activity, forest fires and lightning. In the atmosphere, NO is rapidly oxidised to NO_2 by ozone and HO_2 radicals:

$$NO + O_3 \rightarrow NO_2 + O_2$$

$$NO + HO_2^{\bullet} \rightarrow NO_2 + OH^{\bullet}$$

NO_2 is considerably more toxic than NO, and most of the health concerns have been focused on NO_2. NO_2 can alter lung function and increase susceptibility to respiratory infections. Asthmatics, young children and individuals suffering from chronic respiratory disease may be particularly sensitive to NO_x pollution. Furthermore, NO_x pollution contributes to ground level ozone (O_3), a major component of photochemical smog which affects many urban areas throughout the world (see Section 3.1.3). The red-brown tinge of the air that can be observed in many polluted cities is due to NO_2. NO_x can also damage vegetation and materials. NO_2 contributes to acid rain as it is easily converted to nitric acid (HNO_3) in the atmosphere (see Chapter 2). The contribution of NO_x to acid rain is approximately equal to that of SO_2, and in some regions, such as the western USA, NO_x pollution is the major contributor to rainfall acidity. In view of these concerns most industrialised countries have passed legislation to control emissions of NO_x and the limits are getting more stringent.

Ambient NO_2 is routinely measured at most urban air quality monitoring stations, while source emission monitoring is carried out at electric power plants and nitric acid manufacturing plants. The preferred method is based on the chemiluminescent reaction between NO and O_3, which produces excited NO_2 molecules. These excited molecules give off light which can be monitored. NO_2 is first converted to NO and then reacted with O_3. Classical methods are based on bubbling air into a solution followed by colorimetric analysis. One such method is the still the reference method for NO_2 in the USA and Japan.

3.4.2 Methodology

NO_2 is absorbed into a 0.1% sodium arsenite solution, where it is converted to nitrite ion. The nitrite ion is reacted with a mixture of sulfanilic and acetic acids to give a diazo compound. This is reacted with 1-naphthylamine-7-sulfonic acid (Cleve's acid) to give a purple-pink dye, the absorbance of which is determined spectrophotometrically at 525 nm. The method for NO_2^- determination used here was modified from a British Standard method (British Standard 2690, 1968).

3.4.3 Materials

3.4.3.1 For Absorption Train
- Same as for absorbtion train in Section 3.3
- Membrane filter (0.45 μm), 47 mm diameter, *e.g.* Millipore HAWP and filter holder

3.4.3.2 For Analysis
- Spectrophotometer
- Absorbing reagent. Dissolve 4 g NaOH and 1.0 sodium arsenite in laboratory water and make up to the mark in a 1 L flask
- Reagent A. Sulfanilic acid solution prepared by dissolving 0.5 g of sulfanilic acid in a mixture of 120 mL of laboratory water and 30 mL of acetic acid
- Reagent B. Cleve's acid solution prepared by dissolving 0.2 g of 1-naphthylamine-7-sulfonic acid in a mixture of 120 mL of laboratory water and 30 mL of acetic acid
- Stock NO_2^- solution, 1000 mg L^{-1} prepared by dissolving 1.5 g desiccated $NaNO_2$ in laboratory water and making up to the mark in a 1 L flask
- Working NO_2^- solution, 10 mg L^{-1}, prepared by diluting 10 mL of the 1000 mg L^{-1} stock solution to 1 L with water
- Hydrogen peroxide, 30%

3.4.4 Experimental Procedure

3.4.4.1 Sampling. Place 10 mL of the absorbing reagent in the impinger and assemble the sampling train as follows. Place the membrane prefilter upstream of the impinger, and the moisture trap, consisting of a U-tube containing silica gel, downstream of the impinger. This is followed by the rotameter and the pump. The sampling train should be the same as that illustrated in Figure 3.5 except that the Drechsel bottle is replaced with an impinger. Sample air at a flow rate of 0.2 mL min^{-1} for 24 h. Measure the flow at the beginning and end of the sampling period and take the average. Prior to sampling, mark the level of the absorbing solution in the impinger and after sampling make up to the mark with water to compensate for any loss of solution due to evaporation.

3.4.4.2 Analysis. Transfer absorbing reagent from the impinger to a 25 mL volumetric flask. Add 0.2 mL of H_2O_2 solution, 2 mL of reagent A, mix and allow to stand for 20 min. Add 5 mL of reagent B and dilute to the mark with laboratory water. Allow to stand for a further 20 min and measure the absorbance of the solution at 525 nm in a 1 cm cell. Measure the absorbance of a blank prepared by taking 10 mL of absorbing reagent and adding reagents A and B in the same way as described above. Prepare a series of calibration standards by dilution of the working NO_2^- solution to give the following concentrations: 0.25, 0.5, 1.0, 1.5, 2.0, 3.0, 4.0 and 5.0 $\mu g\,mL^{-1}$ NO_2^-. Place 10 mL of each of these calibration standards in 25 mL flasks and treat in the same way as the samples. Subtract the absorbance of the blank from the sample and standard readings and prepare a calibration graph of absorbance *versus* NO_2^- concentration. Read off the concentration in the sample.

Calculate the concentration of NO_2^- in the air from the following equation:

$$NO_2\ (\mu g\,m^{-3}) = 0.82 \times C \times V_1/V$$

where C is the concentration in the absorbing solution ($\mu g\ NO_2^-\ mL^{-1}$), V_1 is the volume of absorbing solution (mL), V is the volume of air sampled in m^3 and 0.82 is the sampling efficiency.

Notes

1. The sampling efficiency for this method depends on the type of sampling device used. Different impingers and bubblers can have widely varying efficiencies. The above efficiency of 82% applies to a tube drawn out to a 1.0 mm diameter and positioned 6 mm from the bottom of the absorber.
2. You may use a 100 mL Drechsel bottle instead of an impinger. In that case, use 50 mL of absorbing solution and sample at the same rate as indicated above. After sampling, add 1 mL of 30% H_2O_2 followed by 2 mL of reagent A. Mix, allow to stand for 20 min. Add 5 mL of

reagent B. Allow to stand for a further 20 min and measure the absorbance at 525 nm. For this method you should use standards made up in water to 50 mL, and then add the respective reagents in the same way as for the sample in order to keep the conditions constant. This method is less sensitive than the one using impingers.

3. For samples with very low concentrations you may try to add reagents to 10 mL of sample or standard without further dilution to 25 mL. This would increase the sensitivity.
4. Nitric oxide (NO) can give a positive interference (5–15%) in this method.
5. Sulfur dioxide (SO_2) interference is eliminated by oxidising with H_2O_2 to sulfate.

3.4.5 Questions and Problems

1. Convert the concentrations of NO_2 you have determined to volume mixing ratios (ppbv).
2. Compare the concentrations you have determined with the standards quoted in Appendix III and comment.
3. List the anthropogenic and natural sources of NO_x.
4. Discuss the effects of NO_x pollution.
5. The second-order rate constants and concentrations of oxidants for the oxidation of NO by HO_2, CH_3O_2 and O_3 respectively are given below:

Reaction	k (cm^3 molecule^{-1} s^{-1})	Oxidant concentration (molecules cm^{-3})
$NO + HO_2$	8.1×10^{-12}	$HO_2 = 6 \times 10^9$
$NO + CH_3O_2$	6.2×10^{-13}	$CH_3O_2 = 1 \times 10^9$
$NO + O_3$	1.6×10^{-14}	$O_3 = 5 \times 10^{12}$

Calculate the % of NO converted to NO_2 in 1 s by each of these mechanisms and conclude which is the most effective mechanism for the conversion of NO to NO_2.

3.4.6 Suggestions for Projects

1. You may determine the absorption efficiency yourself by placing two, or more, impingers in series and sampling. If no NO_2 is observed in the second impinger, that implies that the absorption efficiency of the first impinger was 100%. If you have sufficient impingers in series until no NO_2 is carried over from the last impinger, you may assume that the total collected in all the impingers represents 100% of the

NO_2. You may then calculate the absorption efficiency of the first impinger and use this in your subsequent determinations.

2. Determine NO_2 concentrations at different locations: urban, rural, residential, industrial. Comment on your results.
3. Determine NO_2 concentrations at different heights by sampling from different floors of a tall building. Plot the vertical profile of NO_2.
4. Measure NO_2 concentrations in the same locality during dry periods and during periods of prolonged rain. What do the results show? Investigate any possible relationship between rainfall amount and NO_2 in air. Sample on as many rainy days as you can and plot the NO_2 concentration *versus* rainfall amount (mm). You may obtain rainfall amounts for the dates when you sample NO_2 from your local meteorological station or you may determine rainfall amounts yourself (see Section 2.2). Calculate the correlation coefficient and determine its level of significance.

3.4.7 Reference

British Standard 2690, "Methods of Testing Water Used in Industry, Part 7. Nitrite, Nitrate and Ammonia (Free, Saline and Albuminoid)", British Standards Institution, London, 1968.

3.4.8 Further Reading

R. M. Harrison and R. Perry (eds.), "Handbook of Air Pollution Analysis", Chapman and Hall, London, 1986, p. 314.

American Society for Testing of Materials (ASTM) Standard test method for nitrogen dioxide content of the atmosphere (Griess–Saltzman reaction), D1607-76, in "Annual Book of ASTM Standards, Section 11, Water and Environmental Technology, Vol 11.03, Atmospheric Analysis: Occupational, Health and Safety", ASTM, Philadelphia, 1987, pp. 38–43.

J. P. Lodge, Jr. (ed.), "Methods of Air Sampling and Analysis", 3rd edn., Lewis, Boca Raton, 1988, pp. 389–394.

J. Mulik, R. Fuerst, M. Guyer, J. Meeker and E. Sawicki, Development and optimization of twenty-four hour manual methods for the collection and colorimetric analysis of atmospheric NO_2, *Int. J. Environ. Anal. Chem.*, 1974, **3**, 333–348.

J. H. Margeson, M. E. Beard and J. C. Suugs, Evaluation of the sodium arsenite method for measurement of NO_2 in ambient air, *J. Air Pollut. Control Assoc.*, 1977, **27**, 553–556.

A. A. Christie, R. G. Lidzey and D. W. F. Redford, Field methods for the determination of nitrogen dioxide in air, *Analyst*, 1970, **95**, 519–524.

B. E. Saltzman, Colorimetric microdetermination of nitrogen dioxide in the atmosphere, *Anal. Chem.*, 1954, **26**, 1949–1955.

3.5 OZONE (O_3)

3.5.1 Introduction

Ozone (O_3) is the most important secondary pollutant found in polluted air and a major constituent of photochemical smog. Although high concentrations of O_3 exist in the stratosphere, where it serves a beneficial purpose by filtering the harmful UV radiation from the sun, its concentration in unpolluted air is fairly low (20–50 ppbv). However, NO_2 and hydrocarbons emitted by motor vehicles and other sources can react in a sunlit atmosphere to produce high concentrations of O_3 and other oxidants. The O_3 concentration is an important measure of urban air quality and it is routinely determined in many urban areas. Levels in excess of 80 ppbv are generally considered to be indicative of a photochemical smog episode. Relevant authorities can decide on what protective measures to prescribe for the public on the basis of O_3 measurements. Ozone is a strong oxidising agent and a highly reactive compound. It is involved in many chemical reactions in the atmosphere and it can have harmful effects on human health, plants and materials. As O_3 is formed by the action of sunlight on predominantly traffic emissions, its concentration exhibits a diurnal variation, with highest concentrations observed around noon (Figure 3.1).

The formation of O_3 is initiated by the photodissociation of NO_2 by sunlight:

$$NO_2 + h\nu \rightarrow NO + O$$

$$O + O_2 \rightarrow O_3$$

Ozone can then oxidise NO to NO_2:

$$NO + O_3 \rightarrow NO_2 + O_2$$

The detailed chemistry is much more complex, involving many free radical reactions. Nevertheless, high O_3 concentrations can build up if a high NO_2/NO ratio exists in the atmosphere (see Example 3.1).

Ozone can have severe impact on the respiratory system: it can cause coughing, wheezing, shortness of breath, chest pain and throat irritation. It can also suppress the immune system's ability to fight off infection. Furthermore, O_3 can cause cracking of rubber and damage textile materials such as polyester and nylon. Ozone, alone or in combination with other pollutants such as SO_2, can cause damage to plants and reduce crop yields. Since O_3 is a secondary pollutant, it cannot be controlled directly. Controls aimed at precursor pollutants, such as catalytic converters on motor vehicles that reduce NO_x, CO and hydrocarbons, should result in lower ambient O_3 levels.

Modern air quality monitoring stations employ instrumental methods for O_3 determination. Ozone can be determined by means of chemiluminescence, IR absorption and UV absorption spectroscopy. Instruments based on UV absorption are the ones most commonly employed.

3.5.2 Methodology

An air sample is collected by bubbling into an absorbing solution of potassium iodide (KI). Ozone converts the iodide to iodine (I_2):

$$O_3 + 2I^- + 2H^+ \rightarrow O_2 + I_2 + H_2O$$

The liberated I_2 is determined by spectroscopy. This method is subject to a positive interference by other oxidants such as H_2O_2, Cl_2 and organic peroxides. If oxidants other than O_3 are present in the air, the method is used to determine "total oxidants". SO_2 causes a negative interference.

3.5.3 Materials

- Same as in Section 3.3. for absorption train
- Spectrophotometer
- Potassium dihydrogen phosphate (KH_2PO_4)
- Disodium hydrogen phosphate (Na_2HPO_4)
- Potassium iodide
- Iodine
- Absorbing solution. Prepare the sampling solution as follows. Dissolve 14.20 g of Na_2HPO_4, 13.61 g of KH_2PO_4 and 10.00 g of KI in laboratory grade water and make up to the mark in a 1 L volumetric flask. Shake the bottle. Store in a glass bottle in a fridge and do not expose to sunlight. This solution can be used for several weeks.

3.5.4 Experimental Procedure

3.5.4.1 Sampling. Place 10 mL of absorbing solution into the glass impinger and assemble a sampling train in the following sequence: inverted funnel, impinger, U-tube containing silica gel, rotameter, pump. The absorption train is the same as that in used in Section 3.4 but an impinger is used instead of a Drechsel bottle. The pre-filter consisting of filter holder and filter paper is not used in ozone sampling. Sample at a rate of $1 \, L \, min^{-1}$ for 30 min. Keep the impinger in the dark while sampling (*e.g.* by wrapping with aluminium foil). Prior to sampling, mark the top of the solution in the impinger and after sampling make up to the mark with water to compensate for any loss due to evaporation. Measure the flow rate at the beginning and at the end of sampling and calculate the average flow rate. Use this flow rate in the calculation.

3.5.4.2 Analysis. Transfer a portion of the exposed reagent to a 1 cm absorption cell and determine the absorbance in a spectrophotometer at 352 nm with water in the reference cell. Prepare a series of calibration standards as follows. Prepare a 0.025 M stock standard I_2 solution by dissolving 3.2 g of KI and 0.635 g of I_2 in laboratory water and making up to the mark in a 100 mL flask. Age for one day at room temperature prior to use. Take 4 mL of this solution and dilute to the mark in a 100 mL volumetric flask with unexposed absorbing reagent to give a 0.001 M I_2 solution. Use this solution to prepare a series of calibration standards in 50 mL volumetric flasks, diluting each time with the absorbing reagent. Take the following volumes of the 0.001 M I_2 solution to give the desired I_2 concentration: 0.5 mL (0.00001 M), 1 mL (0.00002 M), 1.5 mL (0.00003 M), 2 mL (0.00004 M) and 2.5 mL (0.00005 M).

Measure the absorbance of the unexposed absorbing reagent and subtract this value from the sample absorbance and from the absorbance of all calibration standards. Prepare a calibration graph by plotting absorbance *versus* concentration and reading off the concentration in the exposed sample. As the stoichiometry of the $O_3 \rightarrow I_2$ reaction is $1 : 1$ the molarity of O_3 is equal to that of I_2. Calculate the concentration of O_3 in the air in $\mu g \, m^{-3}$ from:

$$O_3 \, (\mu g \, m^{-3}) = 48 \times 10^3 \times V_1 \times C/V$$

where C is the concentration of O_3 in the absorbing solution (mol L^{-1}), V_1 is the volume of absorbing reagent (mL) and V is the volume of air sampled (m^3), calculated as the flow rate times the sampling period. In this experiment $V_1 = 10 \, mL$.

Notes

1. If you expect SO_2 to be present in the air you may eliminate its interference by means of a prefilter. Prepare a 20 mL aqueous solution containing 3 g chromium trioxide and 1 mL concentrated H_2SO_4. Spread this solution uniformly over a 20 cm × 20 cm piece of glass fibre filter paper and dry in an oven at 20 °C for 1 h. Cut the filter

paper into 0.5×1.0 cm strips. Store half of the strips in an air-tight jar for future use. Take the other strips, fold them and pack into a U-tube. Connect to an air pump and condition by drawing dry air overnight. You can obtain dry air by placing a U-tube containing silica gel upstream of the U-tube containing the chromic acid impregnated filter paper. Place the conditioned U-tube upstream of the impinger containing the absorbing solution when sampling for O_3 in order to remove any SO_2 in the air.

2. If your sample is too concentrated you may dilute it with absorbing reagent.
3. You may use sampling rates between 0.5 and 3 L min^{-1}.

3.5.5 Questions and Problems

1. Convert the concentrations of O_3 you have determined to volume mixing ratios (ppmv or ppbv).
2. Compare your measurements to the standards given in Appendix III and comment on the quality of the air.
3. Explain how O_3 is formed in the atmosphere.
4. What are the consequences of O_3 pollution?
5. How can O_3 pollution be controlled?

3.5.6 Suggestions for Projects

1. You may sample air at various times during the day and plot the diurnal variation of O_3 in your locality.
2. You may compare the diurnal variation in O_3 during a weekday (working day) and on a Sunday or public holiday (non-working day). Can you discern the influence of traffic on O_3 concentrations?
3. If you have access to a tall building you may sample simultaneously (if you have sufficient equipment), or quickly in succession, from different floors and plot the vertical profile of O_3 concentration.
4. You could compare O_3 concentrations at the same site, the same time of the day and same day of the week, but during different meteorological conditions, *e.g.* during different seasons (summer and winter), on a sunny and on an overcast day, during a dry period and during a wet period. What does this tell you about the influence of sunlight on O_3 concentrations?
5. You could determine and compare O_3 concentrations at different sites, *e.g.* urban, suburban, rural. Make the measurements at similar times of the day at the different sites.

3.5.7 Further Reading

American Society for Testing of Materials (ASTM) Standard test method for oxidant content of the atmosphere (neutral KI), D2912-76, in "Annual Book of ASTM Standards, Section 11, Water and Environmental Technology, Vol 11.03, Atmospheric Analysis: Occupational, Health and Safety", ASTM, Philadelphia, 1987, pp. 96–101.

R. M. Harrison and R. Perry (eds.), "Handbook of Air Pollution Analysis", 2nd edn., Chapman and Hall, London, 1986, pp. 353–369.

J. P. Lodge, Jr. (ed.), "Methods of Air Sampling and Analysis", 3rd edn., Lewis, Boca Raton, 1988, pp. 403–406.

Interscience Committee, Tentative method for the manual analysis of oxidizing substances in the atmosphere, *Health Lab. Sci.*, 1970, **7**, 152–156.

3.6 AMMONIA (NH_3)

3.6.1 Introduction

Ammonia (NH_3) is not generally considered to be an important pollutant; however, it plays an important role in the biogeochemical cycle of nitrogen. Since NH_3 is very soluble in water it can act as a source of nutrient in the biosphere. Although NH_3 can cause injury to plants at very high concentrations (*ca.* 20 ppmv), typical levels encountered in the atmosphere (5–25 ppbv) are not considered to be harmful. Agriculture is the major source of NH_3 as it is released from animal urine and dung and from volatilisation of applied fertilisers. If present in sufficient amounts, ammonia can neutralise atmospheric acid vapours (sulfuric, nitric and hydrochloric) by forming salt aerosols:

$$2NH_3 + H_2SO_4 \rightarrow (NH_4)_2SO_4$$

$$NH_3 + HNO_3 \rightleftharpoons NH_4NO_3$$

$$NH_3 + HCl \rightleftharpoons NH_4Cl$$

The latter two reactions are reversible and the acid gases can be released under specific atmospheric conditions such as warm weather.

Ammonia may act as a source of H^+ ions in the terrestrial environment and contribute to acidification. For example, ammonium salts deposited in soils are oxidised to produce acids:

$$(NH_4)_2SO_4 + 4O_2 \rightarrow H_2SO_4 + 2HNO_3 + 2H_2O$$

NH_3 is not routinely measured in air quality surveys.

3.6.2 Methodology

Ammonia is bubbled into a solution of dilute H_2SO_4. Ammonium reacts with phenol hypochlorite to give a blue indophenol dye which is determined by spectrophotometry at 625 nm. A pre-filter is required in order to remove NH_4^+ present in suspended particles which may interfere in the determination.

3.6.3 Materials

- Same as in Figure 3.5 for the absorption train
- Whatman 41 filter paper, pre-washed and dried, and filter holder
- Spectrophotometer
- Absorbing solution. 0.0025 M H_2SO_4
- Reagent A, prepared as in Section 2.5.2
- Reagent B, prepared as in Section 2.5.2
- Stock NH_4^+ solution, $1000\,\mu g\,mL^{-1}$, or a $100\,\mu g\,mL^{-1}$ $(NH_4)_2SO_4$ solution prepared as in Section 2.5.2
- Funnel (glass or polypropylene)

3.6.4 Experimental Procedure

3.6.4.1 Sampling.   Place 10 mL of the absorbing reagent in the impinger and assemble the sampling train in the usual manner: inverted funnel, pre-filter (pre-washed Whatman No. 41), impinger, moisture trap (U-tube with silica gel), rotameter and pump. This is the same as the absorption train shown in Figure 3.5 except that the Drechsel bottle is replaced with an impinger. Sample at a flow rate of $1-2\,L\,min^{-1}$ for 1–24 h, depending on the expected concentration. Measure the flow rate at the beginning and at the end of the sampling period and take the average. Mark the level of the sampling reagent in the impinger before sampling and make up to the mark with water after sampling to compensate for any loss due to evaporation.

3.6.4.2 Analysis.   After sampling, place 1.0 mL of the solution into a test tube and add 5 mL of solution A followed by 5 mL of solution B. Cover the mouth of the tube with parafilm and shake vigorously. Leave test tube in rack and place in a warm water bath at 37 °C for exactly 15 min or at room temperature for 30 min. Measure the absorbance of the resulting blue-green colour in a 1 cm cell at 625 nm on a UV/visible spectrophotometer with water in the laboratory reference cell, as described in Section 2.5.2. Similarly, analyse a reagent blank consisting of 1.0 mL of 0.0025 M H_2SO_4, and a series of standard NH_4^+ solutions as described in Section 2.5.2, taking 1.0 mL of each standard for analysis. Determine the absorbance of the blue-green dye at 625 nm in a 1 cm cell. Subtract the absorbance of the blank from the sample and standards, prepare a calibration graph of absorbance *versus* NH_4^+ concentration and read off the concentration in the sample.

The concentration of NH_3 in the air is determined from:

$$NH_3 \; (\mu g \, m^{-3}) = 0.94 \times V_1 \times C/V$$

where C is the concentration of ammonia in the absorbing solution ($\mu g \; NH_4^+ \; mL^{-1}$), V_1 is the volume of absorbing reagent used (10 mL) and V is the volume of sampled air (m^3). The factor of 0.94 is needed to convert from the concentration of NH_4^+, which was measured, to the concentration of NH_3.

Notes

1. You may vary the flow rate and sampling time depending on the expected concentration.
2. You may use a Drecshel bottle and larger volume (*e.g.* 50 mL) of absorbing reagent instead of the impinger. In this case you may sample at higher flow rates (*e.g.* 30 L min^{-1}).
3. If the sample has a low concentration of ammonia, you may use larger volumes of the absorbing reagent, 2–5 mL, in the analysis. In this case, also increase the volume of calibration standard used to correspond to the volume of absorbing reagent analysed. Prepare calibration standards over a lower concentration range.
4. You may determine NH_3 without using the pre-filter. Simply bubble the air into the impinger/bubbler directly. This will eliminate the problem of NH_3 loss on the filter paper, and your result will give you the total NH_3 concentration in the atmosphere (NH_3 in gas + NH_4^+ in particulate matter).
5. If using a pre-filter you will need to heat the filter holder and sampling line upstream of the filter holder in order to prevent losses of NH_3 by condensation in the line or filter assembly. You may do this by directing hot air from a hair dryer onto the assembly. At high humidity, NH_3 may react with acid gases on the filter paper, causing loss of NH_3 from the sample. Keeping the filter paper dry will prevent this.
6. If you are filtering the air with a Whatman no. 41 filter paper you should wash and dry it before use. You may use a Teflon filter paper instead.

3.6.5 Questions and Problems

1. Convert the NH_3 concentrations you have measured to mixing ratios (ppbv).
2. List the various anthropogenic and natural sources of NH_3.
3. Discuss the role of NH_3 in the biogeochemical cycle of nitrogen.
4. Would you expect a positive, or inverse, relationship between NH_3 concentrations and those of SO_2, NO_2 and HCl? Justify your answer in terms of your understanding of atmospheric processes.

3.6.6 Suggestions for Projects

1. Determine NH_3 at various locations: on a farm, at an urban site, an industrial site, *etc.*
2. Determine the vertical profile of NH_3 by sampling from different floors of a tall building.
3. Determine NH_3 during dry and wet periods at the same site and assess the effect of rainfall. You can try to investigate any possible relationship between rainfall amount (mm) and NH_3 concentration by sampling on many rainy days. You can obtain daily rainfall amounts for the dates when you sample NH_3 from your local meteorological station or you may determine this yourself (see Chapter 2). Plot the NH_3 concentration *versus* rainfall amount (mm). Calculate the correlation coefficient and determine its level of significance.
4. Determine NH_3 simultaneously with one or more other gases (SO_2, NO_2, HCl) at various sites. Is there any obvious relationship between ammonia and the other gases? Plot the concentration of NH_3 against that of each of the other gases you determine. Calculate the correlation coefficients and test them for significance. If they are significant, calculate the best-fit line by the method of least squares. Discuss your results on the basis of your understanding of atmospheric processes.
5. Use two sampling trains in parallel to determine NH_3 at the same site; one sampling train should be equipped with a pre-filter and the other should be without a pre-filter. Compare the results and interpret them in terms of the presence of particulate NH_4^+ and possible loss of NH_3 on the pre-filter.

3.6.7 Further Reading

J. P. Lodge, Jr. (ed.), "Methods of Air Sampling and Analysis", Lewis, Boca Raton, 1988, pp. 379–381.

R. M. Harrison and R. Perry (eds.), "Handbook of Air Pollution Analysis", 2nd edn., Chapman and Hall, London, 1986, pp. 324–328.

3.7 HYDROGEN CHLORIDE

3.7.1 Introduction

Although hydrogen chloride (HCl) is a highly corrosive gas, its concentration in the general atmosphere rarely approaches levels which could constitute a significant hazard. This has not, however, always been the case.

Hydrogen chloride pollution has an illustrious history, being:

- The first major air pollutant of industrialised society.
- The pollutant responsible for the development of the first-ever comprehensive air pollution control legislation.
- The pollutant responsible for the development of air pollution control technology.

Throughout the 18th century, enormous quantities of HCl gas were released by industries manufacturing sodium carbonate (Na_2CO_3) to feed the burgeoning *alkali industries*. Sodium carbonate was needed for the production of glass, soap and textiles, and it was produced by the Leblanc process. In this process, sulfuric acid and sodium chloride were mixed and heated to produce sodium sulfate:

$$2NaCl + H_2SO_4 \rightarrow Na_2SO_4 + 2HCl$$

The sodium sulfate was mixed with calcium carbonate and charcoal and heated to give sodium carbonate and calcium sulfide:

$$Na_2SO_4 + CaCO_3 + 4C \rightarrow Na_2CO_3 + CaS + 4CO$$

HCl produced as a by-product in the first stage of the process was released to the atmosphere, causing devastation over large areas around the plants: destroying crops and vegetation, damaging laundry and affecting the health of the local population. Typical symptoms included an irritant cough, smarting of the eyes and breathing difficulties. Complaints about HCl pollution led to the Alkali Act of 1863, which required plant operators to control HCl gas emissions. This was achieved by absorbing HCl gas in water, the first example of an air pollution control technology. Also, higher chimney stacks were built to disperse the pollution. The legislation was extremely effective: by 1865, two years after the passage of the act, almost 99% of HCl produced by 64 alkali works in England was being absorbed in water. Since methods for measuring air pollution were not yet developed in the last century, we have no way of knowing what the concentration of HCl may have been in the air, but from the available records it is obvious that it was sufficiently high to cause most serious acidic and corrosive pollution. Luckily, this is one environmental problem that the present generation does not have to deal with.

Combustion of coal is the largest anthropogenic source of HCl. Coal contains between 0.1% and 1% calcium chloride by weight and during combustion this is converted to HCl gas. Incineration of waste, especially discarded poly(vinyl chloride) (PVC), also emits HCl gas into the atmosphere. Other sources include manufacturing industries producing glass, steel, cement, ceramics and HCl itself. Release of chlorinated hydrocarbons can result in secondary HCl formation in the atmosphere by chemical reactions. Natural sources of HCl include volcanoes, biomass combustion and seasalt aerosols. The latter react with acidic gases in the atmosphere to generate HCl:

$$HNO_3 + NaCl \rightarrow HCl + NaNO_3$$
$$H_2SO_4 + 2NaCl \rightarrow 2HCl + Na_2SO_4$$

HCl pollution may contribute to acid rain, but its contribution is minor compared to that of SO_2 and NO_2; HCl contributes only *ca.* 4% to total acidity generated by anthropogenic sources.

Concentrations of HCl vary from 0.02 to 8 ppbv in urban areas, and from 0.05 to 1.2 ppbv in the marine atmosphere. In urban areas there appears to be a diurnal variation in HCl concentrations, with a maximum around the early afternoon to mid-afternoon, coinciding with maximum production of photochemical smog products.

3.7.2 Methodology

Hydrogen chloride gas is absorbed into a 0.001 M NaOH solution after filtration to remove particulate chloride. The solution is then analysed for chloride ion by ion chromatography (IC).

3.7.3 Materials

- Same as for the absorption train in Figure 3.5.
- Membrane filter (0.45 μm) 4.7 cm diameter, *e.g.* Millipore HAWP, and filter holder.
- Sodium hydroxide solution, 0.001 M.
- Ion chromatograph and associated equipment and reagents (see Section 2.4.6).

- Stock chloride solution, $1000 \, \text{mg} \, \text{L}^{-1}$ prepared by dissolving 1.6485 g NaCl in water and diluting to 1 L. Dry NaCl in an oven at 110 °C for 2 h before weighing.
- Working chloride solution, $10 \, \text{mg} \, \text{L}^{-1}$ prepared by diluting 10 mL of a $1000 \, \text{mg} \, \text{L}^{-1}$ stock solution to 1 L with water (see Section 2.4.6).

3.7.4 Experimental Procedure

3.7.4.1 Sampling. Place 50 mL of the 0.001 M NaOH in the Drechsel bottle and assemble the sampling train as in Figure 3.5. This is followed by the rotameter and the pump. The air is drawn through the inverted funnel. Use polypropylene tubing to connect the various components. Sample air at a flow rate of $20 \, \text{L} \, \text{min}^{-1}$ for 24 h. Measure the flow at the beginning and end of the sampling period and take the average. Prior to sampling, mark the level of the absorbing solution in the impinger and after sampling make up to the mark with water to compensate for any loss of solution due to evaporation.

3.7.4.2 Analysis. After sampling analyse for Cl^- in the absorbing solution by ion chromatography as described in Section 2.4.6. Dilute the $10 \, \text{mg} \, \text{L}^{-1}$ stock solution to prepare a series of calibration standards with chloride concentrations of 0.1, 0.2, 0.5, 0.8 and $1.0 \, \mu\text{g} \, \text{mL}^{-1}$ and analyse them by IC. Prepare a calibration graph of peak height against Cl^- concentration and read off the concentration in the absorbing solution.

Calculate the concentration of HCl in the air from the following equation:

$$\text{HCl} \, (\mu\text{g} \, \text{m}^{-3}) = C \times V_1/V$$

where C is the concentration in the absorbing solution ($\mu\text{g} \, Cl^- \, \text{mL}^{-1}$), V_1 is the volume of absorbing solution (50 mL in this case) and V is the volume of sampled air (m^3).

Notes

1. You may use flow rates between 2 and $30 \, \text{L} \, \text{min}^{-1}$ to achieve absorption efficiencies > 99%.
2. For short-term sampling (*e.g.* 2 h), use 10 mL of 0.001 M NaOH in a 30 mL impinger instead of the Drechsel bottle.
3. If you find that the concentration of HCl in air is too low, you may increase the sensitivity of the method by: (a) increasing the flow rate; (b) using an impinger with a lower volume of absorbing solution

(*e.g.* 10 mL); (c) increasing the sampling period (to 48 h or longer); or (d) by changing the settings on the conductivity detector or integrator.

4. For higher concentrations of HCl, the absorbing solution may be analysed using the precipitation titration method outlined in Section 4.12. This may be suitable for analysing ambient samples at more polluted sites or flue gases.

5. Interference caused by the absorption of gaseous HCl onto the pre-filter is possible. Glass-fibre filter papers give the largest error, while PTFE filters produce the lowest interference. Membrane filters give intermediate errors. If possible, use PTFE or carbonate filters.

6. You may determine HCl without using the pre-filter. Simply bubble the air into the impinger/bubbler directly. This will eliminate the problems of HCl loss or formation on the filter paper, and your result will give you the total chloride concentration in the atmosphere (HCl in gas + Cl^- in particulate matter).

7. If using a pre-filter you will need to heat the filter holder and sampling line upstream of the filter holder in order to prevent losses of HCl by condensation in the line or filter assembly. You may do this by directing hot air from a hair dryer onto the assembly. At high humidity, HCl may condense on the filter paper causing loss of HCl from the sample. Keeping the filter paper dry will prevent this.

8. Interference by aerosol chloride is possible. Although the pre-filter is meant to remove particulate chloride, HCl gas may be produced when NaCl particles retained on the filter react with acids such as HNO_3 and H_2SO_4 according to the chemical equations given earlier.

9. Chlorine gas (Cl_2) may also interfere by dissociating to Cl^- in solution:

$$Cl_2 + H_2O \rightleftharpoons HOCl + H^+ + Cl^-$$

10. You may use a chloride ion selective electrode to analyse the absorbing solution if one is available. Follow the instructions that come with the electrode. Prepare a calibration curve and read off the Cl^- concentration in the extract as shown for fluoride in Section 4.13.

3.7.5 Questions and Problems

1. Convert the concentrations that you determine in your experiments to units of ppbv.

2. Explain why HCl was a major air pollutant in the 19th century and how the problem was solved.
3. What are the major sources of HCl in the atmosphere?
4. What are the potential interferences in the determination of HCl?

3.7.6 Suggestions for Projects

1. Measure HCl at various locations: urban, rural, coastal and comment on your results.
2. Using two sampling trains, one with a pre-filter and the other without, measure HCl simultaneously at the same location. Repeat this at other sites. Compare the results by plotting the results obtained with one sampling train against results obtained from the other. Calculate the best-fit line using the method of least squares. Also, calculate the correlation coefficient and determine the level of significance. What do the results tell you? Which of the two methods would you recommend for HCl sampling: with or without filtration?
3. Determine HCl at a polluted urban site by collecting 2-hourly samples throughout the day. Plot the concentration of HCl against time of day. Do you observe a diurnal cycle?
4. Determine HCl simultaneously with one or more of the other gases (*e.g.* SO_2, NO_2) at various sites. Investigate statistical relationships by means of simple correlation and regression analysis. Are there any significant correlations between different gases? Can you identify sources of pollution on the basis of your analysis.

3.7.7 Further Reading

J. P. Lodge, Jr. (ed.), "Methods of Air Sampling and Analysis", 3rd edn., Lewis, Boca Raton, 1988, pp. 579–580.
R. M. Harrison and R. Perry (eds.), "Handbook of Air Pollution Analysis", 2nd edn., Chapman and Hall, London, pp. 445–450.

3.8 ATMOSPHERIC AEROSOL: HEAVY METALS AND MAJOR CONSTITUENTS

3.8.1 Introduction

Aerosol is a general term used to denote solid particles or liquid droplets suspended in the atmosphere. The definition of related terms is given in Table 3.11.

Table 3.11 *Definition of terms relating to atmospheric aerosol*

Term	Definition
Aerosol	Suspension of small particles or liquid droplets in a gas
Smoke	Ash, soot and grit emitted from combustion processes
Haze	Suspension of small particles (less than a few microns in diameter) causing reduction of visibility
Mist	Suspension of droplets causing reduction of visibility (visibility > 1 km)
Fog	Suspension of water droplets causing great reduction in visibility (visibility < 1 km)

Particles can originate from numerous sources, both natural and industrial. Also, they can be emitted as primary pollutants or formed in the atmosphere as secondary pollutants from precursor gases. Natural sources include wind-blown dust, sea spray, volcanoes and forest fires. Combustion of fossil fuels, whether in power stations or motor vehicle engines, is a major anthropogenic source of dust particles. This produces not only visible smoke, but also finer particles that result from atmospheric reactions of gases emitted by fossil fuel combustion. SO_2 and NO_x can react in the atmosphere to produce sulfate- and nitrate-containing particles. Many other activities also generate particulate pollution: smelting and processing of metals, cement manufacture, agricultural activities, quarrying, construction, *etc.* Particulates can vary considerably in size and chemical composition, depending on their origin. They can extend in diameter from fine particles < 0.1 μm to large particles > 100 μm. Generally, larger particles (> 2 μm) originate from natural sources such as wind-blown dust and sea-spray.

Both particle size and composition play an important role in determining the health effects of particulate matter. Particles > 10 μm in diameter do not penetrate the respiratory system as they are removed in the nasal region. Particles < 10 μm can be deposited in the respiratory system and are called *inhalable*. These particles can penetrate beyond the larynx into the tracheobronchial region. Particles < 2.5 μm can penetrate into the pulmonary region, consisting of bronchioles and alveoli, and they are called *respirable*. Because of the potential health effects of the smaller particles, most air quality monitoring stations nowadays measure the inhalable fraction (< 10 μm), which is termed PM_{10}, using specially designed sampling devices. In the past, most stations measured *total suspended particulates* (TSP), also called *suspended particulate matter* (SPM). Most air quality standards are defined in terms of PM_{10}. The US National Ambient Air Quality Standards (NAAQS) for PM_{10} are

$150\,\mu\mathrm{g\,m}^{-3}$ and $50\,\mu\mathrm{g\,m}^{-3}$ for 24 h and one year averaging periods, respectively (Appendix III). Because particles smaller than $2.5\,\mu\mathrm{m}$ ($PM_{2.5}$) penetrate deeper into the lungs than PM_{10} the US EPA is proposing the introduction of $PM_{2.5}$ standards. The proposed $PM_{2.5}$ standards are $50\,\mu\mathrm{g\,m}^{-3}$ and $15\,\mu\mathrm{g\,m}^{-3}$ for 24 h and one year averaging periods, respectively.

Particles can have harmful effects on their own or they may act synergistically to enhance the toxic effects of other pollutants. High levels of SO_2 and particulates can lead to increased mortality, morbidity and aggravation of respiratory and cardiovascular diseases. Sulfuric acid aerosols have the highest irritant potential. Atmospheric particles also contain many toxic metals (*e.g.* lead, cadmium), that can cause various illnesses, and carcinogenic compounds such as polynuclear aromatic hydrocarbons (PAHs). An association has been found in major US cities between particulate matter and mortality and morbidity at levels of PM_{10} below the prescribed NAAQS. In addition, high concentrations of atmospheric aerosol during smog and haze episodes can significantly reduce visibility. This can lead to increased risk of traffic accidents, closure of airports, *etc.* An increase in the total particulate load in the atmosphere could affect the earth's radiation balance, with consequent effects on weather and climate. Particles can also cause soiling of materials. Toxic substances present in atmospheric particles are washed out and deposited at the earth's surface, and thus introduced into terrestrial and aquatic ecosystems. There they can have harmful effects on living organisms. Humans can be exposed to these compounds through the food chain and the water supply. Major constituents of airborne particles and their sources are given in Table 3.12.

Each of these components may contribute between 5% and 50% of the total mass of the particles, depending on the site. Various transition

Table 3.12 *Major constituents of atmospheric particles and their sources*

Component	Major source
Carbonaceous material	Smoke emissions from combustion
Na^+, K^+, Ca^{2+}, Mg^{2+}	Sea salt, wind-blown dust
SO_4^{2-}	SO_2 emissions from coal combustion
NO_3^-	NO_x emissions from combustion
NH_4^+	NH_3 emissions from animal waste
Cl^-	Sea salt, HCl from coal combustion
Insoluble minerals	Wind-blown dust

Table 3.13 *Typical concentrations of some trace metals in air*

| Element | Concentration range (μg m^{-3}) | |
	Urban	Background/Rural
Fe	0.1–10	0.04–2
Pb	0.1–10	0.02–2
Mn	0.01–0.5	0.001–0.01
Cu	0.05–1	0.001–0.1
Cd	0.0005–0.5	0.0001–0.1
Zn	0.02–2	0.003–0.1
V	0.02–0.2	0.001–0.05

metals (*e.g.* Pb, Mn, Fe, Co, Cu, Zn, Ni, Sn, Cd) are also found in atmospheric particles. Generally, each of the trace metals contributes < 1% of the total particulate mass. Anthropogenic sources of trace metals include metal manufacturing industries, fossil fuel combustion and incineration. Typical ranges of concentrations measured in air are given in Table 3.13. Very few countries have established air quality standards for heavy metals, other than for Pb. The standard for Cd in μg m^{-3} is: Germany 0.05, Yugoslavia 0.04 and Switzerland 0.01.

3.8.2 Sampling and Analysis

The most common method of sampling dust particles in the air is by means of filtration. Both low volume and high volume samplers have been used. The filter used in a gas sampling train may be used to determine the dust component; however, high volume (Hi-Vol) samplers are generally used. Hi-Vol samplers suck air though a large filter paper (203 × 254 mm) at a high flow rate (1–2 m^3 min^{-1}), and this allows for the collection of a larger sample that using low volume sampling. This improves accuracy and allows for the analysis of many components in the dust. The sampler is mounted in a shelter of heavy gauge aluminium so that the filter is parallel to the ground and the gabled roof protects the filter from direct precipitation (Figure 3.11). The filter is generally changed every 24 or 48 h, weighed, extracted and analysed. Longer, or shorter, sampling periods can be used if required. The Hi-Vol sampler is efficient for particles < 100 μm and it is used to determine the concentration of total suspended particulates (TSP). Modified Hi-Vol samplers with a special, size-selective, inlet can be used to sample particles less than 10 μm in diameter (PM$_{10}$) and these are employed by most ambient

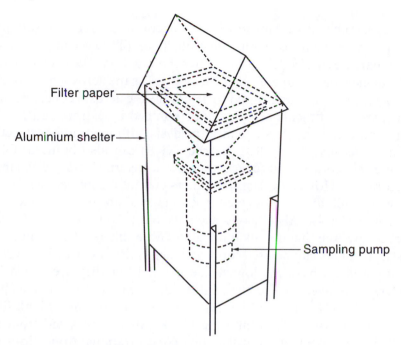

Filter paper

Aluminium shelter

Sampling pump

Figure 3.11 *The high volume (Hi-Vol) aerosol sampler*

air quality monitoring stations. Dichotomous samplers have two inlets and they discriminate particles into two size ranges, one *coarse* (10–25 μm) and one *fine* (< 2.5 μm). Cascade impactors can be used to separate the particulate matter into several size fractions during sampling. The concentration of particulate matter is determined by gravimetry.

Most of the filtration methods determine the long-term average concentration (*e.g.* 24 h). On the other hand, short-term concentrations can be determined using *paper tape samplers*. In these devices the tape moves across a sampling inlet, and new tape can be programmed to advance every 2 h or so. This allows semi-continuous measurements to be made. The β-gauge instrument also uses a roll of tape but it measures the attenuation of β-radiation by dust particles deposited on the filter paper. This is quite a sensitive technique and sampling periods of the order of minutes can be used.

Real-time continuous monitoring of suspended dust particles can be made using instruments operating on the light scattering principle or the piezo-balance. In the piezo-balance method, dust particles are pumped to the surface of a quartz crystal onto which they deposit by electrostatic precipitation. The mass change in the crystal is recorded as a change of its resonant frequency in an oscillator circuit. Another method for making real time measurements of atmospheric aerosol concentrations is the tapered element oscillating microbalance (TEOM) in which the aerosol particles are collected on a filter mounted on a tapered element. The frequency of the tapered element changes as the aerosol mass on the filter increases and it can be used to determine the aerosol concentration.

Chemical constituents are usually determined by extracting the filter papers and carrying out a chemical analysis, although non-destructive techniques such as X-ray fluorescence (XRF) can also be used. Glass fibre filter papers are used for the gravimetric determination of the mass concentration. However, glass fibre filters are not recommended for the analysis of sulfate, nitrate, calcium or magnesium as they have high background levels. Also, gases such as SO_2, NO_x and HNO_3 are absorbed by the highly alkaline glass fibre filters. Glass fibre or polystyrene filters can be used for heavy metals (Pb, Fe, Cd, Zn). Teflon membrane filters have the lowest background but they are expensive. They are recommended for the analysis of major components such as SO_4^{2-}, NO_3^-, NH_4^+ and Cl^- in dust particles. In any case, blank filter papers have to be analysed and the blank values subtracted from the sample measurements when calculating concentrations. Major ions are analysed in aqueous extracts either by ion chromatography (IC) or spectrophotometry. Heavy metals are extracted into an acid solution (usually HNO_3 or a mixture of HNO_3 and some other acid) and determined by flame or graphite furnace atomic absorption spectrometry (AAS), or inductively coupled plasma atomic emission spectrometry (ICP-AES).

A number of processes can lead to errors when utilising filtration to determine the chemical composition of particles. One of these is artifact formation in which gases can react on the surface of the filter. For example, in the presence of high humidity SO_2 may be adsorbed and oxidised to SO_4^{2-}. Also, volatile components may evaporate from the dust particles during sampling.

3.8.3 Sampling Procedure and Mass Determination

3.8.3.1 Methodology. Samples of suspended particulate matter are collected on glass fibre filter papers and weighed on a sensitive analytical balance.

3.8.3.2 Materials
- High volume (Hi-Vol) sampler. Alternatively, a low-volume sampling assembly consisting of filter holder (47 mm), moisture trap

(*e.g.* silica gel in U-tube), flow meter, and pump ($1-30 \, \text{L min}^{-1}$) in series.

- Glass fibre filter paper for Hi-Vol sampler (*e.g.* Whatman GF/A), $203 \times 254 \, \text{mm}$. Alternatively, 47 mm glass fibre filter for low-volume sampling.

3.8.3.3 Experimental Procedure. (*a*) *Hi-Vol Sampling and mass determination.* Leave the glass fibre filter paper ($203 \times 254 \, \text{mm}$) to equilibrate at room conditions for 24 h before and after sampling. The temperature of the room should be in the range $15-30 \, ^{\circ}\text{C}$ and the relative humidity should be in the range $20-45\%$. Weigh on an analytical balance sensitive to 0.1 mg. You may roll the filter paper loosely for weighing if necessary. Place the filter paper in the sampling assembly and sample for 24 or 48 h. The Hi-Vol sampler should be placed either on a roof of a building or in a clearing. Measure the flow rate of the Hi-Vol sampler according to the manufacturer's instructions at the start and at the end of the sampling period, and calculate the mean flow rate. The effective filtration area is $178 \times 229 \, \text{mm}$ (you may have to measure this for your particular case). After sampling, the filter paper should be folded along the long axis with the exposed surface facing inward. Filter papers can be stored in envelopes. After determining the mass concentration, filter papers can be stored for up to 1 year for chemical analysis.

From the difference in weight of the filter paper after and before sampling, the duration of sampling and the average flow rate, determine the concentration of total suspended particles (TSP) in the air:

$$\text{TSP} \, (\mu\text{g m}^{-3}) = \frac{[\text{Final weight (mg)} - \text{initial weight (mg)}] \times 1000}{\text{Flow rate (m}^3 \, \text{min}^{-1}) \times \text{sampling period (min)}}$$

If you are using a special Hi-Vol sampler with a PM_{10} size selective inlet, then your calculation will give you the PM_{10} concentration.

(*b*) *Low-volume sampling.* If a Hi-Vol sampler is not available, you may use a glass fibre filter (acid washed) or membrane filter connected to a low-volume sampling air pump such as that used in the gas sampling train. You may use flow rates between 1 and $20 \, \text{L min}^{-1}$, depending on the pump available. Equilibrate and weigh the filter paper before and after sampling as in the case of the Hi-Vol filter paper and determine the concentration of TSP.

3.8.4 Heavy metals

3.8.4.1 Methodology. Suspended matter collected on a filter is extracted into nitric acid. Heavy metals (Al, Ba, Cd, Cr, Co, Cu, Fe, Pb, Mn, Ni, Si, Sn, Zn and V) are determined by flame AAS.

3.8.4.2 Materials

- Filter paper from Hi-Vol or low-volume sampler
- Filter paper for Hi-Vol sampler (acid-washed glass fibre, Teflon-coated glass fibre, *etc.*)
- Atomic absorption spectrometer
- Stock solutions of heavy metals to be analysed $1000 \, mg \, L^{-1}$ (see Table 4.19)
- Nitric acid, concentrated, high purity
- Nitric acid, 0.25 M
- Hot plate with magnetic stirrer and stirring bar
- Filter paper for low-volume sampling, Whatman 541 (acid-washed) or Millipore HAWP $0.45 \, \mu m$ (47 mm) membrane filter for low-volume sampling.

3.8.4.3 Experimental Procedure. (a) Extraction. Using a stainless steel cork borer with an internal diameter of 16.2 mm, cut out eight circles at random from the Hi-Vol filter paper. You will have to use an area multiplication factor of 24.72 to scale up the analytical results to be representative of the total filtration area. You may use different size cork borers, or simply cut out a piece of the filter paper with scissors. In any case, you will have to calculate the appropriate multiplication factor, so measure the effective sampling area of the filter paper and the area that you have cut out. For rural, unpolluted samples you may have to increase the area of the filter paper that you will extract, while for heavily polluted, industrial samples you may have to reduce the area. Store the unused portion of the filter paper for other analyses.

Carry out the digestion in a fume cupboard. Place the eight discs (or whatever portion of filter you have cut out) and 20 mL of 1 : 1 HNO_3 (10 mL laboratory water + 10 mL HNO_3) in a beaker and cover with a watch glass. Concentrate to about 5 mL on a hot plate at 150–180 °C while stirring with a magnetic stirrer. Add a further 10 mL of 1 : 1 HNO_3 and repeat. Filter the extract through a Whatman 541 filter paper and wash the beaker and filter paper with successive small aliquots of 0.25 M HNO_3. Transfer filtrate and washings into 50 mL volumetric flask and make up to the mark with 0.25 M HNO_3. Repeat this procedure with a number of randomly selected clean filter papers in order to establish the filter blank levels.

In case you are using low-volume rather than Hi-Vol sampling, follow the same procedure but extract the whole filter paper rather than a portion of it. Also extract blank filters in the same way.

(b) Analysis. Determine the concentrations of one or more of the metals using flame AAS. Ensure the appropriate hollow cathode lamp is in place for each metal and use the absorbing lines and type of flame specified in Table 4.20 (see Section 4.16) for the different metals. These are the main resonance lines but other lines may be selected if necessary. Follow the procedure for your spectrometer. For each of the metals, prepare a series (five or six) of calibration standards in 0.25 M HNO_3. These should be prepared daily by dilution from the $1000\,mg\,L^{-1}$ stock solutions over the range of concentrations that you expect for each metal (see Tables 4.19 and 4.20). Draw a calibration graph of absorbance *versus* concentration for each of the metals that you are analysing and read off the concentration in the sample extract.

(c) Calculation. The total atmospheric metal sampled is calculated from:

$$\text{Total metal sampled } (\mu g) = (C \times V \times \text{AMF})_s - (C \times V \times \text{AMF})_b$$

where C is the concentration in extract ($\mu g\,mL^{-1}$), V is the volume of extract (25 mL), AMF is the area multiplication factor (24.72), and subscripts "s" and "b" refer to sample and blank measurements, respectively.

You may have to calculate the AMF for your case, depending on the sampling area of the filter and the size of the filter area extracted. The total mass of metal sampled is divided by the total volume of air sampled (flow rate × sampling period) to give the concentration in the air:

$$\text{Concentration in air } (\mu g\,m^{-3}) = \frac{\text{Total metal sampled } (\mu g)}{\text{Flow rate } (m^3\,min^{-1}) \times \text{sampling period (min)}}$$

In case you are using low-volume rather than Hi-Vol sampling, you should note that your final extract will contain much lower concentrations that in the case of the Hi-Vol method, and you may be operating close to the detection limit of flame AAS. This will lead to greater errors in the final concentration than the Hi-Vol method. You may reduce the errors by extending the sampling period to several days, or even one week. Perform all the calculations as above using the sampling flow rate and sampling period that you employ. In this case there is no need to include the area multiplication factor in the calculation as you are analysing the entire filter paper.

Notes

1. You can use graphite furnace AAS instead, if it is available in your laboratory. This method is more sensitive than flame AAS.

2. You may use ICP-AES instead if it is available in your laboratory.
3. Glass-fibre filters may have high backgrounds of some metals and may be unsuitable for some metal analyses. You may, however, reduce the background content of glass fibre filters by soaking the glass fibre filter paper in dilute HCl and washing extensively with laboratory water. Dry the filter paper and use as normal. As an alternative, you may use low-volume sampling with Whatman 41 filter papers, which have lower background levels of many metals than glass fibre filters. When using low-volume sampling you may have to extend the sampling period to 48 h, or longer, if necessary.
4. If the sample extract is too concentrated, you may dilute it with 0.25 M HNO_3 to fit within the calibration range.
5. If the sample extract is too dilute, you may extract larger portions of the filter paper or you may make up the final extract to a lower final volume (*e.g.* 25 mL) rather than 50 mL.

3.8.5 Major Components

3.8.5.1 Methodology. Particulate samples collected on filter papers are extracted into water and analysed for alkali (Na^+, K^+) and alkaline earth (Mg^{2+}, Ca^{2+}) metals by AAS, chloride (Cl^-), sulfate (SO_4^{2-}) and nitrate (NO_3^-), by IC and ammonium (NH_4^+) by colorimetry.

3.8.5.2 Materials
- Filter papers from Hi-Vol or low volume sampler
- Ion chromatograph
- Atomic absorption spectrometer
- Spectrophotometer
- Hot plate
- Stock solutions of Na, K, Mg, Ca, Cl^-, SO_4^{2-}, NO_3^- and NH_4^+, 1000 mg L^{-1}. If these are not available, prepare by weighing appropriate quantities of soluble salts (*e.g.* NaCl, $NaNO_3$) and diluting to 1 L in a volumetric flask as indicated in Sections 2.4.6, 2.5.1 and 2.5.2.

3.8.5.3 Experimental Procedure. *(a) Extraction.* Place the low-volume filter paper, or a portion of the Hi-Vol filter paper, in a beaker and add 20 mL of laboratory grade water. Warm at 70 °C for 15 min. Pour the liquid into a 50 mL volumetric flask and repeat with another 20 mL of laboratory water. Transfer this to the same volumetric flask and fill to the mark with water. Extract blank filter papers in the same way.

(b) Analysis. (i) Na, K, Ca, Mg. Analyse by flame AAS as described in Section 2.5.1. Prepare a series of standards by diluting the $1000\,mg\,L^{-1}$ stock solutions. You may have to decide on the appropriate range of standards by trial and error, depending on the concentration in your sample extracts. You may also dilute your extracts if they are too concentrated, in order to fit them within your calibration curve.

(ii) Cl^{-}, SO_4^{2-}, NO_3^{-}. Analyse by means of IC in the same way as described for rainwater analysis in Section 2.4.6. You may have to do some optimising in order to choose the most appropriate conditions. Your extracts may be more concentrated than the rainwater and you may decrease the sensitivity of the conductivity detector. You may have to decide on the appropriate range of standards, depending on the levels in your extracts. Prepare the standards by diluting the $1000\,mg\,L^{-1}$ stock solution and, if necessary, dilute your extracts to fit within the calibration curve. You may also analyse these using the methods described in Chapter 4 on water analysis (see Section 4.12 for Cl^{-}, Section 4.14 for SO_4^{2-} and Section 4.10.3 for NO_3^{-}).

(iii) NH_4^{+}. Analyse extracts using the colorimetric phenol–hypochlorite method described in Section 2.5.2 for rainwater. Place 0.5 mL of the filter paper extract into a test tube and add 5 mL of Solution A followed by 5 mL of Solution B. Shake and allow to stand for the recommended period and analyse as instructed in Section 2.5.2.

(c) Calculation. Read off the concentrations in the extract from the calibration graph. The total amount of substance sampled is calculated from:

$$\text{Total mass of substance sampled }(\mu g) = (C_s - C_b) \times V$$

where C_s and C_b are the sample extract and blank extract results in $\mu g\,mL^{-1}$, and V is the volume of extract (50 mL). In the case that you are using a portion of a Hi-Vol filter, you will have to multiply by the area multiplication factor.

The total mass of substance sampled is divided by total volume of air sampled (flow rate × sampling period) to give the concentration in the air:

$$\text{Concentration in air }(\mu g\,m^{-3}) = \frac{\text{Total analyte sampled }(\mu g)}{\text{Flow rate }(m^3\,min^{-1}) \times \text{sampling period }(min)}$$

Notes

1. Ideally, you should use Teflon (PTFE) filters for sampling the major components. Glass fibre filter papers are generally not suitable owing to the very high background content of blank filters. You may,

however, reduce the background content of glass fibre filters by suitable treatment. Soak the glass fibre filter paper in dilute acid (HCl, HNO_3) and wash extensively with laboratory water. Dry the filter paper and use as normal. For low-volume sampling you may use 47 mm Whatman PTFE ($0.5\,\mu m$ pore size) or 47 mm Millipore HAWP ($0.45\,\mu m$ pore size).

2. If the sample extract is too concentrated, you may dilute it to fit within the calibration range.
3. If the sample extract is too dilute, you may extract larger portions of the filter paper and/or reduce the amount of extract (*e.g.* make up the extract into a 25 mL rather than a 50 mL flask).

3.8.6 Questions and Problems

1. Prepare a table summarising all the measurements in your samples (mass concentration, chemical composition).
2. Discuss the potential environmental effects of particulate matter.
3. List all the sources of particulate matter in the atmosphere that you can think of, and classify them as natural or anthropogenic.
4. Explain how particle size affects deposition in the lungs.
5. The detection limit for Mn by flame AAS is given as $0.01\,mg\,L^{-1}$. Calculate the *method detection limit* for Mn in $\mu g\,m^{-3}$ based on sampling air with a Hi-Vol sampler for 24 h at $1.5\,m^3\,min^{-1}$, and extracting and analysing according to the procedure given above in Section 3.8.4.

3.8.7 Suggestions for Projects

1. Simultaneously sample rainwater (see Chapter 2) and aerosol and analyse both for the major components and/or trace metals. Determine the *scavenging ratio* (also called *washout ratio*) for the various species using:

$$\text{Washout ratio, } W = \frac{\text{Contentration in rain } (\mu g\,kg^{-1})}{\text{Concentration in air } (\mu g\,kg^{-1})}$$

Assume the density of rainwater and air to be $1\,kg\,L^{-1}$ and $1.225\,kg\,m^{-3}$, respectively. Inspect the washout ratios for the different components and comment on their respective values. What do they tell you?

2. Sample atmospheric aerosol at different locations: polluted industrial, urban, suburban and rural. Compare the concentrations of major components and heavy metals at the different sites.

3. Measure the concentration of background impurities in different types of filter papers obtained from different manufacturers by analysing blanks. For each type of filter paper, analyse five replicates and calculate the mean and standard deviation. Summarise your results in a table. Which filter paper would you recommend for which analysis?
4. Investigate the effect of different washing techniques on the background impurities in different filter papers. Soak filter papers in water, dilute HCl, dilute HNO_3, *etc.* Summarise your results in a table. Which washing technique would you recommend for which analysis?

3.8.8 Further Reading

R. M. Harrison, Metal analysis, in "Handbook of Air Pollution Analysis", ed. R. M. Harrison and R. Perry, Chapman and Hall, London, 1986, pp. 215–277.

R. W. Shaw, Air pollution by particles, *Sci. Am.*, 1987, **257**, 96–103.

H. Bridgman, "Global Air Pollution: Problems for the 1990s", Wiley, Chichester, 1994, pp. 65–87.

CHAPTER 4

Water Analysis

4.1 INTRODUCTION

4.1.1 The Hydrosphere

Water is one of the most precious commodities, although it is often taken for granted. It has numerous uses, most of them fundamental to life and society. Water is vital for drinking, without which no man or animal could survive, and it is used for maintaining personal hygiene. Agriculture and industry both utilise enormous amounts of water to respectively provide us with food and numerous consumer goods. Also, the fish that we eat depend entirely on water. It is impossible to exhaust the water supplies as water is continuously recycled through the hydrological cycle; however, it is possible to degrade the quality of the water to the point where it is useless, harmful or even deadly. Water pollution has seriously affected many rivers, lakes and even parts of the oceans.

Ninety-seven percent of the world's water is found in the oceans. Only 2.5% of the world's water is non-saline freshwater. However, 75% of all freshwater is bound up in glaciers and ice caps. Only 1% of freshwater is found in lakes, rivers and soils and 24% is present as groundwater. The use of water increases with the growing population, putting increasing strain on the water resources. In 1975, total global use of water was just under 4000 km^3 per year and this is expected to increase to about 6000 km^3 per year by the year 2000. Finding adequate supplies of freshwater to meet our ever-increasing needs, and maintaining its quality, is becoming a problem. Although water availability is not a problem on a global scale, it may be a problem finding high quality freshwater at the required place in the required quantity.

4.1.2 Water Pollution

Human settlements and industries have long been concentrated along rivers, estuaries and coastal zones owing to the predominance of water-borne trade. The human use of water leads to degradation of water quality by introducing various polluting substances. Historically, water pollution was not a major problem until large centres of human population developed. Rivers and seas have their own self-purification ability, but water pollution becomes a problem when it exceeds the self-purification capacity of a water body. A water pollutant is any biological, physical or chemical substance present at excessive levels capable of causing harm to living organisms, including man. Water quality determines its potential use. For example, the quality of water required for industrial use depends on the industrial process involved. Polluted water may be of use for some processes but not for others.

Water pollution can be a serious health hazard. People can become sick from drinking, washing or swimming in polluted water. Biological hazards of water include pathogenic bacteria, viruses and parasites, while chemical hazards include nitrates, lead, arsenic and radioactive substances. Of greatest concern is the quality of water intended for domestic water supplies as it must be free from substances harmful to human health. From a public health point of view it is most important that domestic water supplies are free from disease carrying microorganisms. Pathogens which can be transmitted through water include those responsible for intestinal tract infections (dysentery, cholera, typhoid fever) and those responsible for infectious hepatitis and polio. In developed countries of Europe and North America, water supplies are generally free from these organisms, but this cannot be said of developing countries where outbreaks of water-borne diseases such as typhoid fever and cholera still occur. Nevertheless, even in developed countries, water-borne disease is still a potential threat and water supplies are routinely screened for faecal coliform bacteria, the presence of which indicates the presence of faecal discharges in the water. Water supplies should also be free from toxic chemicals such as pesticides, insecticides, heavy metals and radioactive substances.

A common form of physical water pollution is thermal pollution. Industry utilises water for cooling purposes and this water is then returned to water bodies at higher than ambient temperatures. Major culprits in this respect are power stations, which use enormous amounts of surface waters for condensing steam which drives the turbines that produce electricity. The heated water is then discharged back to rivers, lakes, estuaries and seas. Fish are extremely sensitive to temperture and an increase of only a few degrees may have devastating consequences on life in a water body. For this reason, the temperature of water is reduced inside cooling towers at power stations before discharge. As this book

Table 4.1 *Major chemical pollutants in the hydrosphere*

Pollutant	*Typical sources*	*Comments*
Radioactivity	Accidental discharges from nuclear industry, transport of nuclear materials and nuclear tests.	Emotive subject. Health effects are frequently debated.
Organic chemicals	Agricultural use of herbicides and pesticides. Industrial and domestic wastes. Marine oil pollution from oil-tanker accidents.	A wide variety of chemicals are discharged (petroleum hydrocarbons, pesticides, detergents, *etc.*). Potentially harmful effects on human health and aquatic ecosystems.
Heavy metals	Wastes from industry, agriculture, urban drainage and domestic households.	Many metal pollutants can have potentially harmful effects on human health and aquatic ecosystems (*e.g.* mercury, lead, cadmium).
Acids	Drainage from mines, wastes from industry and acid deposition from the atmosphere.	Can harm aquatic ecosystems (*e.g.* sulfuric acid, nitric acid) by mobilising toxic metals.
Nutrients	Agricultural use of fertilisers and sewage.	May cause eutrophication (*e.g.* phosphorus and nitrogen compounds). Nitrates may affect human health.

deals with chemical analysis, physical and biological pollution, although important, will not be discussed in detail.

Primary sources of water pollution are: sewage, industrial waste-waters, land drainage and oil spills. Major categories of water pollutants and their sources are listed in Table 4.1. Many of these pollutants are continuously discharged into water bodies on a daily basis. Occasional releases of large quantities of pollutants attract the greatest public concern. Typical examples are accidental releases of radioactive substances from the Sellafield nuclear reprocessing plant in the UK into the Irish Sea, French nuclear tests in the Pacific Ocean and oil-tanker accidents such as the Exxon Valdez.

4.1.3 Biochemical Oxygen Demand (BOD)

Wastewater entering into a river or ocean from an effluent pipe outlet can be considered as a point source of pollution, the concentration of pollutant decreasing with increasing distance downstream from the source owing to dilution. Organic pollutants in wastewater present a specific pollution problem. Bacteria in the water degrade the organic pollution, utilising *dissolved oxygen* (DO) in the process. The amount of oxygen necessary to decompose organic matter in a unit volume of water is called the *biochemical oxygen demand* (BOD). The BOD can serve as a measure of organic pollution and it is routinely determined in waste-water treatment plants and water quality laboratories. If there is too much organic pollution present and consequently the BOD is too high, the DO may become too low to support aquatic life. The decomposition of organic pollution begins as soon as the wastewater enters the river and close to the point source of the pollution there is a *pollution zone*, with high BOD and low DO. Further downstream, there is an *active decomposition zone*, where the DO is at a minimum due to the biochemical decomposition of organic pollutants. Finally, there is a *recovery zone* in which the DO increases while the BOD decreases because most of the organic pollution has decomposed. Most water bodies have some ability to degrade organic waste; however, problems arise when the receiving water is overloaded with organic, oxygen-demanding wastes. If the water body's natural cleansing capability is overpowered, all the DO may be used up, resulting in fish deaths. BOD is discussed in more detail in Section 4.7.

4.1.4 Eutrophication

The increase in the concentration of nutrients in a water body is called eutrophication. Nutrients are elements essential for the growth of living organisms (C, N, P, K, S and some trace metals). Eutrophication is a natural and desirable process; however, human activities can greatly increase the presence of nutrients, leading to *cultural eutrophication*, a major problem in many areas. Sewage and fertiliser-rich farmland drainage, containing high levels of nitrogen and phosphorus, discharged into water bodies can lead to an increase in the phytoplankton population, the so-called *algal bloom*. The algae become so thick that light cannot penetrate the water and algae beneath the surface die owing to the absence of light. As the algae decompose the dissolved oxygen is used up, and fish, starved of oxygen, also start to die.

Problems of eutrophication are usually encountered in lakes receiving nutrient-rich wastes. Recently, concern has been expressed about possible eutrophication of coastal waters in the tropics by sewage and its potentially detrimental effects on coral reefs. Eutrophication problems can be solved by ensuring that wastewaters discharged into water bodies do not contain excessive nutrient levels. This can be achieved by employing advanced methods of wastewater treatment capable of

Table 4.2 *Eutrophication criteria for lakes and reservoirs*

Parameter	Oligotrophic	Mesotrophic	Eutrophic
Total N (μg L^{-1})	<200	200–500	>500
Total P (μg L^{-1})	<10	10–20	>20
DO in hypolimnion (% saturation)	>80	10–80	<10
Chlorophyll a (μg L^{-1})	<4	4–10	>10
Phytoplankton production (g C m^{-2} d^{-1})	7–25	75–250	350–700

Adapted from C. F. Mason (1990).

removing these substances. It would be more effective to control phosphorus rather than nitrogen, because phosphorus is usually the limiting agent in algal growth and much of it is released from point sources such as sewage works. On the other hand, most of the nitrogen comes from diffuse sources such as land drainage. Tertiary wastewater treatment involving chemical flocculation is capable of removing phosphorus from the effluent, but most sewage treatment plants do not employ tertiary treatment (see Section 4.1.9). It is possible, but expensive, to reverse eutrophication of affected water bodies.

Table 4.2 gives the relationship between the trophic status of a water body and several measurable parameters.

4.1.5 Acidification

Acidification of freshwaters has been going on for over a century, but it has only been recognised as a major water pollution problem since the late 1960s. Investigations of pH trends in rivers and lakes in Sweden, Norway, Canada and the US show a decrease in pH over the last 40 years and the pH in some poorly buffered lakes can be as low as 4–4.5. The cause of acidification is the deposition of acid rain and this is discussed in Section 2.1. Acidification can cause toxic metals, such as aluminium, to be leached into solution. One method to reverse acidification of a lake is by liming, but this is only a temporary solution.

Another source of acidity in surface waters is mine drainage from sulfur-bearing deposits of coal, iron, lead, zinc and copper. Drainage waters from underground and surface coal mines are especially acidic due to the presence of pyrite (FeS_2) in coal seams. Pyrite reacts with water and air in the presence of certain bacteria to produce sulfuric acid:

$$2FeS_2 + 7O_2 + 2H_2O \rightarrow 2FeSO_4 + 2H_2SO_4$$

Example 4.1

A river water sample was analysed and the concentrations of Ca^{2+} and CO_3^{2-} found to be 30 and 0.25 mg L^{-1}, respectively. Calculate the degree of saturation with respect to calcite. $K_{sp} = 8.7 \times 10^{-9}$ mol^2 L^{-2} for calcite.

The equilibrium between a salt and the dissolved ions is defined by the solubility product, K_{sp}. For mineral calcite the equilibrium can be described by:

$$CaCO_3(s) \rightleftharpoons Ca^{2+} + CO_3^{2-} \qquad K_{sp} = [Ca^{2+}][CO_3^{2-}]$$

where (s) denotes the solid phase. The product of dissolved ions actually measured in solution is called the *ion activity product* (IAP). For the case above, the concentrations are first converted to units of mol L^{-1} by dividing by the respective ionic masses and then the IAP is calculated:

$$[Ca^{2+}] = 30 \times 10^{-3}/40.08 = 7.5 \times 10^{-4}\ mol\ L^{-1}$$
$$[CO_3^{2-}] = 0.25 \times 10^{-3}/60.01 = 4.2 \times 10^{-6}\ mol\ L^{-1}$$
$$IAP = (7.5 \times 10^{-4}) \times (4.2 \times 10^{-6}) = 3.19 \times 10^{-9}\ mol^2\ L^{-2}$$

The relationship between IAP and K_{sp} can be used to evaluate the state of a solution:

IAP $> K_{sp}$, the solution is supersaturated and the salt will precipitate.
IAP $= K_{sp}$, there is equilibrium between the saturated solution and the salt.
IAP $< K_{sp}$, the solution is undersaturated and the salt will dissolve.

In the above case, IAP (3.19×19^{-9}) $< K_{sp}$ ($K_{sp} = 8.7 \times 10^{-9}$) and the solution is undersaturated. Calcite will therefore dissolve.

$$\text{Degree of saturation} = IAP/K_{sp} = 3.19 \times 10^{-9}/8.7 \times 10^{-9} = 0.37$$

The oxidation of ferrous (Fe^{2+}) iron to ferric (Fe^{3+}) iron produces yet more sulfuric acid:

$$4FeSO_4 + 10H_2O + O_2 \rightarrow 4Fe(OH)_3 + 4H_2SO_4$$

Mine drainage is controlled by sealing abandoned mines, drainage control and chemical treatment involving liming.

Acid sulfate soils can also act as a source of acidity under appropriate environmental conditions. Acid sulfate soils are rich in pyrite (FeS_2) and they tend to occur in inter-tidal swamps and coastal plains. When drainage brings oxygen into these soils the pyrite is oxidised to sulfuric acid according to the same reactions as shown above for acid mine drainage. The pH of the water can fall to well below 4 and toxic levels of aluminium are released into solution, posing a considerable hazard to aquatic life. Acid sulfate soils are found mainly in the tropics, the best known example being the Bangkok Plain in Thailand where acid sulfate soils occupy some 600 000 ha.

Effects of acidic pollution include:

- Destruction of aquatic life. Below pH 4, most life forms in surface waters die.
- Increased corrosion. This can have an effect on boats, structures erected in the water body (*e.g.* piers) and plumbing systems.
- Damage to agricultural crops. If the pH of irrigation water drops below 4.5, metals toxic to plants can be leached into solution.

4.1.6 Salinity

While most of the world's water is saline (*i.e.* seawater), in freshwaters salinity is undesirable. However, freshwater can become saline due to:

- Industrial wastewaters which tend to contain high levels of inorganic salts formed by various industrial processes.
- Road drainage waters which contain salt used to melt snow and ice on highways.
- Irrigation which dissolves salts from the soil.
- Seawater penetration into rivers during high tides and low runoff.
- Brines from oil wells and mines which are sometimes released into fresh water.

Salinity in freshwaters can cause a number of problems:

- Saline waters are not suitable as a source of drinking water.
- Salinity can adversely affect aquatic organisms in fresh waters because, unlike marine life forms, they are generally not adapted to high salinity.
- Salinity in irrigation water can adversely affect plant growth and crop production.

4.1.7 Coastal Pollution

The composition of seawater is quite different from that of fresh waters (see Table 4.4); seawater is a concentrated electrolyte solution containing high concentrations of several ions. Furthermore, seawater is more resistant to changes in pH upon addition of acids or alkalis than freshwaters owing to its greater buffering capacity. Despite being by far the largest reservoir in the hydrosphere, the oceans' capacity for absorbing wastes is not limitless and there is evidence of seawater pollution throughout the world. It is, however, the coastal areas that are the worst affected. Coastal waters receive both direct discharges of effluents and contaminated river outflows. Pollution of coastal zones has long been a major issue throughout the world and the problem is growing. Coastal pollution is especially severe in developing countries, where sewage and other wastes are discharged to estuarine and coastal waters without any prior treatment. Often, these effluents are loaded with organic waste that is high in nutrient content. Population densities in coastal zones, even in areas that until quite recently could have been considered as remote or pristine, have been rising due to the expansion of tourism as a major earner of foreign currency in many developing nations. The effect of coastal pollution on public health is a serious concern as seafood is the primary and preferred source of nutrition in these countries. Increasing pollution of coastal waters threatens their use for both tourism and the production of seafood. Coastal pollution carries with it the following threats:

- *Eutrophication*. Discharge of effluents high in nutrients can produce algal blooms and lead to deoxygenation of coastal waters. Decay of organic matter uses up oxygen and depressed DO levels can result in the death of fish and other organisms both in coastal fish farms and capture fisheries, reducing overall marine productivity and decreasing food supplies.
- *Bioaccumulation of toxic metals*. Toxic metals discharged in effluents can be bioconcentrated in seafood, especially in shellfish such as oysters, mussels and cockles, to levels in excess of public health standards, thus presenting a health hazard to those eating seafood.
- *Microbial contamination*.The discharge of raw sewage to coastal water carries with it the threat of microbial contamination of the water and seafood. Increasing levels of *E. coli* in coastal waters have been correlated with symptoms of gastrointestinal and skin complaints among bathers. Consumption of contaminated seafood is also a significant public health problem in developing countries. However, outbreaks of food poisoning by seafood, especially shellfish, are not uncommon even in developed countries, as evidenced by public warnings given out from time to time.
- *Red tide*. The potential contribution of coastal pollution to toxic

algal blooms, the so-called "red tides", is a grave concern in coastal zones of many developing countries. Red tides are due to the presence of a specific dinoflagellate species; one species of particular importance in east Asia is *Pyrodinium bahamense* var. *compressa*. Red tides have been responsible for many deaths over the last two decades, mainly among those eating shellfish from waters supporting blooms of toxic algae. One possible explanation which has been proposed for the growing occurrence of red tides is that the shift in the nitrogen to phosphorus ratio in sewage and the increasing nutrient loads may have contributed to a shift in local phytoplankton species dominance, with more toxic dinoflagallates displacing the diatoms.

4.1.8 Groundwater Pollution

Groundwater has long served as a source of drinking water and it is still important today. In the US, about 50% of the population depends on groundwater as a source of drinking water. As groundwater is isolated from the surface, most people take it for granted that groundwater should be relatively pure and free from pollutants. Although most groundwaters are still of high quality, at some locations it is becoming increasingly difficult to maintain the purity of groundwaters. Sources of groundwater contamination include:

- *Saltwater intrusion.* Intensive pumping of groundwater may cause water levels to decline, allowing seawater to infiltrate inland and contaminate the aquifer.
- *Landfills.* Seepage of chemicals from municipal and industrial landfills can penetrate through the soil and contaminate underlying groundwater. Methods to control this pollution include employing liners to prevent leaks and a system of collecting the leachate that seeps from the landfill.
- *Underground storage tanks.* In North America alone there are some 1.4 million underground storage tanks containing petrol and other hazardous substances. Leakage of only a few litres of these compounds can have drastic consequences on groundwater quality.
- *Agriculture.* Several agricultural practices contribute to groundwater pollution: fertiliser and pesticide application, irrigation and animal feeding operations. Agricultural leachate can seep through the soil and contaminate groundwater supplies.
- *Septic tanks.* Septic tanks are generally used in rural communities not connected to sewers. Even in a developed country such as the

US there are over 20 million septic tanks. Leakage from badly designed and poorly operated septic tanks may contaminate groundwaters with nutrients, toxic substances and bacteria.

- *Oil wells.* Abandoned oil wells may contaminate groundwaters with brine when well casings rupture.
- *Road drainage.* Salts applied on roads for snow control, chemicals from accidental spillages and other substances that end up on roads are washed away as road drainage water which may seep through to the aquifer.

Still other sources include radioactive disposal sites, mining wastes and construction excavations.

Groundwaters differ in several ways from surface waters. Bacterial breakdown of pollutants is extremely slow compared to surface waters owing to the absence of aerobic bacteria in groundwater because of the lack of oxygen. Also, the rate of flow of groundwater though the permeable rock is very slow compared to surface water flow rates and pollution cannot be diluted or dispersed so easily. Nevertheless, rocks and soils may filter some of the pollution.

4.1.9 Wastewater Treatment

The prevention of water-borne disease was the main reason for introducing water pollution control methods. In the Middle Ages in Europe, sewage was discharged into the streets and left there. In England, the discharge of sewage into rivers was started around 1810. During the Industrial Revolution, many rivers were extremely polluted and cholera epidemics were common. The world's first sewer system was installed in Hamburg, Germany, in 1843 and the first sewer system in the US was constructed in 1855. Filters were first introduced in England in the early 19th century but this did not get rid of pathogens. In 1875 the first Public Health Act was passed in the UK, making local authorities responsible for sewage disposal. This act stated that sewage should be free from dangerous bacteria, but the technology necessary for achieving this was not yet available. The Rivers Pollution Prevention Act was passed in the UK in 1876, prohibiting direct discharge of industrial wastes into rivers. Industries were required to discharge their wastes into sewers and town councils were responsible for sewage treatment. In the US, sewage treatment plants were first built in the 1870s and by 1910 about 10% of the sewage in the US was treated. Disinfection techniques were first introduced in the early 20th century in England and their widespread use has greatly reduced the occurrence of water-borne diseases in many countries. Today, in many developed countries, close to 100% of the sewage is treated and many countries have legally enacted water quality standards (see Appendix III). However, in many developing countries, sewage treatment is almost non-existent and raw sewage is discharged into water bodies, causing serious water pollution problems.

Sewage treatment takes place at specially designed plants that receive sewage from households and industries. Some industries may have their own wastewater treatment plants designed to reduce specific pollutants produced at the site. After treatment, the wastewater is discharged into a water body (river, lake or ocean). Wastewater treatment is classified into three categories: primary, secondary and tertiary, in order of increasing purification capability.

- *Primary treatment.* This is the most rudimentary treatment which can remove most of the solids in the water and moderately lower the BOD. The sewage enters the plant through a series of screens that remove large floating objects. Next, the sewage enters a grit chamber where sand, stones and grit settle by gravity to the bottom of the chamber. The grit is disposed of at a landfill site. The sewage then enters a sedimentation tank where suspended solids settle out to form a raw sludge which is collected and disposed. Chemical additives can be used to help the sedimentation process. Primary treatment removes about 60% of the solids and approximately 35% of the BOD. This is the most common type of treatment and many plants employ primary treatment only.
- *Secondary treatment.* This utilises biological processes to remove additional suspended material and further lower the BOD. Secondary treatment generally employs the *activated sludge process*, although *trickling filters* are also used. Sewage is introduced into an aeration tank, where it is mixed with air and with activated sludge recycled from the subsequent sedimentation tank. The sewage remains in the tank for several hours and aerobic bacteria, naturally present in the sludge, break down organic pollutants in the wastewater. The sewage then enters a final sedimentation tank where the sludge settles out. Some of this sludge is recycled into the aeration tank. Most of the sludge is taken to a sludge digester where it is treated with anaerobic bacteria, which further degrade the sludge. Methane is produced in the digester and it is either burned off or used as a fuel in the plant. The sludge is then dried and disposed of by landfill. In trickling filters, wastewater is sprayed onto the surface of tanks filled with large-pored material (rocks, brick fragments, *etc.*). Microorganisms form a film on the material

and consume nutrients from the wastewater trickling through to the bottom of the tank. Approximately 90% of the solids and BOD can be reduced by secondary treatment.

- *Tertiary or advanced treatment.* Tertiary treatment is designed to reduce further the concentration of specific pollutants in the wastewater. Many processes are available depending on the pollutant to be removed. Tertiary treatment can remove suspended solids, dissolved organic compounds, dissolved nutrients (phosphorus and nitrogen) and heavy metals. Methods involve sand filters, carbon filters, electrodialysis, reverse osmosis, ion exchange and the use of chemical additives such as coagulants (alum) and oxidants (ozone, hydrogen peroxide). Some of these processes can be quite expensive. Tertiary treatment can remove more than 95% of the pollutants in the wastewater, but it is not widely used in sewage treatment plants.

Regardless of what type of treatment is used, the sewage is finally treated with chlorine gas before being discharged into the environment. This destroys disease-causing bacteria. The effectiveness of conventional primary and secondary treatment processes that are widely used in sewage treatment plants is summarised in Table 4.3.

The water entering and exiting the sewage treatment plant is analysed for some of the parameters discussed later on in this chapter. Many of the treatment processes have to be designed specifically for the type of wastewater in question and therefore it is necessary to know the type and concentration of pollutants in the raw sewage that is to be treated. Also, it is necessary to ascertain that the treated wastewater meets the legally enacted standards and can be discharged safely into the environment. Many of the processes require careful control of operating conditions (*e.g.* pH) and it may be necessary to perform analysis at several stages

Table 4.3 *Efficiency of removal of pollutants present in raw sewage by conventional treatment processes in sewage treatment plants*

	Removal efficiency (%)	
Pollutant/parameter	*Primary treatment only*	*Primary + secondary treatment*
Suspended solids	60	90
Total nitrogen	20	50
Total phosphorus	10	30
Dissolved minerals	0	5
Biochemical oxygen demand	35	90
Chemical oxygen demand	30	80

during the treatment process. A modern wastewater treatment plant may have continuous, in-line, instrumental analysers installed.

4.1.10 Potable Water Treatment

On average, each person in the US utilises between 300 and 400 litres of water every day for domestic use (drinking, washing, cooking, flushing toilets, watering gardens). Approximately 8% of the global freshwater resources is utilised for domestic purposes. Water for domestic use is mainly drawn from groundwater and surface waters. Groundwater, extracted from aquifers below the water table, has long been an important source of domestic water. Wells are still used to pump groundwater in many rural communities. Although groundwaters are generally of higher quality than surface waters, they, too, are prone to pollution (see Section 4.1.8). Water intended for domestic use generally needs to be treated to conform to national drinking water standards (see Appendix III). Approximately 60% of the water used for domestic purposes is returned to rivers as wastewater. In regions with large populations and considerable industrial activity, the contribution of sewage to total river flow can be high, especially during dry periods. Hence, wastewater discharged into a river may be extracted further downstream and serve as a source of domestic water for another community. In other words, what was one person's waste may be another person's drinking water and it is of utmost importance that high quality is maintained in order to protect the public health. Generally, water intended for domestic use is first stored in reservoirs, where fine sediment may settle out, improving the clarity of the water. The water is then treated in a water treatment plant before entering the municipal water supply. Treatment varies depending on the quality of the water, but it generally involves filtration and chemical treatment (*e.g.* chlorination, fluoridation). In some locations where there are few sources of freshwater, seawater desalination is practiced. Water analysis is performed routinely at water treatment plants to ensure that legally enacted standards are met.

Owing to public concerns (often deemed unfounded) over the quality of domestic water, and objections to the taste of chlorinated water, many people prefer to drink bottled mineral water. Water analysis is performed routinely at mineral water bottling plants, and suppliers are required to specify the composition of the water. Although both bottled mineral waters and domestic waters in developed countries are generally safe to drink, there have been isolated cases of contamination which went unnoticed in the quality control laboratories. This is not surprising, considering that approximately 1000 substances may be present in public water supplies, and laboratories generally screen the water for the more commonly encountered substances. Water may also become contaminated during distribution. For example, considerable concern has been expressed in the past over lead pollution in drinking water. The source of lead was lead pipes between water mains and homes, lead pipes in buildings, leaded solder in copper pipes and brass taps. Nowadays, lead pollution in drinking water is mainly a problem in older buildings. Water may also be inadvertently contaminated during treatment as a result of negligence, and such incidents have been reported for both bottled mineral waters and domestic waters.

4.1.11 Water Analysis

Water analysis can involve any of the following samples: surface water from lakes, rivers and seas, groundwater, drinking water, industrial and municipal wastewater and boiler water. Concentrations of major ions in some surface waters are listed in Table 4.4. Water quality is routinely monitored in sewage and water treatment plants. Methods have been developed for a wide range of inorganic and organic analytes utilising a variety of analytical techniques. Those methods which have been adopted by the water industry are called *standard methods*. Methods which can be readily carried out using basic equipment are given in the present book, and some of these were adapted from standard methods.

There are many water analysis tests kits available on the market. These are mainly based on colorimetric and electrochemical methods, and they are generally intended for use in the field. Most of these test methods involve the addition of reagents in tablet or powder form, the latter being

Table 4.4 *Concentrations of major ions in seawater* (mg kg^{-1}), *salt lakes* (mg kg^{-1}) *and riverwater* (mg L^{-1})

Constituent	Seawater	Salt lake water	Riverwater
HCO_3^-	140	180–17 400	17.9–183
SO_4^{2-}	2649	264–13 590	0.44–289
Cl^-	18 980	1960–112 900	2.6–113
NO_3^-	2	1.2–1.6	0.3–1.9
Ca^{2+}	400	3.9–15 800	5.4–94
Mg^{2+}	1272	23–41 960	0.5–30
Na^+	10 556	1630–67 500	1.6–124
K^+	380	112–7560	0.0–11.8

supplied in sachets. These reagents are frequently the same as those used in standard methods but are supplied in pre-weighed quantities. The procedures developed for these test kits are convenient and easy to follow. One example is the Hach test kit (see Hach reference in reading list below). Some kits also include portable probes for pH, temperature, conductivity and dissolved oxygen. Portable probes based on ion selective electrodes are becoming increasingly available for a variety of other chemical components.

The methods which will be discussed here are suitable for any type of water sample, but it is assumed that most samples will be surface water samples collected from rivers, lakes, coastal waters or wastewaters. You may also apply these methods to the analysis of tapwaters and bottled mineral waters.

4.1.12 Sampling and Storage

The volume of sample to be collected will depend on the number of analyses that will be carried out. If a comprehensive analysis of the water is to be carried out, 2 L of sample should be collected. If one or few analytes are to be determined, smaller volumes may be satisfactory depending on the volume required for the test. Plastic or glass bottles can be used, depending on the analyte. In some instances it may be necessary to collect samples in several different bottles. Analysis of dissolved oxygen requires separate bottles.

You should remember that the sample you collect is only representative of the place and time of sampling. A detailed survey of a water body may require an investigation of spatial and temporal variability. The composition of the sample may also change during transport and storage, mainly due to biochemical and surface reactions. Be sure to follow the recommended procedure for each analyte.

Surface water samples are the easiest to collect. Collecting samples at depth requires special collectors, of which there are several types (Ruttner, Kemmerer, Dussart, Valas, Watt, *etc.*). The most common sampler is the Van Dorn sampler, shown in Figure 4.1. The sampler consists of a hollow cylinder of PVC and two rubber valves at the open ends. The two valves are interconnected by a rubber tube and also have a string attached to them. Prior to immersion in water the two valves are pulled out and the strings are attached to a lock fitted outside the cylinder. The cylinder, open at both ends, is lowered to the desired depth by means of graduated rope. A metallic messenger is released along the rope. The messenger strikes the lock and releases the two rubber valves to close the cylinder. The sampler is then pulled out of the water and the

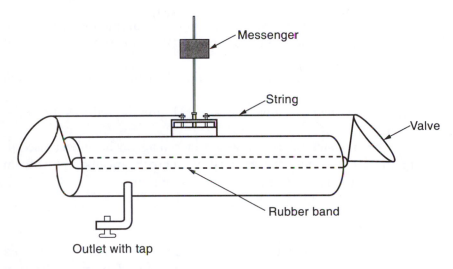

Figure 4.1 *The Van Dorn sampler*

sample transferred to a bottle for storage. It is also possible to take samples from different depths by pumping water through a plastic tube lowered to the desired depth.

You will generally be collecting samples at, or close to, the water surface. Take clean sampling bottles, filled with pure laboratory water to the site. Empty the bottle and then rinse several times with the water to be collected. Each time, empty the bottle downstream. Finally fill the sample bottle, seal and label it, and take it to the laboratory for analysis. The bottle should be filled slowly to avoid turbulence and air bubbles. When sampling water from wells and taps, the pump or tap should be left to run long enough to flush the pipe sufficiently to give a representative sample.

Recommended bottle materials, preservatives and storage times are given in Table 1.4. The quoted storage times should be considered as maximum recommended times. Preferably, the samples should be analysed as soon as possible, but this may not always be practical. Follow any preservation, storage or treatment procedures that may be required for the specific analysis. For example, dissolved oxygen determination requires addition of a preserving reagent on site. Nutrient analysis (*e.g.* nitrogen, phosphorus) should be carried out as soon as possible after sampling since metabolic conversion of the components will lead to unrepresentative results if analysis is delayed. Several preservation procedures have been recommended for nutrient samples when they cannot be analysed promptly:

- *Refrigeration* — this slows down, but does not eliminate, bacterial and chemical reactions; reaction rates at 4 °C are about a quarter of the rates at 25 °C.
- *Freezing* — this reduces bacterial and chemical reactions even further.
- *Addition of acid* — lowering the pH greatly reduces bacterial activity.
- *Addition of a bactericide* — many bacterial agents have been used to eliminate completely the activity of microorganisms; chloroform and mercury(II) chloride are the most common.

4.1.13 Filtration

Filtration if often employed in water analysis to separate dissolved components from those present in suspended matter. However, filtration can present several problems:

- *Penetration of insoluble material through the filter.* The standard procedure is to filter the sample through a filter with a nominal pore size of 0.45 μm, and components passing through the filter are deemed to be "dissolved" or "soluble". This operational definition, while practical, is arbitrary and inaccurate as colloids and polymers can penetrate through the filter together with truly dissolved substances.
- *Contamination.* Substances present in the filter material may dissolve into the sample during filtration. This is especially a problem when analysing trace metals in relatively clean samples. Filters for trace metal analysis should be washed with acid before use, so as to remove any metal contaminants from the filter material.
- *Adsorption.* Some analytes may adsorb onto the filter material or filtration unit. For example, certain metals may be adsorbed onto the walls of glass filtration units.

Instructions for filtration are given in the appropriate experiments. The filters should be pre-conditioned by rinsing, first with laboratory water and then with some sample which should be discarded. Only then should the sample proper be filtered. Filter units made of glass and plastic are available. All-glass Millipore filter holder and vacuum flask are the most suitable for trace analysis as no adsorptive losses have been observed. However, other apparatus may be suitable for general anlyses (*e.g* Buchner funnel and filtering flask). A typical filtration unit is shown

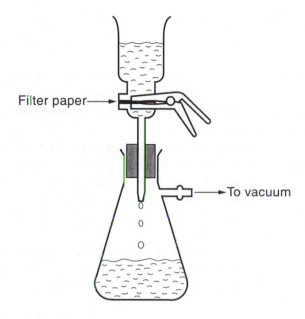

Figure 4.2 *Vacuum filtration unit*

in Figure 4.2. Vacuum is applied by connecting the flask to an air pump or a water aspirator and turning on the pump or water before connecting the flask. When filtration is complete, the connecting tube should be disconnected from the water aspirator or pump before the pump or water are turned off.

With metal analysis, samples should be acidified only after filtration. Acidifying before filtration could dissolve some metals which were not in true solution in the sample, and hence the results of the analysis would not be representative of field conditions. Filtration is also recommended when it is necessary to store samples for longer periods as it removes larger biological materials which could cause an interference by reacting with the analyte during storage.

4.1.14 Questions and Problems

1. Although water is one of the most abundant resources, concerns are being expressed about its availability and quality in the future. Discuss.
2. Describe the various uses of water and explain how water quality may affect these uses.
3. Outline the different types of water pollutants and their sources.
4. Describe the process of eutrophication.

5. Explain how DO and BOD vary with distance downstream from a point source of wastewater containing high levels of organic matter.
6. Describe the different types of wastewater treatment (primary, secondary, tertiary).
7. Discuss the problems of obtaining a representative water sample.
8. Calculate the average residence times of Na^+, Cl^- and Ca^{2+} in the oceans, taking the volume of the oceans as 1.23×10^{21} L, the river flow to the oceans as 3.6×10^{16} L a^{-1}, and the average river water concentrations of Na^+, Cl^- and Ca^{2+} as 6.3, 7.8 and 15 mg L^{-1}. The average seawater composition is given in Table 4.4 and the density of seawater is 1.025 kg L^{-1}. Ignore other inputs to the oceans (*e.g.* rainfall) and assume steady state.
9. Calculate the ionic strength of a river water sample with the following composition (in mg L^{-1}): 108 HCO_3^-, 19 SO_4^{2-}, 0.5 F^-, 4.9 Cl^-, 0.3 NO_3^-, 23 Ca^{2+}, 6.2 Mg^{2+}, 16 Na^+ and 0.1 K^+ (see Example 5.1).

4.1.15 Suggestions for Projects

1. Compile as much information on wastewater and its treatment in your area (town/city) to include:
   - Wastewater composition
   - Wastewater volume discharge rates
   - Wastewater treatment technologies
   - National/local effluent standards

   Classify, and if possible quantify, the wastewater according to its source, *i.e.* domestic, industrial, agricultural. You may be able to obtain useful information from your local sewage treatment works and the pollution control department of your local government authority (*e.g.* town council). If sufficient information is available you may be able to calculate the discharge rates of individual pollutants (concentration × volume discharge rate). You could arrange to visit the local sewage works and see for yourself the various wastewater treatment processes at work. Write a critical review of wastewater treatment technologies, standards and management strategies employed in your community. Compile the data you have gathered into visually appealing and comprehensible formats (tables, graphs, pie-charts, bar-charts, *etc.*) and prepare a report on your findings. You may present the results of your study in a seminar and discuss them further with your fellow students.
2. Carry out a comprehensive survey of surface water quality in your

area by collecting samples at various sites and analysing them using the procedures outlined in this chapter. You may consider analysing the samples two (duplicate) or three (triplicate) times each and reporting the mean of the measurements. This will give you some idea about how well you are performing your analysis. Enter your results in a table. You should also record the colour, odour and any other relevant observations (proximity of sewage outlet, industrial wastewater outlet, *etc.*).

3. Carry out a detailed analysis of a sewage effluent using the methods described in this chapter and enter your results in a table.
4. Sample and analyse tapwater from several houses in your area. Also, analyse various bottled mineral waters on the market. Compare the quality of municipal tapwater and bottled mineral waters.

4.1.16 Reference

C. F. Mason, Biological aspects of freshwater pollution, in "Pollution: Causes, Effects and Control", 2nd edn., ed. R. M. Harrison, Royal Society of Chemistry, Cambridge, 1990, pp. 99–125.

4.1.17 Further Reading

H. H. Rump and H. Krist "Laboratory Manual for the Examination of Water, Wastewater and Soil", 2nd edn., VCH, Weinheim, 1992.

APHA, "Standard Methods for the Examination of Water and Wastewater", 16th edn., American Public Health Association, Washington, 1985.

C. N. Sawyer, P. L. McCarty and G. F. Parkin, "Chemistry for Environmental Engineering", 4th edn., McGraw-Hill, New York, 1994.

H. L. Golterman, R. S. Clymo and M. A. M. Ohnstad, "Methods for Physical & Chemical Analysis of Fresh Waters", Blackwell, Oxford, 1978.

Hach Company, "Hach Water Analysis Handbook", Hach, Loveland, 1992.

W. Fresenius, K. E. Quentin and W. Schneider (eds.), "Water Analysis: A Practical Guide to Physico-Chemical, Chemical and Microbiological Water Examination and Quality Assurance", Springer, Berlin, 1988.

H. Fish, Freshwaters, in "Understanding our Environment: An Introduction to Environmental Chemistry and Pollution", ed. R. M. Harrison, Royal Society of Chemistry, Cambridge, 1994, pp. 53–91.

R. F. Packham, Water quality and health, in "Pollution: Causes, Effects and Control", 2nd edn., ed. R. M. Harrison, Royal Society of Chemistry, Cambridge, 1990, pp. 83–97.

J. N. Lester, Sewage and sewage sludge treatment, in "Pollution: Causes, Effects and Control", 2nd edn., ed. R. M. Harrison, Royal Society of Chemistry, Cambridge, 1990, pp. 33–62.

4.2 SOLIDS IN WATERS

4.2.1 Introduction

The term *solids* refers to the quantity of solid matter remaining in a water sample after drying or igniting at a specified temperature. Several categories of solids can be defined: total, dissolved, suspended, settleable, fixed and volatile. Differentiation between dissolved and suspended solids in water is accomplished by means of filtration. Solids in waters are undesirable for many reasons. They degrade the quality of drinking water and they also reduce the utility of water for irrigation and industrial purposes. Waters with high solids content require additional mechanical and chemical treatment, and the cleaning process becomes more expensive. Furthermore, high levels of solids in water increase the water density, affect osmoregulation of freshwater organisms and reduce the solubility of gases (*e.g.* O_2). Suspended solids in untreated wastewater can lead to sludge deposits and anaerobic conditions in receiving surface waters.

The main sources of solids in natural waters are water, rain and wind erosion of soil surfaces. Water erosion is more pronounced in humid and semi-humid regions with heavy storms or prolonged wet seasons (*e.g.* in monsoon climates). Wind erosion is very common in arid regions (*e.g.* American prairies, tropical savannas). Solids, originating from soil, decrease water transparency, inhibit photosynthetic processes and eventually lead to increase of bottom sediments and decrease of water depth in lakes, ponds and rivers. The latter can increase the frequency and severity of flooding.

Municipal and industrial wastewaters are polluted by solids originating from domestic wastes, road runoff and industrial processes. Unlike the above-mentioned diffuse sources of solids in natural waters, these solids may be removed from wastewater in sewage treatment plants. The amount of suspended solids increases with the degree of water pollution. In sludges, most of the solid matter is in suspended form and the fraction of dissolved solids is of minor importance. The concentration of various types of solids in untreated domestic wastewaters is given in Table 4.5. The total dissolved solids content of drinking water varies between 20 and 1000 mg L^{-1}, and it consists mainly of inorganic salts, some organic substances and dissolved gases.

Suspended solids determination is extremely useful in the assessment

Table 4.5 *Solids content of domestic wastewater with different degrees of contamination*

Solids content (mg L^{-1})	Degree of contamination		
	Weak	Medium	Heavy
Total solids (TS)	300	700	1000
Total dissolved solids (TDS)	200	500	700
Fixed dissolved solids (FDS)	120	300	400
Volatile dissolved solids (VDS)	80	200	300
Suspended solids (SS)	100	200	300
Fixed suspended solids (FSS)	20	50	60
Volatile suspended solids (VSS)	80	150	240
Settleable solids (SETS)	4	8	15

of polluted waters. It is widely used to determine the strength of domestic wastewaters and to evaluate the efficiency of water treatment processes. The determination of settleable solids is also important in the analysis of wastewaters. It serves as the basis for deciding whether primary sedimentation is required for waste treatment. In the past, total solids determination was used as a measure of the pollution present in wastewater. Nowadays, this has been largely replaced by BOD and COD tests, which are used to evaluate the strength of pollutants in wastewater more accurately. Information on total dissolved solids can easily be obtained from conductivity measurements. Total dissolved concentrations in some waters are given in Table 4.6. Because fixed and volatile solids are related to the amount of inorganic and organic matter present in the solid fraction of wastewater, activated sludge and industrial wastes, their determination is useful in the control of wastewater treatment plant operations. However, BOD, COD and organic carbon determinations provide better measures of organic matter.

Table 4.6 *Total dissolved solids (TDS) in some water samples*

Water type	TDS (mg L^{-1})
Sea water	35 000
Recommended for potable water	< 1000
Irrigation water	< 500 no restriction on use
	500–2000 moderate restriction on use
	> 2000 severe restriction on use

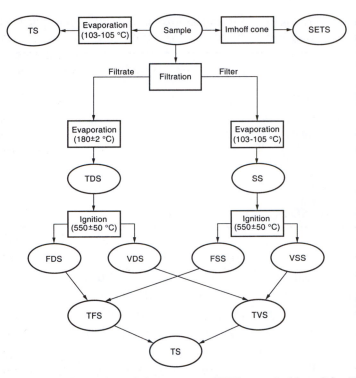

Figure 4.3 *Analysis scheme for solids in water. SETS = settleable solids, TS = total solids, TDS = total dissolved solids, SS = suspended solids, FDS = fixed dissolved solids, VDS = volatile dissolved solids, FSS = fixed suspended solids, VSS = volatile suspended solids, TFS = total fixed solids, TVS = total volatile solids*
(Adapted from "Water Quality Characteristics, Modelling, Modification", G. Tchobaoglous and E. D. Schroeder, Addison-Wesley, Reading, MA, 1985)

4.2.2 Methodology

The analysis scheme for determining solids is illustrated in Figure 4.3 and it involves the heating and evaporating of samples to constant weight at a definite temperature. The temperature at which the water residues are dried or ignited determines the general classification of solid types:

- Total solids dried at 103–105 °C
- Total dissolved solids dried at 180 ± 2 °C
- Total suspended solids dried at 103–105 °C
- Fixed and volatile solids in wastewater ignited at 550 ± 50 °C

The first three methods are suitable for the determination of solids in potable, surface and saline waters, as well as domestic and industrial wastewater in the range up to 20 000 mg L^{-1}. The last method is suitable for the determination of solids in sediments, as well as solid and semi-solid materials produced during water and wastewater treatment. Settle-able solids are determined volumetrically using an Imhoff cone.

4.2.2.1 Sample Handling and Preservation. Use plastic or glass bottles to collect the sample and analyse as soon as possible after collection. If samples must be stored before analysis can be carried out, store in a refrigerator at 4 °C to minimise microbiological decomposition of solids.

4.2.3 Total Solids (TS)

4.2.3.1 Methodology. A water sample is evaporated in a weighed dish and dried at 105 °C. The increase in weight over that of the empty dish represents the total solids.

4.2.3.2 Materials
- Evaporating dishes: porcelain, platinum, high-silica glass (*e.g.* Vycor), stainless steel or aluminium. Platinum or Vycor are prefer-able. Porcelain dishes are not recommended owing to a tendency to lose weight. However, they may be used if the other materials are not available. Platinum dishes are not available in many labora-tories owing to their high cost.
- Desiccator
- Drying oven
- Analytical balance, capable of weighing to 0.1 mg

4.2.3.3 Experimental Procedure. Clean evaporating dish with labora-tory water and heat in an oven at 105 °C for 1 h. Store in desiccator and weigh immediately before use. Measure accurately a volume (100–500 mL) of well-mixed sample and transfer into a pre-weighed evaporating dish. Evaporate to dryness in a preheated drying oven at 105 °C. This usually takes about 6 h but highly mineralised water may require longer. If sample volume exceeds dish capacity, add successive portions to the same dish after evaporation. Take the dish out of the oven and cool in a desiccator. Weigh dish on an analytical balance to the nearest 0.1 mg. Calculate the total solids as follows:

$$TS \ (mg \ L^{-1}) = 1000 \times (M_t - M_d)/V$$

where M_t is the weight of the dish + dried residue (mg), M_d is the weight of the dish (mg) and V is the sample volume (mL).

Notes

1. Highly mineralised waters from arid and semi-arid regions with considerable calcium, magnesium, chloride and/or sulfate content may be hygroscopic and require prolonged drying, efficient desiccation and rapid weighing.
2. For accurate results repeat the drying and weighing cycle until constant weight is obtained (± 0.5 mg).

4.2.4 Total Dissolved Solids (TDS)

4.2.4.1 Methodology. The water sample is filtered through a standard glass fibre filter, and the filtrate is evaporated to dryness in a weighed dish and dried at 180 °C. The increase in weight over that of the empty dish represents the total dissolved solids.

4.2.4.2 Materials
- Same as in Section 4.2.3
- Buchner funnel and suction flask or Millipore filtration unit
- Glass fibre filter paper, Whatman GF/C, or similar
- Hot water bath

4.2.4.3 Experimental Procedure. Heat evaporating dish of appropriate size in oven at 180 °C for 1 h. Cool in a desiccator and weigh. Measure accurately a volume (100–500 mL) of well-mixed sample and pass through the filter under slight suction. Wash any remaining solids from the measuring cylinder with three successive 10 mL portions of laboratory water and pass the washings through the filter. Transfer filtrate to a pre-weighed evaporating dish and evaporate to dryness on a hot water bath. If filtrate volume exceeds dish capacity, add successive portions to the same dish after evaporation. Dry for at least 1 h in an oven at 180 °C, cool in a desiccator and weigh. Calculate the total dissolved solids (TDS) using the same equation as in Section 4.2.3.

Notes

1. Residues dried at 180 °C will lose almost all mechanically occluded water.
2. Highly mineralised waters from arid and semi-arid regions containing high levels of calcium, magnesium, chloride and/or sulfate content may be hygroscopic and require prolonged drying, efficient desiccation and rapid weighing. Samples high in bicarbonate may require prolonged drying at 180 °C to ensure complete conversion of bicarbonate to carbonate.

3. For accurate work, repeat the drying, desiccating and weighing cycle as for total solids.

4.2.5 Suspended Solids (SS)

4.2.5.1 Methodology. A water sample is filtered through a pre-weighed glass fibre filter. The filter is dried at 105 °C and re-weighed. The amount of suspended solids is determined from the increase in weight of the filter paper.

4.2.5.2 Materials
- Same as in Section 4.2.4

4.2.5.3 Experimental Procedure. Wash filter in filter holder under suction with successive small volumes of laboratory water. Remove filter paper from assembly, place in the aluminium or stainless steel dish and dry in oven at 105 °C for 1 h. Cool in desiccator and weigh. Repeat the drying, desiccating and weighing until a constant weight is obtained or until drift is less than 0.5 mg. Replace the filter in the filtration assembly and dampen with laboratory water. Measure accurately a volume of well-mixed water sample (100–500 mL) and filter under slight suction. Wash any remaining solids from the measuring cylinder with three successive 10 mL portions of laboratory water and pass the washings through the filter. Remove the filter and dry in oven at 105 °C for 1 h. Cool in a desiccator and weigh. Calculate the suspended solids (SS) as follows:

$$SS \ (\text{mg L}^{-1}) = 1000 \times (M_t - M_b)/V$$

where M_t is the weight of the filter paper after sample filtration (mg), M_b is the weight of the filter paper prior to sample filtration (mg) and V is the volume of the sample (mL).

You may also determine suspended solids by difference from the measurements of total solids in unfiltered (*i.e.* total solids) and filtered samples (*i.e.* total dissolved solids) obtained in Sections 4.2.3 and 4.2.4, respectively:

$$\text{Total suspended solids} = \text{Total solids} - \text{Total dissolved solids}$$

4.2.6 Fixed and Volatile Solids

4.2.6.1 Methodology. The residue from the determination of total solids, dissolved solids and suspended solids is ignited at 550 °C in a muffle furnace. Solids remaining after ignition are the fixed solids while the weight lost on ignition represents the volatile solids. Loss in weight is due to conversion of organic matter to CO_2 and H_2O.

4.2.6.2 Materials
- Same as in Sections 4.2.3, 4.2.4 and 4.2.5
- Muffle furnace

4.2.6.3 Experimental Procedure. Heat muffle furnace to 550 °C. Place dish in furnace for 1 h. Store in a desiccator and weigh immediately before use (this gives value of M in equation below when determining fixed total solids or fixed dissolved solids). If determining fixed and volatile suspended solids, ignite filter paper at 550 °C in muffle furnace for 15 min. Cool in desiccator and weigh (this also gives value of M in equation below when determining fixed suspended solids). Proceed to determine total solids and dissolved solids as described in Sections 4.2.3 and 4.2.4, respectively, using dishes prepared according to the above procedure, and determine suspended solids according to Section 4.2.5 using the filter paper prepared as above. These determinations yield the value of M_b in the equation below.

Ignite dishes with residue obtained from total solids (Section 4.2.3) and total dissolved solids (Section 4.2.4) determinations and the filter paper from suspended solids determination (Section 4.2.5) in the muffle furnace for 1 h. Cool partially in laboratory air and transfer to a desiccator for further cooling. Weigh as soon as the dish has cooled to room temperature. Perform the calculations as follows:

$$\text{Volatile solids (mg L}^{-1}) = 1000 \times (M_b - M_a)/V$$

$$\text{Fixed solids (mg L}^{-1}) = 1000 \times (M_a - M)/V$$

where M_b is the weight of dish (or filter) + weight of residue prior to ignition (mg), M_a is the weight of dish (or filter) + weight of residue after ignition (mg), M is the weight of dish (or filter) (mg) and V is the sample volume (mL).

Negative errors may be produced in the volatile solids determination owing to loss of volatile matter during drying. Considerable errors may arise when determining low concentrations of volatile solids in the presence of high concentrations of fixed solids.

4.2.7 Settleable Solids

4.2.7.1 Methodology. Settleable solids are determined volumetrically after allowing solids to settle under the influence of gravity.

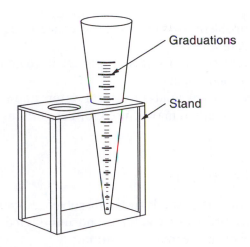

Figure 4.4 *Imhoff sedimentation cone*

4.2.7.2 Materials
- Imhoff cone (see Figure 4.4) or measuring cylinder (500 mL or 1 L)

4.2.7.3 Experimental Procedure.
Shake sample and place 1 L into the Imhoff cone. Allow solids to settle for 45 min and gently spin the cone to dislodge particles adhering to the walls of the cone. Leave undisturbed for a further 15 min and read off the volume of settled solids directly in mL L^{-1} from the graduated scale.

4.2.7.4 Alternative Procedure.
If Imhoff cones are not available in your laboratory you may use the following alternative procedure. Shake sample and pour about 500 mL into a 1 L measuring cylinder. Allow to stand undisturbed for 1 h. Thereafter, carefully siphon off 250 mL of liquid from halfway between any settled and floating material; make sure that you do not disturb the settled material while siphoning. Determine the suspended solids in the portion of liquid siphoned off according to the procedure in Section 4.2.5 above. This should give you SS_s in the equation below. Calculate the amount of settleable solids (in mg L^{-1}) from the difference between the suspended solids (SS) normally determined in the well-mixed sample prior to settling and the amount determined after settling (SS_s):

$$\text{Settleable solids (mg L}^{-1}) = SS - SS_s$$

4.2.8 Questions and Problems

1. What categories of solids can be determined in different waters?
2. What are the sources of error and variability in solids analysis?

3. What are the problems of environmental concern regarding solids in natural and waste waters?
4. Why are different types of solids of interest in the control of waste-water treatment plant operations?
5. What are the sources of solids in natural waters and wastewaters?
6. What alternative method may be used for determining total dissolved solids (TDS)?

4.2.9 Suggestions for Projects

1. Compare the content of solids (total, dissolved, suspended, fixed and volatile) in various waters (polluted and unpolluted). Discuss your results in terms of the degree of pollution.
2. Measure the content of solids in river water during dry and wet seasons (in tropics) or winter and spring periods (in temperate regions). Explain your results.
3. Measure the solids in riverwater sampled at an inlet to a lake and in the lake water. Explain your results, taking into account the role of stream flow and sedimentation.

4.2.10 Further Reading

APHA, "Standard Methods for the Examination of Water and Wastewater", 16th edn., American Public Health Association, Washington, 1985, pp. 92–100.

V. P. Sokoloff, Water of crystallization in total solids of water analysis, *Ind. Eng. Chem. Anal. Ed.*, 1933, 5–336.

C. N. Sawyer, P. L. McCarty and G. F. Parkin, "Chemistry for Environmental Engineering", 4th edn., McGraw-Hill, New York, 1994, pp. 567–576.

4.3 ELECTROCHEMICAL MEASUREMENTS

pH, conductivity and redox potential can be grouped together because they are all based on electrochemical measurements involving electrodes. Conductivity and pH are normally the first determination to be carried out on a water sample, either in the field or in the laboratory.

4.3.1 pH

The pH is one of the most important measurements commonly carried out in natural waters and wastewaters. The pH is a way of expressing the H^+ concentration in water (see Section 2.1) and it is used to express the acidic or alkaline nature of a solution. The content of H^+ ions in natural waters is mainly related to the quantitative ratio of carbonic acid and its ions. This acid dissociates in waters according to:

$$H_2CO_3 \rightleftharpoons H^+ + HCO_3^-$$

and consequently waters containing large quantities of dissolved carbon dioxide (CO_2) are acidic (pH < 7.0). Hard waters are slightly alkaline due to the dissolution of limestone and other minerals containing Ca and Mg:

$$CaCO_3 \rightleftharpoons Ca^{2+} + CO_3^{2-}$$
$$CO_3^{2-} + H_2O \rightleftharpoons HCO_3^- + OH^-$$

The changes in pH values in water are closely related to photosynthetic processes (due to CO_2 uptake by water plants) and consequently to the decomposition of organic matter.

Many industrial wastewaters contain mineral acids. Sulfuric acid and sulfuric acid salts are found in mine drainage waters and wastewaters from the metallurgical industries. If sulfur, sulfides or iron pyrite are present, these can be converted to sulfuric acid and sulfates by sulfur-oxidising bacteria under aerobic conditions:

$$2S + 3O_2 + 2H_2O \rightarrow 2H_2SO_4$$
$$FeS_2 + 3.5O_2 + H_2O \rightarrow FeSO_4 + H_2SO_4$$

In mine drainage waters the hydrolysis of heavy metals salt also plays a significant role in determining the pH due to hydrolysis:

$$Me^{2+} + 2H_2O \rightleftharpoons Me(OH)_2 + 2H^+$$

where Me^{2+} stands for heavy metals such as Fe, Zn, Cd, Cu, Ni, *etc.* Similar processes determine the pH of other industrial and municipal wastes. Mineral acids can also be found in some natural waters. Acid rain, containing nitric and sulfuric acid (see Chapter 2), can significantly reduce the pH of poorly buffered lakes and other receiving waters. Typical pH values of different waters are shown in Table 4.7.

In surface waters, pH is one of the most important water quality indices. The concentration of H^+ ions has great significance for chemical and biological processes occurring in natural waters as it can influence the growth of water biota, especially fish population, as shown in

Table 4.7 *Typical pH ranges in different waters*

Water type	pH
Unpolluted surface waters	6.5–8.5
Polluted surface waters	3.0–12.0
Unpolluted rainwater	4.6–6.1
Acidic rainwater	2.0–4.5
Seawater	7.9–8.3
Swamp waters	5.5–6.0
Ground waters	$\leqslant 2.0$–12.0
Mine drainage waters	1.5–3.5
Industrial and municipal wastewaters	$\leqslant 1.0$–$\geqslant 12.0$

Table 4.8 *Minimum pH values in surface water required for survival of fauna*

pH_{min}	Species lost at $< pH_{min}$
6.0	Death of snails and crustaceans
5.5	Death of salmon, rainbow trout and whitefish
5.0	Death of perch and pike
4.5	Death of eels and brook trout

Table 4.8. The speciation of toxic metals is greatly influenced by the pH and this can be a major concern, especially in acidified waters. Fish kills in acidified rivers and lakes have not been directly attributed to the acidity but to the mobilisation of toxic aluminium forms. Corrosion of metals and concrete is also influenced by the pH. The pH is important in the control of water and wastewater treatment processes such as water softening, disinfection, chemical coagulation, biological wastewater treatment processes, *etc.*

4.3.1.1 Experimental Procedure. Follow exactly the procedure outlined in Section 2.4.5.

4.3.2 Conductivity

Electrical conductivity (EC), also called specific conductance, is a measure of the ability of a water sample to convey an electrical current and it is related to the concentration of ionised substances in water. Conductivity can be used as an approximate measure of the total concentration of inorganic substances in water. Ions that have a major influence on the conductivity of water are H^+, Na^+, K^+, Mg^{2+}, Ca^{2+},

Table 4.9 *Conductivities of some types of water*

Water type	Conductivity (μmho cm^{-1})
Freshly distilled water	0.5–2
Aged distilled water	2–4
Rainwater	3–60
Drinking water	50–1500
Wastewater	>10 000

Cl^-, SO_4^{2-} and HCO_3^-. Other ions such as Fe^{2+}, Fe^{3+}, Mn^{2+}, Al^{3+}, NO_3^-, HPO_4^{2-}, $H_2PO_4^-$ and dissolved gases have a minor influence on the conductivity. Compounds that do not dissociate in aqueous solution are poor conductors of electricity; hence organic compounds have little influence on the conductivity. The water conductivity increases with temperature owing to a decrease in viscosity and increasing dissociation.

Conductivity increases with increasing mineral content. Typical conductivity values for some types of water are given in Table 4.9. Surface waters tend to have conductivities in tens to hundreds of μmho cm^{-1}, while salt lake waters may have conductivities as high as 600 000 μmho cm^{-1}.

Conductivity is often used to express the mineral content of a water sample. It is an important measurement in waters destined for various uses: irrigation, drinking, food industry and industrial boilers. Conductivity is also used to monitor the operation of desalination plants. Measurement of conductivity is widely used as a substitute for total dissolved solids (TDS) and salinity determinations. These parameters can be calculated from conductivity measurements by means of empirical equations. Many instruments are equipped to give direct readings of salinity and TDS on the basis of conductivity measurements.

4.3.2.1 Experimental. Determine the conductivity by following the procedure in Section 2.4.4.

4.3.3 Redox Potential

In addition to the pH, the reduction–oxidation (*redox*) potential (Eh) is one of the most important characteristics of natural waters. The redox potential is a measure of the oxidising or reducing power of the water. During oxidation processes, substances lose electrons, whereas during reduction they take up electrons. Both of these processes take place simultaneously in any given system. The redox potential measures the ability of the aquatic system to supply electrons to an oxidising agent and take up electrons from a reducing agent. Low values indicate high reduction whereas high values indicate high oxidation. Natural waters contain different ions and neutral molecules of the same element and these ions and molecules create the individual reduction–oxidation couples. Consequently, the reduction–oxidation state of water is a complex, dynamic equilibrium of many such co-existing redox couples. The Eh value influences geochemical speciation of many polyvalent elements such as Mn, Fe, Cr, As, *etc.*

Natural redox systems vary considerably, depending on the presence of chemical species. The main systems are formed by O_2, Fe, S and some organic compounds. Oxygen is a universal oxidant, and even at low concentrations it can significantly influence the Eh value. Under increasing O_2 content in water the Eh value also increases and can reach up to $+700$ mV. Owing to the existence of different oxidation states, from S^{2-} to S^{6+}, sulfur creates a number of intermediate forms. The presence of H_2S leads to a low redox potential ($\leqslant -100$ mV) and reducing conditions for the majority of natural compounds. The oxidative, trivalent form of iron (Fe^{3+}) easily undergoes hydrolysis. Under typical pH conditions, the content of Fe^{3+} is very low ($10^{-1}-10^{-2}$ mg L^{-1}) while the content of Fe^{2+} can be as high as several mg L^{-1}. Hence the value of the redox potential for the iron system ($Fe^{3+} + e^- \rightleftharpoons Fe^{2+}$) depends to a great degree on the pH, the Eh decreasing sharply with increasing pH. Since the behaviour of chemical substances is greatly influenced by both the pH and the Eh, it is common to illustrate this behaviour by means of Eh–pH diagrams (see Figure 4.5).

Rapidly mixing waters with low quantities of organic matter have Eh values determined by the reduction of O_2. If reducing substances that use up oxygen are present in still waters, the water may become depleted of oxygen. This can lead to the reduction of species intermediate in the redox scale (*e.g* Fe^{3+}, NO_3^- and MnO_2). Eventually, sulfur species are reduced to sulfide. Anaerobic biological processes in surface waters and wastewaters can decrease the Eh down to very low values. Under natural conditions, the redox potential may vary from -500 to $+500$ mV, as shown in Table 4.10. In seawater, measured redox potentials vary from

Table 4.10 *Redox potential values in different natural waters*

Water type	Eh (mV)
Surface waters with free O_2	$+100$ to $+500$
Surface and ground waters with free H_2S	$\leqslant -100$
Ground waters in contact with oil deposits	up to -500

-600 mV, for water from the bottom sediments containing organic matter, to $+300$ mV in aerated surface water.

The redox potential is a very important chemical parameter in environmental analysis. It characterises the chemical state of elements in water and, together with the pH, controls chemical processes occurring in nature. The Eh values determine the microbiological activity of waters and consequently the degree of water self-purification from various organic and inorganic pollutants. The redox potential is also important in wastewater treatment. Effective aerobic treatment of wastewaters requires high Eh values ($+200$ to $+600$ mV), while anaerobic treatment requires low Eh (-100 to -200 mV). See Example 4.2.

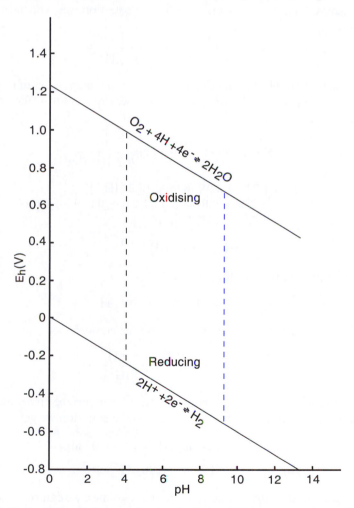

Figure 4.5 *Typical Eh–pH diagram showing the limits of Eh in nature and typical pH limits in soil and water*

Example 4.2

Calculate the theoretical limits of Eh in water as a function of pH and plot
these on an Eh–pH diagram.

Oxygen is the strongest oxidising agent commonly found in natural waters
and its reduction sets the upper limit of the redox potential:

$$O_2 \text{ (g)} + 4H^+ + 4e^- \rightleftharpoons 2H_2O \qquad E° = 1.229 \text{ V}$$

The lower limit is set by the reduction of hydrogen:

$$2H^+ + 2e^- \rightleftharpoons H_2 \qquad E° = 0.00 \text{ V}$$

For oxygen reduction we can write the Nernst equation (see Appendix II) as:

$$E = E° - \frac{0.0591}{n} \log_{10} \frac{[H_2O]^2}{pO_2[H^+]^4}$$

Since the activity of H_2O in solution is 1, and the partial pressure of oxygen in
the atmosphere $pO_2 = 0.21$ atm, and since 4 electrons are transferred in the
reaction, we can write:

$$E = 1.229 - \frac{0.0591}{4} \log_{10} \frac{1}{0.21[H^+]^4}$$

or:
$$\begin{aligned} E &= 1.229 + 0.0148 \log_{10} (0.21 \, [H^+]^4) \\ &= 1.229 + 0.0148 \log_{10} 0.21 + 0.059 \log_{10} [H^+] \\ &= 1.22 + 0.059 \log_{10} [H^+] \end{aligned}$$

Since
$$pH = -\log_{10}[H^+]$$

the above equation reduces to:

$$Eh = 1.229 - 0.059 \text{ pH}$$

For hydrogen reduction we can write the Nernst equation as:

$$E = E° - \frac{0.0591}{2} \log_{10} \frac{pH_2}{[H^+]^2}$$

The maximum pressure that H_2 could reach in near surface environments is
1 atm and substituting this extreme value into the equation we get:

$$E = 0.00 + 0.03 \log_{10} [H^+]^2 - 0.03 \log_{10} 1$$

$$Eh = -0.06 \text{ pH}$$

Since both equations that we derived are straight lines, we can substitute two
extreme pH values encountered in nature and draw straight lines between

them to define the theoretical Eh limits. The pH of natural systems can vary between 0 and 14 but it is generally between 3 and 10. Substituting these values into the derived equations we get:

	pH	
Eh (V)	3	10
for O_2	1.052	0.639
for H_2	-0.18	-0.6

A plot of the Eh limits is shown in Figure 4.5. Measured Eh values in natural waters are always below the theoretical maximum calculated for oxygen. Chlorine gas (Cl_2), widely used in water treatment, is a stronger oxidising agent than oxygen with a higher oxidation potential ($E^\circ = 1.359$ V). Eh values in nature are usually above the calculated theoretical minimum. Stronger reducing reactions may be encountered under local conditions within organic materials out of contact with water (*e.g.* coal and oil deposits).

4.3.3.1 Methodology. The redox potential is determined by measuring the potential difference between a platinum electrode and a reference electrode.

4.3.3.2 Materials
- pH meter with a millivolt display
- Platinum electrode
- Saturated calomel electrode

4.3.3.3 Experimental Procedure. Immerse the platinum and calomel electrodes in the sample held inside a beaker. Record the potential in mV after the reading has stabilised, which should take no more than several minutes. Rinse the electrodes with water between measurements.

Notes

1. If the reading does not stabilise after 5 min this implies that the platinum electrode is contaminated and it should be cleaned. You may clean the electrode by immersing in hot 1:1 H_2SO_4 and leaving overnight (1:1 HNO_3 may be used instead). Rinse the electrode with water before use.
2. If you are measuring the Eh of a sample you should also determine the pH, as Eh values on their own are not of much use.

4.3.4 Questions and Problems

1. Discuss the main anthropogenic factors which influence the pH, Eh and conductivity in natural waters and wastewaters.
2. What does water conductivity mean and how is this parameter related to the mineral content of water?
3. Discuss the possible chemical reactions in water which determine the pH and Eh of waters.
4. To what extent do the pH and Eh values in wastewaters differ from those in natural waters?
5. Calculate the H^+ and OH^- concentrations of solutions having the following pH values: (a) 5.5, (b) 8.2 and (c) 9.5.
6. If the pH increases by 2 units, by how much would the OH^- concentration change?
7. If the H^+ concentration is 5.0×10^{-4} mol L^{-1}, what is the OH^- concentration?
8. Calculate the pH of the following solutions: (a) 0.01 M NaOH, (b) 1 M HCl and (c) 10^{-4} mol L^{-1} H_2SO_4.

4.3.5 Suggestions for Projects

1. Measure pH and Eh in different natural waters of your locality (rivers, lakes, ponds, streams, swamps, groundwater from wells, tapwater). Plot the values of pH against Eh and calculate the correlation coefficient. If this is significant, calculate the regression line and plot it on the graph.
2. Measure pH, Eh, dissolved oxygen, Fe^{2+}, Fe^{3+}, SO_4^{2-}, NO_3^-, NH_3, *etc.*, in wastewaters and evaluate the main redox systems.
3. Measure the conductivity and mineral content of surface and ground water and tabulate the results for your region.
4. Carry out a survey of pH in various surface waters and evaluate the potential danger to fish, snail and crustacean populations.

4.3.6 Further Reading

APHA, "Standard Methods for the Examination of Water and Wastewater", 16th edn., American Public Health Association, Washington, 1985, pp. 76–80 and 429–437.

B. Moldan and J. Cherny (eds.), "Biogeochemistry of Small Catchments. SCOPE 51", Wiley, New York, 1994, pp. 164–188 and 207–223.

N. G. Bunce, "Environmental Chemistry", 2nd edn., Wuerz, Winnipeg, 1994, pp. 131–198.

C. N. Sawyer, P. L. McCarty and G. F. Parkin, "Chemistry for Environmental Engineering", 4th edn., McGraw-Hill, New York, 1994, pp. 458–463.

4.4 HARDNESS

4.4.1 Introduction

The term *hardness* refers to the ability of water to precipitate soap. Hard waters are undesirable for two reasons: they require considerable amounts of soap to produce a lather, and they produce scale in industrial boilers, heaters and hot water pipes. The former is of concern to domestic users of water, whereas the latter is a problem for engineers. On the other hand, soft waters have been linked with an increased incidence of cardiovascular disease. Other advantages of hard water include the neutralisation of acid deposition and the reduction of the solubility of toxic metals.

The major contributors to water hardness are dissolved calcium and magnesium ions. These ions combine with soap to form insoluble precipitates. Other polyvalent cations such as those of iron, zinc, manganese, aluminium and strontium may also contribute to hardness, but their contribution is usually insignificant due to low concentrations of these metals in water. The hardness of water derives largely from the weathering of minerals such as limestone ($CaCO_3$), dolomite ($CaCO_3 \cdot MgCO_3$) and gypsum ($CaSO_4 \cdot 2H_2O$) and it varies considerably from place to place, depending on the nature of geological formations. Groundwaters are generally harder than surface waters. *Total hardness* is defined as the sum of Ca and Mg concentrations expressed as calcium carbonate in mg L^{-1} or ppm. Waters are classified according to the hardness scale shown in Table 4.11.

Table 4.11 *Classification used for water hardness*

Degree of hardness	Hardness (mg equivalent $CaCO_3$ L^{-1})
Soft	< 50
Moderately soft	50–100
Slightly hard	100–150
Moderately hard	150–200
Hard	200–300
Very hard	> 300

Scaling problems tend to occur with moderately hard and very hard waters. Calcium is associated with bicarbonate ions in solution and upon heating these are converted to calcium carbonate, which forms a thick scale on the surface of domestic and industrial boilers, water heating appliances, kettles, pipes, *etc.*:

$$Ca^{2+} + 2HCO_3^- \rightarrow CaCO_3(s) + CO_2 + H_2O$$

This hardness, which can be removed by heating, is called *temporary* or *carbonate hardness*. Temporary hardness is derived from contact with carbonates (limestone and dolomite). Hardness which cannot be removed by boiling is called *permanent* or *non-carbonate hardness* and it is due to anions such as chloride, nitrate, sulfate and silicate. This hardness does not contribute to scale formation. Contact with gypsum would result in permanent hardness. *Calcium hardness* is that due to Ca only, while *magnesium hardness* is due to Mg only. Magnesium hardness can be calculated from a determination of total and calcium hardness:

Magnesium hardness = total hardness − calcium hardness

4.4.2 Methodology

A simple, rapid and inexpensive method commonly used in the water industry for hardness determination is the direct complexation titration with ethylenediaminetetraacetic acid (EDTA):

$$\text{HOOCCH}_2 \diagdown \atop \text{HOOCCH}_2 \diagup \text{N}-\text{CH}_2\text{CH}_2-\text{N} {\diagup \text{CH}_2\text{COOH} \atop \diagdown \text{CH}_2\text{COOH}}$$

EDTA forms 1:1 complexes with divalent metals such as calcium:

$$Ca^{2+} + EDTA \rightarrow \{Ca \cdot EDTA\}_{complex}$$

Eriochrome Black T or Calmagite can be used as indicators. If a small quantity of indicator is added to a water sample containing Ca and Mg ions at pH 10, the solution becomes wine red. The indicator forms complexes with Ca and Mg ions which give the solution a wine-red colour:

$$Ca^{2+} + \text{Eriochrome Black T} \rightarrow \{Ca \cdot \text{Eriochrome Black T}\}_{complex}$$
$$\text{wine red}$$

As EDTA is added it displaces the cations from the cation–indicator complex by forming more stable complexes with the cations. When all of the Ca and Mg is complexed with EDTA, (at the end point), the solution turns from wine red to blue due to the free Eriochrome Black T indicator. In order to obtain a sharp end point a small amount of magnesium ions must be present. This is generally not a problem with natural water samples which tend to contain some Mg, but it is a problem when standardising EDTA solutions with pure $CaCO_3$. A small quantity of $MgCl_2$ is added to the standard EDTA solution to ensure the presence of Mg ions.

You may use the method outlined here to determine hardness, as well as Ca and Mg concentrations in surface waters, effluents, tapwater and bottled mineral water.

4.4.3 Materials

- pH meter and combination electrode or glass + calomel electrode
- Primary standard $CaCO_3$ solution, 0.004 M. Dry $CaCO_3$ reagent at 110 °C for 2 h. Weigh accurately 100 mg of dried $CaCO_3$ in a 250 mL conical flask. Add approximately 5 mL of concentrated HCl and after dissolution is complete add 50 mL of water. Boil the solution for 5 min to expel CO_2, cool and transfer to a 250 mL volumetric flask. Make up to the volume with laboratory grade water and mix well. Calculate the exact concentration of $CaCO_3$.
- Standard disodium ethylenediaminetetraacetate dihydrate ($Na_2EDTA \cdot 2H_2O$) solution (formula weight = 372.2), 0.004 M. Dissolve about 1.6 g of the salt in 200 mL of water containing a pellet of NaOH and 80 mg of $MgCl_2 \cdot 6H_2O$. Make up to 1 L and store in a plastic bottle.
- 0.1 M NaOH
- Eriochrome Black T indicator solution, 0.5% (w/v) in ethanol, freshly prepared
- Patton & Reeder's indicator (mix with 100 times its weight of Na_2SO_4)
- Buffer solution. Dissolve 7.0 g NH_4Cl in 57.0 mL of concentrated ammonia solution and dilute to 1 L.
- Hydrochloric acid, concentrated

4.4.4 Experimental Procedure

4.4.4.1 Standardisation of EDTA Solution. Pipette four 20 mL aliquots of the primary standard $CaCO_3$ (0.004 M) solution into 250 mL conical flasks. Titrate each with the 0.004 M EDTA solution in the following manner. Add 15 mL of the EDTA solution from the burette, then add 10 mL of the buffer solution followed by 5 drops of the Eriochrome Black T indicator solution. Continue the titration while stirring continuously until the colour changes sharply to blue. Generally the wine-red colour turns slowly to purple and then sharply to blue. Discard the result of the first titration, which should serve as a rough guide to the end-point location. Take and average of the last three titrations to calculate the exact concentration of the EDTA solution and the number of milligrams of $CaCO_3$ per one millilitre of EDTA solution.

4.4.4.2 Determination of Total Hardness. Pipette 20 mL of the water sample into a 250 mL conical flask and add 10 mL of buffer solution and 5 drops of indicator solution. Carry out one rough titration to get an idea of the location of the end point. The colour change is the same as in the standardisation experiment. Repeat the titration using another 20 mL aliquot of the water sample but this time add 3/4 of the expected titrant before adding the buffer and indicator. If sufficient sample is available, repeat the titration several times and take the average. Calculate the total hardness as mg L^{-1} $CaCO_3$ equivalents from the following expression:

$$\text{Hardness as mg } CaCO_3 \text{ L}^{-1} = 1000 \times V_t \times M/V_s$$

where V_t is the amount of titrant needed to reach the end point (mL), M is the mg $CaCO_3$ equivalent to 1 mL of EDTA titrant (determined in the standardisation procedure) and V_s is the volume of sample analysed (mL).

4.4.4.3 Determination of Calcium Hardness. Adjust the pH of 20 mL water sample to between 12 and 12.25 using 1 M NaOH. Do not add buffer. Add 0.2 g of Patton & Reeder's indicator mixture to the solution and titrate as before until the colour changes from wine red to blue. Repeat the titration several times on different 20 mL aliquots of sample and take the average. Calculate the calcium hardness in the water sample as mg $CaCO_3$ L^{-1} using the same expression as for total hardness. You may calculate the calcium concentration directly from this titration as:

$$Ca(\text{mg L}^{-1}) = 401 \times V_t \times M/V_s$$

where the terms V_t, M and V_s are the same as in the equation for total hardness given above. Calculate the magnesium hardness as the difference between the total and calcium hardness, and calculate the magnesium concentration.

Notes

1. You may use higher volumes of sample if required (*e.g.* 50 mL).
2. Calmagite may be used instead of Eriochrome Black T as the indicator for total hardness determination.
3. Murexide may be used instead of Patton & Reeder's indicator for the Ca determination but the end point may not be as sharp. Mix about 200 mg of the Murexide powder with 100 g dry NaCl and grind to a fine powder. Use about 0.2 g for each titration.
4. Ca and Mg may be determined using flame AAS, flame emission spectroscopy (see Section 2.5.1) or flame photometry. Total hardness may then be calculated as:

$$\text{Total hardness (mg equivalent CaCO}_3 \text{ L}^{-1}) = 2.497\,[\text{Ca}] + 4.118\,[\text{Mg}]$$

where the concentrations of Ca and Mg are expressed in mg L^{-1}.

4.4.5 Questions and Problems

1. Discuss the advantages and disadvantages of hard waters.
2. Differentiate between temporary and permanent hardness.
3. Which metals would you expect to interfere in the determination of hardness?
4. What is the meaning of the term "standardisation" and why is it important to standardise reagents?
5. Why is it necessary to add Mg when standardising EDTA solution with primary standard $CaCO_3$ solution?
6. Calculate the pH at which 100 mg L^{-1} $Mg(OH)_2$ would begin to precipitate (K_{sp} for $Mg(OH)_2 = 9 \times 10^{-12}$ mol^3 L^{-3}).
7. Why is it necessary to raise the pH to >12 when determining Ca hardness?
8. Convert the following to total hardness in mg equivalent $CaCO_3$ L^{-1}:
   - (a) 82 mg L^{-1} $CaSO_4$
   - (b) 65 mg L^{-1} $MgCl_2$
   - (c) 37 mg L^{-1} $MgCO_3$
   - (d) 75 mg L^{-1} $CaCl_2$ + 32 mg L^{-1} $MgCO_3$
9. Permanent hardness can be removed by adding washing soda, $Na_2CO_3 \cdot 10H_2O$, to the water to precipitate calcium carbonate which can then be removed by filtration. Calculate the mass of washing soda required to soften 1000 litres of hard water containing 250 mg L^{-1} Ca^{2+}.

4.4.6 Suggestions for Projects

1. Determine the total hardness and Ca and Mg concentrations for waters from different sources. Classify the waters according to your results.
2. Determine Ca and Mg in different samples using AAS and total hardness using the titration method. Calculate the total hardness from the AAS measurements and plot it against the results of the titration. Calculate the correlation coefficient and obtain the best fit line by the method of least squares. Comment on your results.
3. Determine Ca and Mg several times in the same sample (*i.e.* replicate analysis) using AAS and also determine total hardness several times using the titration method. Calculate the total hardness from the AAS measurements. Carry out all the measurements using one method in one day. Calculate the mean, standard deviation and coefficient of variation for the two methods. Comment on the repeatability of the two methods.
4. Do the same as in the previous experiment but perform only one analysis on one day using either or both methods. Complete the analysis of all the replicates over a longer period (weeks or months) by analysing on separate days, as you find convenient. Carry out the same kind of statistical analysis of your data as indicated above and comment on the reproducibility of the two methods.
5. Determine the hardness in groundwaters at different locations by taking samples from wells. Explain your results in terms of the geological formations with which the water has been in contact.

4.4.7 Further Reading

APHA, "Standard Methods for the Examination of Water and Wastewater", 16th edn., American Public Health Association, Washington, 1985, pp. 209–214.

C. N. Sawyer, P. L. McCarty and G. F. Parkin, "Chemistry for Environmental Engineering", 4th edn., McGraw-Hill, New York, 1994, pp. 485–492.

4.5 ALKALINITY

4.5.1 Introduction

The *alkalinity* of a water sample is a measure of its capacity to neutralise acids. In other words, it is the sum of all the titratable bases. Alkalinity is an operationally defined concept, *i.e.* it is defined in terms of the amount of acid required to react with the sample to a designated pH. Hence, experimentally determined values vary considerably with the end-point pH used in the titrimetric analysis. Alkalinity is commonly expressed in mg $CaCO_3$ L^{-1}.

Alkalinity of waters is mainly due to salts of strong bases and salts of weak acids. Alkalinity of many surface waters is primarily a function of hydroxide, carbonate and bicarbonate concentrations. Bicarbonate is the major contributor to alkalinity and it arises from the action of CO_2 in percolating water on basic minerals in soils and rocks:

$$CO_2 + CaCO_3 + H_2O \rightarrow Ca^{2+} + 2HCO_3^-$$

Other compounds present in natural and waste waters may also make a minor contribution to alkalinity. For example, ammonia and salts of weak inorganic acids such as borates, silicates and phosphates may contribute to alkalinity, as may salts of organic acids (*e.g.* humic, acetic, propionic).

Alkalinity is widely used as a measure of buffer capacity of natural waters and wastewaters. Alkalinity measurements are used in the control and interpretation of water and wastewater treatment processes such as chemical coagulation, aeration, anaerobic digestion, ammonia stripping and water softening. Domestic wastewater has an alkalinity that differs slightly from that of the water supply, owing to the addition of materials during domestic use. Untreated domestic wastewaters are normally alkaline with alkalinities between 50 and 2000 mg $CaCO_3$ L^{-1}. Many regulatory and municipal authorities prohibit the discharge of industrial wastes containing hydroxide (caustic) alkalinity to sewers and receiving waters. Alkalinity in excess of Ca and Mg concentrations is important in determining the applicability of water for irrigation. Boiler waters, surface waters with thriving algal populations and waters softened by means of lime or lime-soda treatment can contain significant hydroxide and carbonate alkalinity. At present, an important environmental aspect of alkalinity in natural waters is the capacity to neutralise acidity originating from atmospheric deposition.

Although alkalinity has little public health significance, highly alkaline waters are unpalatable and are not used for domestic water supply.

4.5.2 Methodology

Alkalinity is determined by titrating a measured volume of sample with H_2SO_4. If the pH of the sample is greater than 8.3, the titration is carried out in two stages. In the first stage the titration is carried out to a pH of

8.3, the phenolphthalein end point. By the time the pH has been reduced to 10, all the hydroxides would have been neutralised:

$$2OH^- + H_2SO_4 \rightarrow SO_4^{2-} + 2H_2O$$

The phenolphthalein end point corresponds to the equivalence point for the following reaction:

$$2CO_3^{2-} + H_2SO_4 \rightarrow SO_4^{2-} + 2HCO_3^-$$

This is known as *phenolphthalein alkalinity*. Subsequently, the titration is carried out to a pH of about 4.5, the methyl orange end point which corresponds approximately to the equivalence point for the following reaction:

$$2HCO_3^- + H_2SO_4 \rightarrow SO_4^{2-} + 2H_2CO_3$$

This is referred to as *total alkalinity*. If a sample has a pH less than 8.3, a single titration to the methyl orange end point is carried out. Alternatively, a potentiometeric titration can be carried out, measuring the pH with an electrode and plotting the titration curve. A typical potentiometric titration curve is shown in Figure 4.6, illustrating the location of

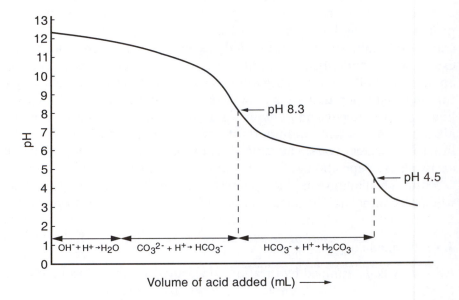

Figure 4.6 *Alkalinity titration curve*

the stoichiometric end points and the contribution of various species to the alkalinity.

4.5.3 Materials

- Sodium carbonate solution, 0.025 M. Weigh about 4 g of primary standard Na_2CO_3 and dry at 110 °C for 2 h. Cool in a desiccator. Weigh accurately 2.66 g to the nearest 0.1 mg. Dissolve in water and dilute to 1 L in a volumetric flask. This solution can be used for up to 1 week. Thereafter, discard and prepare fresh reagent. Calculate the exact concentration of this solution on the basis of the weight and use it to standardise the sulfuric acid titrant.
- Standard sulfuric acid, 0.05 M and 0.01 M. You may use hydrochloric acid, 0.1 and 0.02 M, instead.
- Phenolphthalein indicator. Weigh 5 g of reagent and dissolve in 500 mL of ethanol. Add 500 mL of water while stirring. Filter the solution.
- Methyl orange indicator. Weigh 0.5 g of the free acid form and dissolve in 1 L of water. Filter. If using the sodium salt, dissolve 0.5 g of reagent into 1 L of water. Add 15 mL of 0.1 mol L^{-1} HCl and filter.
- pH meter with combination electrode or glass and reference electrode.

4.5.4 Experimental Procedure

4.5.4.1 Sampling. Collect samples using glass or plastic (polyethylene or polypropylene) bottles. Fill bottles to the brim and seal tightly. Do not leave any air pockets or bubbles in the bottle. Analyse as soon as possible and preferably within 24 h of collection.

4.5.4.2 Standardisation. Standardise the titrant as follows. Measure 40 mL of the 0.025 M Na_2CO_3 solution in a conical flask, and add 60 mL of water and 3 drops of methyl orange indicator. Titrate with 0.05 M H_2SO_4 (or 0.1 M HCl) until the colour changes from yellow to orange. Record the volume of titrant used. Repeat with two more 40 mL aliquots of the Na_2CO_3 solution. Calculate the exact molarity of the H_2SO_4 titrant from the average of the three readings from the following equation:

$$Na_2CO_3 + H_2SO_4 \rightarrow Na_2SO_4 + CO_2 + H_2O$$

For example, if 20 mL of H_2SO_4 is required for the titration, the concentration of H_2SO_4 is exactly 0.05 M. Since the molecular mass of Na_2CO_3 is 106, 1 mL of the 0.05 M H_2SO_4 solution is equivalent to 5.30 mg of Na_2CO_3. Similarly, 1 mL of 0.05 M H_2SO_4 is equivalent to 5.00 mg of $CaCO_3$, the molecular mass of $CaCO_3$ being 100.1. Standardise the 0.01 M H_2SO_4 (or 0.02 M HCl) against 20 mL of 0.025 M Na_2CO_3.

4.5.4.3 Analysis by Indicator.   Measure the pH of the water sample. If the pH of the sample is <8.3, determine only the total alkalinity (part B). Otherwise determine both the phenolphthalein and total alkalinity (parts A and B). Measure 100 mL (or 50 mL) of the water sample into a conical flask.

(A) Add 4 drops of phenolphthalein indicator to the sample in the flask. The colour of the solution should turn magenta. If it does not, proceed to stage B. Titrate with 0.05 M H_2SO_4 (or 0.1 M HCl) until the pink colour just disappears and record the amount of titrant used. This value will be used to calculate the phenolphthalein alkalinity. Continue by proceeding to stage B.

(B) Add 3 drops of methyl orange indicator. The sample should turn yellow. Titrate with 0.05 M H_2SO_4 (or 0.1 M HCl) until the colour just turns orange. Record the volume of titrant used.

4.5.4.4 Analysis by Potentiometric Titration.   Measure 100 mL (or 50 mL) of the sample and insert a combination pH electrode (or glass electrode and reference electrode) into the solution. Monitor the pH while adding standard 0.05 M H_2SO_4 (or 0.1 M HCl) from a burette. Record the volumes of acid required to reach pH 8.3 (phenolphthalein alkalinity) and pH 4.5 (total alkalinity).

Calculation. Calculate the alkalinity from the titration results by using the following expression:

$$\text{Alkalinity (mg CaCO}_3 \text{ L}^{-1}) = 1000 \times V_t \times M/V_s$$

where V_t is the volume of standard acid used (mL), M is mass (in mg) of $CaCO_3$ equivalent to 1 mL of titrant (5.00 mg mL^{-1} for 0.05 M H_2SO_4) and V_s = volume of sample (mL).

When calculating the phenolphthalein alkalinity, V_t is the volume of acid used to reach the phenolphthalein end point, or pH 8.3 in the potentiometric titration. When calculating the total alkalinity, V_t is the volume of acid used to reach the phenolphthalein end point (part A) plus the amount of acid used to reach the methyl orange end point (part B). In

Table 4.12 *Alkalinity relationships*[a]

Condition	Titration result	Hydroxide alkalinity (HA) as $CaCO_3$	Carbonate alkalinity (CA) as $CaCO_3$	Bicarbonate alkalinity (BA) as $CaCO_3$
(a)	PA = TA	TA	0	0
(b)	PA > 0.5TA	2PA − TA	2(TA − PA)	0
(c)	PA = 0.5TA	0	2PA	0
(d)	PA < 0.5TA	0	2PA	TA − 2PA
(e)	PA = 0	0	0	TA

[a] PA = phenolphthalein alkalinity; TA = total alkalinity.

the potentiometric titration, V_t is the total volume of titrant required to reach a pH of 4.5. The pH of the end point should be quoted when reporting the alkalinity.

4.5.4.5 Calculation of Contributing Anions from Alkalinity Measurements. The three principal contributors to alkalinity are hydroxide (OH^-), carbonate (CO_3^{2-}) and bicarbonate (HCO_3^-) ions. It is possible to calculate the contribution of each of these to the alkalinity on the basis of the alkalinity determination if it is assumed that other species do not contribute to the alkalinity and if it is assumed that OH^- and HCO_3^- cannot coexist. The following alkalinity conditions are then possible in the sample: (a) hydroxide alone, (b) hydroxide and carbonate, (c) carbonate alone, (d) carbonate and bicarbonate and (e) bicarbonate alone. The results of the alkalinity determination can be used to work out the alkalinity relationships from Table 4.12. See also Example 4.3.

4.5.4.6 Determination of Contributing Anions from pH and Alkalinity Measurements. Given that $[OH^-] = K_w/[H^+]$, the hydroxide alkalinity can be calculated from the sample pH. Given that 1 mol L^{-1} of OH^- is equivalent to 50 000 mg L^{-1} of alkalinity as $CaCO_3$, the hydroxide alkalinity (HA) can be determined from:

$$HA \text{ (mg } CaCO_3 \text{ } L^{-1}) = 50\,000 \times 10^{(pH - pK_w)}$$

Carbonate alkalinity (CA) and bicarbonate (BA) alkalinity (in mg $CaCO_3$ L^{-1}) can then be determined from the following expressions:

$$CA = 2(PA - HA)$$

$$BA = TA - (CA + HA)$$

Example 4.3

The results of alkalinity determinations in a water sample were as follows: phenolphthalein alkalinity (PA) = 10 mg $CaCO_3$ L^{-1}; total alkalinity (TA) = 120 mg $CaCO_3$ L^{-1}. Calculate the concentrations of the ions contributing to the alkalinity.

From Table 4.12 it is apparent that the analysis corresponds to condition (d) (*i.e.* PA < 0.5TA).

Hydroxide alkalinity (HA) = 0 mg $CaCO_3$ L^{-1}

Carbonate alkalinity (CA) = 2PA = 2 × 10 = 20 mg $CaCO_3$ L^{-1}

Bicarbonate alkalinity (BA) = TA − 2PA = 120 − 20 = 100 mg $CaCO_3$ L^{-1}

Concentrations of the individual ions can be obtained by using conversion factors obtained from the ratio of the equivalent weight of the anion and the equivalent weight of $CaCO_3$. The equivalent weight of $CaCO_3$ is 100/2 = 50. The conversion factors for the different species are given below:

Anion	Equivalent weight	Conversion factor
OH^-	17/1 = 17	17/50 = 0.34
CO_3^{2-}	60/2 = 30	30/50 = 0.6
HCO_3^-	61/1 = 61	61/50 = 1.22

Therefore, for the example above, the concentrations of contributing ions are:

$$CO_3^{2-} = 20 \times 0.6 = 12 \text{ mg L}^{-1}$$

$$HCO_3^- = 100 \times 1.22 = 122 \text{ mg L}^{-1}$$

Notes

1. For samples with high alkalinity, use 0.05 M H_2SO_4 or 0.1 M HCl. For samples with low alkalinity, use 0.01 M H_2SO_4 or 0.02 M HCl.
2. Instead of the phenolphthalein indicator you may use metacresol purple indicator (dissolve 100 mg reagent in 100 mL of water).
3. Instead of the methyl orange indicator, various mixed indicators can be used (*e.g.* bromocresol green–methyl red indicator, or methyl orange–indigo carmine–methyl red indicator) for a more distinct end-point colour change.
4. Most surface water samples, including seawater, have pH values less than pH 8.3. Therefore, the determination of phenolphthalein alkali-

nity can be omitted. Wastewater samples and some natural samples (*e.g.* hard water lakes) may have higher pH values and both phe-nolphthalein and total alkalinity should be determined. Measure the pH of the sample in order to determine which procedure to follow.

4.5.5 Questions and Problems

1. What are the sources of alkalinity in natural water?
2. What are the total alkalinity and phenolphthalein alkalinity, and why do these values depend on the end-point pH used?
3. Illustrate the speciation of H_2CO_3, HCO_3^- and CO_3^{2-} with pH. Why are the pH values not a good guide to alkalinity?
4. How can one use the alkalinity values in interpretation and control of water and wastewater treatment processes?
5. Why is alkalinity expressed as mg $CaCO_3$ L^{-1}?
6. Calculate the alkalinity of a water sample if 12.64 mL of 2.05×10^{-2} M HCl are needed to titrate a 250 mL water sample to pH 4.5.
7. Calculate the alkalinity of a water sample if 8.62 mL of 4.60×10^{-3} M HCl are needed to titrate a 500 mL sample to a methyl orange endpoint.
8. A 100 mL sample required 15 mL of 0.04 M H_2SO_4 to reach the phenolphthalein end point and 30.2 mL to reach the methyl orange end point (the latter includes the amount required to reach the phenolphthalein end point). Calculate the phenolphthalein and total alkalinity of the sample and the concentrations of contributing anions.
9. A 100 mL sample of water with a pH of 10.5 was titrated with 0.05 M H_2SO_4. The phenolphthalein end point was reached after addition of 12.3 mL of titrant, and the methyl orange end point was reached after addition of 36.8 mL of titrant (the latter includes the amount required to reach the phenolphthalein end point). Calculate the contributions of hydroxide, carbonate and bicarbonate alkalinity using (a) alkalinity measurements only and (b) using pH and alkalinity measurements.

4.5.6 Suggestions for Projects

1. Carry out a survey of alkalinity values in different natural waters and wastewaters. Interpret the results in terms of the level of pollution.
2. On the basis of alkalinity determinations in various waters, calculate the concentrations of the anions contributing to the alkalinity in each sample.

4.5.7 Further Reading

N. J. Bunce, "Environmental Chemistry", 2nd edn., Wuerz, Winnipeg, 1994,
    pp. 131–158.
APHA, "Standard Methods for the Examination of Water and Wastewater",
    16th edn., American Public Health Association, Washington, 1985, pp. 269–
    273.
C. N. Sawyer, P. L. McCarty and G. F. Parkin, "Chemistry for Environmental
    Engineering", 4th edn., McGraw-Hill, New York, 1994, pp. 471–484.

4.6 DISSOLVED OXYGEN (DO)

4.6.1 Introduction

All aerobic life forms, including aquatic ones, require oxygen for respiration. A warm-water aquatic ecosystem should have a dissolved oxygen (DO) concentration of at least 5 mg L^{-1} in order to support a diversified biota, including fish. DO concentration is a function of temperature, pressure, salinity and the biological activity in the water body. The solubility of oxygen in fresh water at sea-level and at 25 °C is 8.3 mg L^{-1}. In seawater, at the same conditions, the solubility of oxygen is 6.7 mg L^{-1}. Photosynthesis by plants can produce O_2 during daylight hours:

$$6CO_2 + 6H_2O + h\nu \rightarrow C_6H_{12}O_6 + 6O_2$$

Respiration by plants and animals, and aerobic bacteria, can consume O_2. Chemical oxidation reactions can also cause minor depletion of O_2.

Pollution can cause DO concentration to drop below the level necessary for maintaining a healthy biota. Waste products flowing into lakes and rivers can quickly lead to oxygen depletion. Wastes provide nutrients for microorganisms which grow and multiply rapidly, consuming oxygen in the process. Normally, these microorganisms act to purify the lake or river by breaking down wastes, but if an excess of polluting waste is present the oxygen may be completely used up. Most pollutants contain carbon as the most abundant element. Carbon bearing wastes may be converted to carbon dioxide with the help of bacteria:

$$C + O_2 \rightarrow CO_2$$

According to the stoichiometry of this reaction, 32 g of oxygen is needed to oxidise 12 g of carbon, *i.e.* 8 mg L^{-1} of oxygen is required to oxidise 3 mg L^{-1} of dissolved carbon. This implies that a small drop of oil could use up all the oxygen in 5 L of water, illustrating how readily oxygen may be depleted. When the demand for oxygen exceeds the supply, the microorganisms and fish start to die. The body of water can no longer purify itself and anaerobic processes take over. Anaerobic decomposition leads to the formation of obnoxious products such as sulfides, methane and ammonia. Such reducing conditions are said to be "*anoxic*". As a consequence, the river or lake may become lifeless.

In order to protect aquatic ecosystems, liquid wastes discharged into surface waters are required to have a high level of dissolved oxygen (5–8 mg L^{-1}). This is often a legal requirement enforced by means of effluent standards. If wastewaters do not fulfill this requirement, they have to be treated. Aeration is an important process for treating wastewaters, ensuring that there is sufficient dissolved oxygen for the effluent standard to be met. Continuous exposure of fish to DO levels between 1 and 5 mg L^{-1} can slow down their growth, while levels below 1 mg L^{-1} are lethal to fish if exposure lasts more than a few hours.

In temperate climates, DO levels tend to be more critical during summer months because the rate of biochemical reactions increases with increasing temperature and stream flows tend to be lower. In thermally stratified lakes, DO exhibits characteristic vertical profiles which can vary with the season (Figure 4.7). According to the temperature profile, three regions of the lake can be defined: the *epilimnion* (warm upper layer), the *thermocline* (transition zone) and the *hypolimnion* (the colder, deeper region). The epilimnion is well oxygenated because of the dissolution of oxygen at the surface and the production of oxygen by photosynthesis. If the level of productivity in a lake is high, nutrient-rich detritus will sink through the hypolimnion. The detritus will deplete DO from the hypolimnion while releasing nutrients. It is not uncommon for the DO concentration to be 0 at the bottom of the lake, in which case anaerobic decomposition predominates [Figure 4.7(b)]. On the other hand, if productivity of a lake is low, the amount of detritus will be substantially lower and little DO will be depleted from the hypolimnion. In this case the water will be saturated with oxygen from top to bottom [Figure 4.7(a)]. Some lakes may exhibit unusual profiles with a maximum DO concentration in the thermocline [Figure 4.7(c)].

The presence of DO in wastewater is desirable because it prevents the formation of noxious odours. Very high levels of DO are to be avoided in domestic and industrial water supplies in order to minimise corrosion of iron and steel pipes in the distribution system and in boilers. Most countries do not set a DO standard for drinking water with the exception of Russia, which sets a maximum allowable concentration (MAC) of 4.0 mg L^{-1} for drinking water. For fisheries and aquatic life

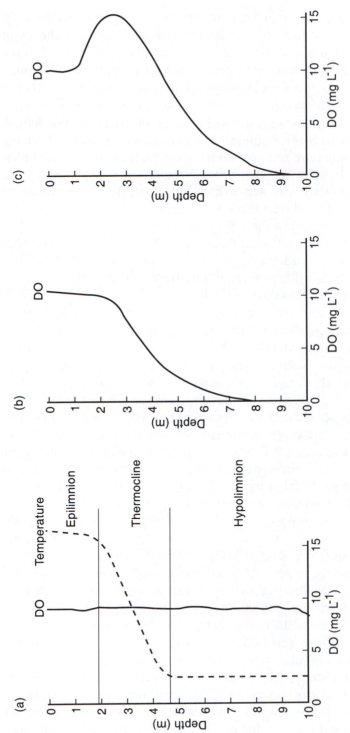

Figure 4.7 *Typical dissolved oxygen (DO) profiles in thermally stratified lakes*

the following DO standards (in mg L^{-1}) have been adopted: EU 5.0–9.0, Canada 5.0–9.5, Russia 4.0–6.0 (with a lower level acceptable under ice).

In view of the importance of DO in maintaining life in aquatic ecosystems, it is necessary to monitor its concentration both in natural waters and in wastewaters. DO is also monitored in domestic and industrial water supplies. Although oxygen probes for use in the field have been developed, direct titration on water samples is still widely used. When sampling for DO it is important to add a preservative immediately after sample collection. DO determination is used in the biochemical oxygen demand (BOD) test. See also Example 4.4.

Example 4.4

The DO in a water sample was determined and found to be 3.8 mg L^{-1}. The temperature of the sample was 15 °C and the pressure was 760 mmHg. Calculate the percent saturation of DO in the water sample, given that Henry's law constant for O_2 at 15 °C is 1.5×10^{-3} mol L^{-1} atm^{-1}. Assume that the salinity is less than 100 mg L^{-1} and that it does not affect the DO concentration.

Henry's law constant is defined as:

$$K_H = [O_2 \cdot H_2O]/pO_2 = 1.5 \times 10^{-3} \text{ mol } L^{-1} \text{ atm}^{-1}$$

where $[O_2 \cdot H_2O]$ is the concentration of dissolved oxygen in equilibrium with atmospheric oxygen (*i.e.* DO at saturation) and pO_2 is the partial pressure of oxygen in the atmosphere.

It follows that:

$$[O_2 \cdot H_2O] = K_H \times pO_2$$

Given that the total pressure is 760 mmHg or 1 atm, and the oxygen concentration in air is 20.95%, the partial pressure of oxygen is 0.2095 atm. Therefore:

$$[O_2 \cdot H_2O] = (1.5 \times 10^{-3}) \times 0.2095 = 3.14 \times 10^{-4} \text{ mol } L^{-1}$$

Since the molecular mass of O_2 is 32:

$$O_2 \text{ at saturation} = (32 \times 3.14 \times 10^{-4}) \times 1000 = 10.05 \text{ mg } L^{-1}$$

Therefore, % saturation of the water sample $= 100 \times 3.8/10.05 = 37.8\%$

4.6.2 Methodology

The iodometric titration method for DO determination was developed by Winkler and it is sometimes referred to as the *Winkler* method. Manganese(II) solution is added to the sample to fix the dissolved oxygen in the form of a brown precipitate of manganese(III) hydroxide:

$$4Mn(OH)_2 + O_2 + 2H_2O \rightarrow 4Mn(OH)_3 \downarrow$$

The brown precipitate is dissolved by addition of an acid and reacted with iodide to produce iodine:

$$Mn(OH)_3 + I^- + 3H^+ \rightarrow Mn^{2+} + 0.5I_2 + 3H_2O$$

The iodine is then titrated with sodium thiosulfate:

$$2S_2O_3^{2-} + I_2 \rightarrow S_4O_6^{2-} + 2I^-$$

Four moles of thiosulfate corresponds to one mole of dissolved oxygen. If a 0.025 M $Na_2S_2O_3$ solution is used for the titration of a 200mL sample, then:

$$1 \text{ mL } 0.025 \text{ M } Na_2S_2O_3 = 1 \text{ mg DO L}^{-1}$$

Nitrites, which are especially present in treated sewage, can interfere in the method. This interference is eliminated by adding sodium azide to the sample which destroys the nitrite:

$$HNO_2 + HN_3 \rightarrow N_2 + N_2O + H_2O$$

4.6.3 Materials

- 250 mL BOD bottles (Erlenmeyer flasks may be used as alternatives)
- Thermometer
- Manganese(II) sulfate solution. Dissolve 36.4 g $MnSO_4$ in water, filter and dilute to 100 mL. This solution should not give a colour with starch when added to acidified KI solution.
- Alkali–iodide–azide reagent. Dissolve 50.0 g NaOH, 13.5 g NaI and 1 g of NaN_3 in water and dilute to 100 mL.
- Sulfuric acid, concentrated
- Starch solution

- Standard sodium thiosulfate titrant, 0.025 M. Dissolve 6.205 g $Na_2S_2O_3 \cdot 5H_2O$ in laboratory grade water. Add 1.5 mL 6 M NaOH (or 0.4 g solid NaOH) and dilute to 1 L. Standardise with potassium iodate solution.
- Standard potassium iodate solution, 4.167×10^{-3} M. Dissolve 900.7 mg of dry KIO_3 in water and dilute to 1 L. KIO_3 has a purity of 99% and it should be dried at 120 °C. Considering the purity, 900.7 mg of the salt is equivalent to 891.7 mg of pure KIO_3.
- Potassium iodide

4.6.4 Experimental Procedure

4.6.4.1 Standardisation of Thiosulfate Solution. Dissolve approximately 2 g KI, free from iodate, in a conical flask with 100 mL of water. Add a few drops of concentrated sulfuric acid and 20 mL of the standard iodate solution prepared above. Dilute to 200 mL. This reaction produces iodine:

$$IO_3^- + 5I^- + 6H^+ \rightarrow 3I_2 + 3H_2O$$

Titrate liberated iodine with thiosulfate solution while swirling the flask. When the solution becomes pale straw yellow in colour, add a few drops of starch solution. Continue titrating until the colour changes from blue to colourless. Record the volume of titrant used up. When the solutions are of equal strength, 20.0 mL 0.025 M $Na_2S_2O_3$ should be required. Repeat the standardisation two more times with fresh portions of potassium iodate solution.

4.6.4.2 Sampling. Record the temperature of the water in the field. Record the location, the date and the time of sampling. Remember to make a note of any observation which may be of significance (proximity of sewage outlet, presence of oil films, depth of sampling, *etc.*). At each site, collect three replicate samples (*i.e.* fill three sampling bottles). Fill the 250 mL BOD bottle (or Erlenmeyer flask) to the brim and stopper it while it is below the surface. Perform the sampling carefully. Do not agitate the sample or allow it to come into contact with air. You must avoid introducing air into the sample.

Either in the field, or in the laboratory (but no more than two hours after sample collection), add 1 mL of the manganese(II) sulfate solution, followed by 1 mL of the alkali–iodide–azide solution. This must be carried out very carefully in order to avoid introducing any air into your sample. Dip the pipette well below the surface of the water sample. Make sure that you do not aerate the sample while pipetting the solutions (*i.e.* remove the pipette as soon as the reagent has been delivered). Rinse the pipettes between addition of reagents. After addition of the reagents, stopper the bottle carefully. This will displace some of the solution but do not worry about it. While holding the stopper, invert the bottle several times to mix the sample and the reagents. You should observe the appearance of a brown precipitate of manganese(III) hydroxide. If the precipitate is white, it indicates that DO is absent. Allow the precipitate to settle completely for 15 min. The precipitate should settle in the lower half of the bottle, leaving a clear solution above. At this point, add 1 mL concentrated H_2SO_4 and let it run down the inner walls of the bottle. Stopper the bottle and invert it several times to dissolve the precipitate.

4.6.4.3 Analysis.   Titrate a volume corresponding to 200 mL of the original sample. You must allow for loss of sample due to displacement by reagents. What volume of solution should you therefore titrate?

Titrate with 0.025 M $Na_2S_2O_3$ solution to a pale yellow colour. Only then add two drops of starch indicator and swirl to mix. Continue titrating until the solution changes from dark blue to colourless. Record the number of millilitres of titrant required to titrate the sample. Repeat the procedure with the two other replicate samples from each site. Report the dissolved oxygen concentration in mg L^{-1} (1 mL 0.025 M $Na_2S_2O_3$ = 1 mg O_2 L^{-1} for a 200 mL sample). Calculate the mean for each site.

Note. You may use $MnSO_4 \cdot H_2O$ or $MnSO_4 \cdot 2H_2O$ instead of $MnSO_4$.

4.6.5 Questions and Problems

1. Discuss the significance of DO in surface waters.
2. Outline the basic chemical principles underlying the Winkler method.
3. The azide modification is employed to eliminate interference by nitrite. What other substances found in water may interfere in the Winkler method? Describe some other modifications of the Winkler method and explain how they are used to eliminate these interferences.
4. Explain why 20 mL of 0.025 M $Na_2S_2O_3$ solution is equivalent to 20 mL of a 4.167×10^{-3} M KIO_3 solution.
5. DO can be determined with a precision, expressed as a standard deviation, of about 0.02 mg L^{-1} in distilled water and about 0.06 mg L^{-1} in wastewater. In the presence of appreciable interferences the standard deviation may be as high as 0.1 mg L^{-1}. In the light of this information, discuss the precision of your own measurements.

6. Why is it important to fix the DO in the field?
7. The following temperatures and DO concentrations were determined at different depths in a lake:

Depth (m)	0	2	4	6	8	10
T (°C)	24	22	7	4	4	4
DO (mg L^{-1})	8.6	7.5	9.0	8.4	4.7	0

Calculate the DO at saturation for each temperature and the % saturation at each depth. Plot the T, DO, and DO at saturation as a function of depth. Explain these profiles.

4.6.6 Suggestions for Projects

1. Determine DO in different types of waters (*e.g.* rivers, lakes, ponds, sea) at different sites (near sewage outlets, far from industrial and urban areas, *etc.*) and relate to the degree of pollution of the water.
2. Measure DO at varying distances downstream from a sewage effluent outlet. Plot the DO as a function of distance from the outlet.
3. Investigate the diurnal and seasonal variation of DO in a lake or pond. Measure the DO at the same site during day and night (*e.g.* you could take turns with other students to sample every hour during a 24 hour period), or at weekly or monthly intervals throughout the year. Record the temperature of the water each time you sample. Plot the diurnal and/or seasonal variation in DO. Calculate the DO at saturation for each sample from the temperature and plot this on the same graph as the measured DO. Compare the DO at saturation with the measured DO and comment on the conditions prevailing in the water body.
4. If you have appropriate sampling equipment, determine DO concentration at different depths in a lake or pond during different seasons. Plot the concentration of DO *versus* depth and comment on your results. You could also determine the concentrations of nitrogen and phosphorus at various depths and relate these to the DO concentrations. Interpret your results in terms of the productivity of the lake or pond.

4.6.7 Further Reading

APHA, "Standard Methods for the Examination of Water and Wastewater", 16th edn., American Public Health Association, Washington, 1985, pp. 413–426.

C. N. Sawyer, P. L. McCarty and G. F. Parkin, "Chemistry for Environmental Engineering", 4th edn., McGraw-Hill, New York, 1994, pp. 515–527.

R. W. Raiswell, P. Brimblecombe, D. L. Dent and P. S. Liss, "Environmental Chemistry, the Earth–Air–Water–Factory", Arnold, London, 1980, pp. 84–91.

W. B. Clapham, Jr., "Natural Ecosystems", Macmillan, London, 1973, pp. 139–141 and 161–166.

4.7 BIOCHEMICAL OXYGEN DEMAND (BOD)

4.7.1 Introduction

The oxygen content of water samples stored in a full, airtight bottle decreases with time owing to the oxidation of organic matter by microorganisms. The sources of this organic matter are the excretions of aquatic biota, water-soluble humus compounds and industrial, domestic and agricultural effluents. The biochemical oxygen demand (BOD) is an empirical determination of the amount of oxygen required to oxidise the organic matter in the sample. In addition to measuring the oxygen required for the biochemical degradation of organic matter, termed *carbonaceous biochemical oxygen demand* (CBOD), the BOD test also measures the amount of oxygen used to oxidise reduced forms of nitrogen, the so-called *nitrogenous biochemical oxygen demand* (NBOD), as well as inorganic substances such as ferrous iron and sulfides. BOD is determined by incubating the water sample with aerobic microorganisms under specific conditions of time and temperature. The most widely used test, BOD_5, is based on a 5-day incubation period at 20 °C. The dissolved oxygen (DO) is measured at the beginning and end of the incubation period, and the BOD represents the difference between the initial and final DO:

$$BOD = (\text{initial DO} - \text{final DO}) \text{ mg L}^{-1}$$

The BOD_5 value does not represent the total BOD, since the biological oxidation of organic matter takes a lot longer than 5 days to go to completion. About 95–99% of the reaction is complete after 20 days, but as this is too long to wait for results, a 5-day incubation period is usually used. For most domestic and industrial wastewaters the BOD_5 value represents between 60% and 80% of the total BOD.

Since most wastewaters contain more oxygen-demanding materials than the amount of DO available in air-saturated water, it is necessary to dilute the samples before incubation. Nutrients such as nitrogen, phosphorus and trace metals are added to the dilution water, which is also buffered to ensure that the pH of the incubated sample will be in a

Table 4.13 *Classification of surface water quality based on BOD values*

Degree of pollution	BOD_5 (mg O_2 L^{-1})
Very clean	<1.0
Clean	1.1–1.9
Moderately polluted	2.0–2.9
Polluted	3.0–3.9
Very polluted	4.0–10.0
Extremely polluted	>10

range suitable for bacterial growth. For samples that do not contain a large population of microorganisms, such as disinfected effluents, it is necessary to "seed" the sample with a mixed bacterial culture, usually obtained from a sewage treatment plant. Most surface waters and wastewaters contain sufficient numbers of bacteria and seeding is not required. If only carbonaceous demand is to be determined, a nitrification inhibitor is added to the sample.

The BOD test is widely used to determine the pollution strength of wastewaters and the quality of receiving surface waters. It is a very important measurement in water treatment plants, where it is used to determine the approximate quantity of oxygen required to stabilise organic matter biologically in wastewater, the size of the wastewater treatment facility, the efficiency of different treatment processes and compliance with regulatory discharge standards. On the basis of BOD values the degree of water pollution may be estimated (Table 4.13). A river can self-purify itself if the BOD is below 4, but not if it is greater than 4 mg O_2 L^{-1}.

Most countries do not set standards for BOD in drinking water. One exception is Russia, where the maximum allowable BOD for drinking water is 2.9 mg O_2 L^{-1}. For fisheries and aquatic life the EU sets a BOD standard of 3–6 mg O_2 L^{-1}, while Russia sets a standard of 3 mg O_2 L^{-1}. Typical BOD values of wastewaters are given in Table 4.14.

Table 4.14 *BOD_5 values for wastewaters*

Type of wastewater	BOD_5 (mg O_2 L^{-1})
Very well treated effluent	3–5
Standard effluent (after primary and secondary treatment)	10–30
Badly treated sewage	40–80
Strong sewage	400–600
Trade effluents (animal and vegetable waste)	>1000

As mentioned previously, oxidation of nitrogen compounds also contributes to the BOD. This is accomplished by two groups of nitrifying bacteria, one group converting ammonia to nitrite, and the other converting nitrite to nitrate:

$$2NH_3 + 3O_2 \rightarrow 2NO_2^- + 2H^+ + 2H_2O$$

$$2NO_2^- + O_2 \rightarrow 2NO_3^-$$

The population of nitrifying bacteria in domestic wastewaters is generally low, and their reproductive rate is slow at 20 °C, the temperature of the BOD test. Since it takes between 8 and 10 days for the population of nitrifying bacteria to reach significant numbers, the contribution of nitrification to the 5-day BOD is insignificant, and it can generally be assumed that the BOD_5 value is representative of CBOD. When nitrifying bacteria are present in significant numbers, as in effluents from biological treatment units such as the activated sludge process or trickling filters, nitrification may contribute significantly to the 5-day BOD. Wastewater engineers are interested in CBOD rather than total BOD, as CBOD is used to measure plant efficiency. In this case, the interference caused by nitrification can be eliminated by sample pre-treatment (chlorination, acidification and pasteurisation) or by adding an inhibitor (methylene blue, 2-chloro-6-(trichloromethyl)pyridine, *etc.*). River samples may also contain significant populations of nitrifying bacteria, but a specific CBOD test has not been proposed for surface water samples.

There are many other potential sources of error in the BOD test, some leading to an overestimation and others leading to an underestimation of the true BOD value. Toxic metals may depress the BOD. Anaerobic bacteria which may be present in septic sewage and river mud may give a low BOD value. Also, there are differences between conditions under which the BOD test is carried out and those prevailing in the natural environment. For example, the BOD test is carried out in the dark, while in natural environments the reaction proceeds in sunlight for half of the time (daytime). Algae that thrive in sunlight would die in the dark, contributing to the BOD. Slight variations in the temperature during incubation can also cause errors; ± 1 °C change in the temperature can produce a $\pm 4.7\%$ change in the BOD. Residual chlorine, if present, should be removed from samples. The reproducibility of BOD determinations is generally much lower than that of typical chemical analyses.

Many variations of the BOD method are in use. These include using longer or shorter incubation periods, determination of oxygen consumption rates, continuous oxygen uptake measurements by respirometric

Example 4.5

Calculate, from theory, the BOD, CBOD and NBOD of a water sample containing 200 mg L^{-1} of alanine ($C_3H_7NO_2$).

$$\text{Molecular mass of alanine} = 89$$

$$\text{Molecular mass of } O_2 = 32$$

Assume that the biological oxidation of an organic compound takes place in three steps:

1. Nitrogen and carbon are converted to NH_3 and CO_2, respectively.
2. Ammonia is oxidised to nitrite.
3. Nitrite is oxidised to nitrate.

Write balanced equations for each of these steps:

1. $C_3H_7NO_2 + 3O_2 \rightarrow NH_3 + 3CO_2 + 2H_2O$ CBOD
2. $NH_3 + 1.5O_2 \rightarrow HNO_2 + H_2O$ $\left.\vphantom{\begin{array}{c}a\\b\end{array}}\right\}$ NBOD
3. $HNO_2 + 0.5O_2 \rightarrow HNO_3$

Total BOD = the sum of the oxygen demands of steps 1, 2 and 3
$$= 3 + 1.5 + 0.5 = 5 \text{ mol } O_2 \text{ mol}^{-1} \text{ alanine}$$
$$= 5 \times 32 = 160 \text{ g } O_2 \text{ mol}^{-1} \text{ alanine}$$

$$200 \text{ mg } L^{-1} \text{ alanine} = 0.2/89 \text{ mol } L^{-1} = 2.25 \times 10^{-3} \text{ mol } L^{-1}$$

$$BOD = 160 \times (2.25 \times 10^{-3}) = 0.360 \text{ g } O_2 \text{ L}^{-1} = 360 \text{ mg } O_2 \text{ L}^{-1}$$

CBOD = oxygen demand of step 1
$$= 3 \text{ mol } O_2 \text{ mol}^{-1} \text{ alanine} = 3 \times 32 = 96 \text{ g } O_2 \text{ mol}^{-1} \text{ alanine}$$
$$= 96 \times (2.25 \times 10^{-3}) = 0.216 \text{ g } O_2 \text{ L}^{-1} = 216 \text{ mg } O_2 \text{ L}^{-1}$$

NBOD = sum of the oxygen demands of steps 2 and 3
$$= 1.5 + 0.5 = 2 \text{ mol } O_2 \text{ mol}^{-1} \text{ alanine} = 2 \times 32 = 64 \text{ g } O_2 \text{ mol}^{-1} \text{ alanine}$$
$$= 64 \times (2.25 \times 10^{-3}) = 0.144 \text{ g } O_2 \text{ L}^{-1} = 144 \text{ mg } O_2 \text{ L}^{-1}$$

techniques, *etc*. Only the 5-day BOD test is described here. See also Example 4.5.

4.7.2 Methodology

The water sample is diluted with a suitable volume of oxygen-saturated water containing nutrients and incubated in the dark for 5 days at 20 °C.

The DO concentration is determined before and after incubation and BOD_5 is calculated by difference. The BOD is reported as mg O_2 L^{-1} consumed by 1 L of undiluted sample. No more than 70% of the oxygen should be consumed during incubation.

4.7.3 Materials

- All the reagents required for the dissolved oxygen (DO) test by the Winkler method (see Section 4.6)
- BOD incubation bottles: 250 mL capacity, with ground-glass stoppers ground to a point to prevent the trapping of air when stoppers are inserted. Bottles should be cleaned with chromic acid solution or a good quality detergent, and then rinsed with water to remove traces of cleaning agent. In order to avoid drawing air into the bottles during incubation, use a water seal. You can obtain satisfactory water seals by inverting bottles in a water bath. Place a paper or plastic cup, or foil cap, over the mouth of the bottle to reduce evaporation of the water seal during incubation.
- Air incubator or water bath operated at $20 \pm 1\,°C$. Place in the dark.
- Phosphate buffer solution prepared by dissolving 21.75 g K_2HPO_4, 8.5 g KH_2PO_4, 33.4 g $Na_2HPO_4 \cdot 7H_2O$ and 1.7 g NH_4Cl in water and diluting to 1 L. This solution should have a pH of 7.2.
- Calcium chloride solution prepared by dissolving 27.5 g $CaCl_2$ in water and diluting to 1 L
- Magnesium sulfate solution prepared by dissolving 22.5 g $MgSO_4 \cdot 7H_2O$ in water and diluting to 1 L
- Ferric chloride solution, prepared by dissolving 0.25 g $FeCl_3 \cdot 6H_2O$ in distilled water and diluting to 1 L
- Dilution water. Prepare daily by adding 1 mL of each of the phosphate buffer, magnesium sulfate, calcium chloride and ferric chloride solutions per litre of water and saturate with air. This can be achieved by bubbling filtered laboratory air using a pump or compressed air into the solution through an air stone or sintered glass frit.
- Hydrochloric acid, 1:1 (prepared by mixing equal volumes of concentrated HCl and water)
- Sodium hydroxide, 6 M, prepared by dissolving 240 g NaOH in water and diluting to 1 L
- pH meter and electrode

N.B. Discard any of the reagent solutions if they show signs of bacterial growth.

4.7.4 Experimental Procedure

4.7.4.1 Sampling and Storage. Collect 500 mL–1 L of sample. The test should be started as soon as possible after sampling. If this is not possible, store at 3–5 °C in a refrigerator for up to 24 h. Leaving the sample at room temperature for 8 h could reduce the BOD by as much as 40%. Bring the samples to 20 °C before starting the test.

4.7.4.2 Analysis. Measure the pH of the sample, and adjust to pH 7–8 using HCl or NaOH. If the pH is between 6.5 and 8.2, do not adjust. The following dilutions are recommended depending on the sample:

Raw sewage:	30–100 times dilution
Settled sewage:	10–50 times dilution
Purified effluents:	3–10 times dilution
Polluted river water:	1–5 times dilution

Ideally, the dilution should be carried out in such a way that after 5 days of incubation at least 2 mg O_2 L^{-1} is consumed and at least 2 mg O_2 L^{-1} remains in the solution. You may have to prepare several dilutions to cover the range of possible values. Place about 400 mL of dilution water in a 1 L measuring cylinder. Pour the water slowly down the wall of the cylinder to prevent entrainment of air. Add the required volume of the sample from another measuring cylinder or pipette and dilute to 1 L. Mix carefully with a mixing rod so as not to entrain air bubbles.

Completely fill three 250 mL BOD bottles with the diluted solution, ensuring that no air bubbles are trapped. This may be done by siphoning. Determine the dissolved oxygen (DO) in one bottle immediately using the Winkler method described in Section 4.6.4. Stopper the other two bottles, water-seal, place in the water bath or air incubator and incubate in the dark for 5 days at 20 °C. After 5 days, determine DO using the Winkler method.

Fill three bottles with dilution water in the same way. Determine DO in one bottle immediately and in the other two bottles after incubating for 5 days. This dilution water blank serves as a rough check on the quality of the dilution water and it should not exceed 0.2 mg L^{-1}.

Calculate the BOD from the following:

$$BOD \ (mg \ L^{-1}) = F(D_i - D_f)$$

where D_i is the initial DO in the diluted sample, D_f is the DO in the diluted sample after 5 days of incubation and F is the dilution factor:

F = total volume after dilution (mL)/volume of undiluted sample (mL)

Report the average of the BOD values measured in the two incubated bottles. If the dilution water blanks do not exceed 0.2 mg L^{-1}, it is not necessary to correct for the DO uptake by the dilution water.

Notes

1. For samples whose BOD_5 does not exceed 7 mg L^{-1} dilution is not necessary, and you may perform the test directly on the undiluted samples. However, these samples should be aerated to bring the DO content to saturation.
2. Analyse samples containing residual chlorine only after dissipating the chlorine and seeding the dilution water (see below). Standing the sample in the light for several hours will dissipate the chlorine. Chlorine most often dissipates during transport of samples to the laboratory. You can avoid taking chlorinated samples of wastewater by sampling ahead of the chlorination unit.
3. If seeding of dilution water is required, proceed as follows. Obtain an appropriate seed containing a high population of microorganisms. The preferred seed is undisinfected effluent from a biological treatment process such as a trickling filter or activated sludge process obtained at a sewage works. Otherwise, you may seed the dilution water with a sample of surface water collected downstream (up to several km) from a wastewater effluent discharge point. Another alternative is to allow domestic wastewater to settle for 1–36 h at 20 °C and use the supernatant as the seed. Add 5 mL of the seed solution to 1 L of the dilution water. Use the seeded dilution water to dilute the sample exactly as described in the procedure and carry out the incubation and analysis. Also determine the BOD of the seed in the same way that you would for a sample. Prepare a *seed control* by diluting the seed solution with diluting solution. Measure the initial DO and the DO after 5 days incubation and use these values to correct the sample BOD measurement as shown in the equation below. It is not necessary to measure the oxygen uptake of seeded dilution water blanks. When seeding is employed the BOD of the sample is calculated as follows:

$$BOD \ (mg \ L^{-1}) = F[(D_i - D_f) - f(C_i - C_f)]$$

where D_i is the initial DO in diluted sample, D_f is the DO in diluted sample after 5 days of incubation, F is the dilution factor (calculated as shown above), C_i is the initial DO in diluted seed solution (seed control), C_f is the DO in diluted seed solution (seed control) after 5 days of incubation and f is defined as:

$$f = \frac{\% \text{ seed in diluted sample}}{\% \text{ seed in diluted seed solution (seed control)}}$$

4. When analysing samples supersaturated with DO, such as cold waters or those with a high photosynthetic activity, aerate the sample in the same way as described for the dilution water. This will bring the DO level down to the saturated value.

5. You may use an oxygen electrode instead of the Winkler method to determine the DO concentrations.

4.7.5 Questions and Problems

1. Define biochemical oxygen demand. Compare and contrast biochemical oxygen demand and chemical oxygen demand.
2. Discuss the limitations of the BOD test.
3. Which samples require seeding, and what would you use as a seed material?
4. What are the applications of BOD data in environmental engineering?
5. Calculate the theoretical BOD, CBOD and NBOD of wastewater samples containing: (a) 130 mg L^{-1} glycine ($C_2H_5NO_2$), (b) 250 mg L^{-1} aniline (C_6H_7N) and (c) a mixture of 200 mg L^{-1} glycine and 130 mg L^{-1} alanine ($C_3H_7NO_2$).

4.7.6 Suggestions for Projects

1. Carry out a survey of BOD values in surface waters of your region. Perform sampling at the same time of the day to avoid the influence of diurnal variations in O_2 concentration. Apply the data from Table 4.13 to rank water quality and compare the results with national standards for BOD. Comment on the sources of pollution.
2. Measure BOD values at several locations downstream from an effluent source. Comment on your results.
3. Measure BOD at several locations inside a sewage treatment works. Discuss the results in terms of the treatment technology used.
4. Carry out any of the above three projects together with measurements of other water quality parameters (pH, DO, nitrogen compounds,

etc.) and investigate any possible relationship between the BOD and other parameters. Discuss the results on the basis of your knowledge of aquatic processes.

4.7.7 Further Reading

APHA, "Standard Methods for the Examination of Water and Wastewater", 16th edn., American Public Health Association, Washington, 1985, pp. 525–532.
N. J. Bunce, "Environmental Chemistry", 2nd edn., Wuerz, Winnipeg, 1994, pp. 131–158.
C. N. Sawyer, P. L. McCarty and G. F. Parkin, "Chemistry for Environmental Engineering", 4th edn., McGraw-Hill, New York, 1994, pp. 527–544.

4.8 CHEMICAL OXYGEN DEMAND (COD)

4.8.1 Introduction

The composition and concentration of organic compounds in natural waters are related to various biotic and abiotic processes, the most important among them being: excretion of hydrobiota; inputs from atmospheric deposition and surface runoff; and inputs of industrial, municipal and agricultural wastes. The COD is a measure of the oxygen equivalent of the organic matter in a sample that is susceptible to oxidation by a strong oxidising agent. This test is widely used to measure the organic strength of domestic and industrial waste-waters, and it can be related empirically to BOD, organic matter and organic carbon. The main difference between COD and BOD tests is that while BOD determines compounds that can be biologically oxidised, COD measures those substances which can be chemically oxidised.

COD determination involves reacting the sample with an excess of an oxidising agent for a specified period of time, after which the concentration of unreacted oxidising agent is usually determined by a redox back-titration. The quantity of oxidising agent used up is determined by difference from the initial oxidant concentration. COD is reported in terms of oxygen equivalent.

There are a number of different experimental procedures for measuring COD. Nowadays, the most widely used method involves refluxing the sample with an excess of potassium dichromate ($K_2Cr_2O_7$) (see Example 4.6). The dichromate remaining after reaction is determined by titration with ferrous ammonium sulfate. Dichromate is preferred to other oxidants because of its superior oxidizing ability; with few exceptions,

---

Example 4.6

What is the equivalence between dichromate and oxygen?

The dichromate method for COD involves oxidation–reduction (redox) reactions. Transfer of electrons and changes in oxidation state take place during redox reactions. The reaction of dichromate in acidic solution may be represented as:

$$Cr_2O_7^{2-} + 14H^+ + 6e^- \rightarrow 2Cr^{3+} + 7H_2O$$

while oxidation by oxygen can be represented as:

$$O_2 + 4H^+ + 4e^- \rightarrow 2H_2O$$

Since each $Cr_2O_7^{2-}$ ion consumes 6 electrons and since each O_2 molecule consumes 4 electrons, then 1 mole of $Cr_2O_7^{2-}$ is equivalent to $6/4 = 1.5$ moles of O_2.

---

organic compounds are almost completely oxidised (95–100%). Pyridine and volatile organic compounds are not oxidised by dichromate. High concentrations of chloride can interfere with the chromate method:

$$6Cl^- + Cr_2O_7^{2-} + 14H^+ \rightarrow 3Cl_2 + 2Cr^{3+} + 7H_2O$$

Addition of mercury(II) sulfate to the sample prior to analysis eliminates this interference:

$$Hg^{2+} + 2Cl^- \rightleftharpoons HgCl_2$$

The oxidation of organic compounds by dichromate during digestion can be represented by the following unbalanced equation (see also Example 4.7):

$$(C_aH_bO_cN_d) + xCr_2O_7^{2-} + H^+ \rightarrow CO_2 + H_2O + NH_4^+ + Cr^{3+}$$

The theoretical amount of dichromate required to oxidise a particular organic compound can be computed from:

$$x = 2a/3 + b/6 + c/3 + d/2$$

An older method involves reacting the sample with an excess of potassium permanganate. This test is also referred to as *permanganate*

Example 4.7

Calculate the theoretical COD of a wastewater sample containing 250 mg L^{-1} of glucose ($C_6H_{12}O_6$).

The theoretical number of moles of $Cr_2O_7^{2-}$, x, required to oxidise one mole of glucose can be calculated from the previous equation as:

$$x = (2 \times 6/3) + 12/6 + 6/3 = 8$$

Since 1 mol $Cr_2O_7^{2-}$ = 1.5 mol O_2 (see Example 4.6), 8 mol $Cr_2O_7^{2-}$ = 12 mol O_2.

Molecular mass of glucose = 180
Molecular mass of O_2 = 16

250 mg L^{-1} glucose = 0.250 g L^{-1} glucose = 0.25/180 mol L^{-1} glucose

No. of mol L^{-1} of $Cr_2O_7^{2-}$ required = 8 × (0.25/180) = 0.011 mol L^{-1}

No. of mol L^{-1} of O_2 required = 12 × (0.25/180) = 0.017 mol L^{-1} = 32 × 0.017
$$= 0.544 \text{ g } O_2 \text{ L}^{-1}$$
$$= 544 \text{ mg } O_2 \text{ L}^{-1}$$

COD of wastewater containing 250 mg L^{-1} glucose is 544 mg O_2 L^{-1}

---

value (PV) or the *oxygen absorbed* test. The unreacted permanganate can be determined by reacting with iodide to liberate iodine:

$$2MnO_4^- + 16H^+ + 10I^- \rightarrow 2Mn^{2+} + 8H_2O + 5I_2$$

The liberated iodine is titrated with sodium thiosulfate. The permanganate method is rapid and relatively easy to carry out, and it is useful with less polluted surface waters and tapwaters. The main disadvantage of this method is that most organic compounds are only partially oxidised (up to 80%). There are several variations of this test involving different reaction times. Only the dichromate method is described in detail here.

In addition to organic compounds, various inorganic substances such as nitrite, sulfite and ferrous iron are also oxidised during COD determination. The main advantage of the COD method is that it is rapid; it takes only a few hours to complete compared to 5 days for BOD. The COD of wastewater is generally higher than the BOD because more substances can be chemically oxidised than can be biologically oxidised. It is possible to correlate COD with BOD for many types of wastes, and the COD test can then be used as a quick substitute for BOD. Also, wastewaters containing compounds toxic to microorganisms can be analysed for COD. Furthermore, the COD method yields more reproducible results than the BOD method. One disadvantage of the COD test is that it cannot differentiate between biologically oxidisable and biologically inert organic matter. It is therefore less relevant to natural processes than the BOD test.

COD values can characterise the degree of pollution and self-purification of various waters. National and international standards and guidelines have been set for COD. The COD test is useful for monitoring and control after correlation with sources of organic matter has been established. COD test kits containing ampoules with pre-measured reagents are commercially available. These are supplied with instructions for use.

4.8.2 Methodology

The sample is refluxed with excess potassium dichromate in concentrated sulfuric acid for 2 h. This oxidises most of the organic matter in the sample. Silver sulfate is included as a catalyst to speed up the oxidation process. After digestion, the unreacted dichromate remaining in solution is titrated with ferrous ammonium sulfate:

$$Cr_2O_7^{2-} + 6Fe^{2+} + 14H^+ \rightarrow 2Cr^{3+} + 6Fe^{3+} + 7H_2O$$

The amount of dichromate consumed is calculated and the oxidisable organic matter is reported in terms of oxygen equivalents.

4.8.3 Materials

- Reflux apparatus, consisting of a flask (round-bottomed or Erlenmeyer) with ground glass neck, reflux condenser and a hot plate or heating mantle (Figure 4.8).
- Standard potassium dichromate solution, 0.0417 M. Dry some $K_2Cr_2O_7$, primary standard grade, in an oven at 105 °C for 2 h. Dissolve 12.259 g in water and dilute to 1 L. This solution is stable indefinitely.
- Sulfuric acid reagent prepared by dissolving 15 g of Ag_2SO_4 in 1 L concentrated sulfuric acid.
- Ferroin indicator solution. Purchase ready-made indicator solution or prepare by dissolving 0.7 g $FeSO_4 \cdot 7H_2O$ and 1.485 g 1,10-phenanthroline monohydrate ($C_{12}H_8N_2 \cdot H_2O$) in water and diluting to 100 mL.

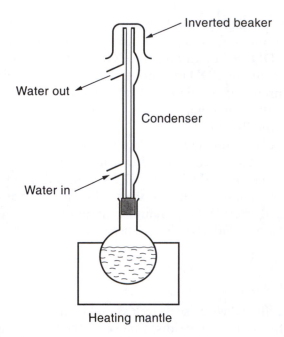

Figure 4.8 *Reflux apparatus*

- Standard ferrous ammonium sulfate solution, *ca.* 0.25 M. Dissolve 98 g ferrous ammonium sulfate, $Fe(NH_4)_2(SO_4)_2 \cdot 6H_2O$, in water. Add 20 mL concentrated H_2SO_4, cool and dilute to 1 L. Standardise this solution daily against standard $K_2Cr_2O_7$ solution as outlined below.

4.8.4 Experimental Procedure

4.8.4.1 Standardisation of Ferrous Ammonium Sulfate Solution. As ferrous ion is slowly oxidised to ferric ion by dissolved oxygen, it is necessary to standardise the ferrous ammonium sulfate solution each time that it is used. Dilute 10.0 mL of the standard potassium dichromate solution to about 100 mL with water and add 30 mL concentrated H_2SO_4. Cool and titrate with the standard ferrous ammonium sulfate solution after adding 3 drops of ferroin indicator to show the end point, at which the colour changes from blue-green to violet-red. Calculate the molarity of the standard ferrous ammonium sulfate solution from:

$$\text{Molarity} = 0.25 \times \frac{\text{Volume of standard potassium dichromate solution (mL)}}{\text{Volume of standard ferrous ammonium titrant used up (mL)}}$$

Use this value when calculating the COD of samples.

4.8.4.2 Sampling and Storage. Samples should be collected in glass bottles and the test performed as soon as possible after sample collection. If this is not possible, preserve samples by acidifying to pH ≤ 2 with concentrated H_2SO_4 and store in a refrigerator at $4\,°C$. Wastewater samples are left to stand for 2 h to remove solids by sedimentation.

4.8.4.3 Analysis. Place 50 mL of the water sample in a 500 mL refluxing flask (Figure 4.8). Add 25 mL of the standard potassium dichromate solution and some anti-bump granules or glass beads. Very slowly add 75 mL sulfuric acid reagent while mixing. Let the solution cool and mix thoroughly. Attach flask to condenser, and turn on cooling water. Place a small beaker over the open end of the condenser. Turn on the heater and carefully boil. After refluxing for 2 h, turn off the heater and cool the solution. Rinse the condenser with approximately 150 mL of water before disconnecting the flask. Let the solution cool to room temperature and add 3 drops of ferroin indicator solution. Titrate with standard ferrous ammonium sulfate solution until the colour changes sharply from blue-green to violet-red. Also reflux a reagent blank consisting of 50 mL of water, 25 mL standard potassium dichromate solution and 75 mL sulfuric acid reagent in the same way as the sample and titrate. Calculate the COD as follows:

$$\text{COD (mg } O_2 \text{ L}^{-1}) = 8000 \times M \times (V_1 - V_2)/V_s$$

where V_1 is the volume of ferrous ammonium sulfate titrant used to titrate the blank (mL), V_2 is the volume of ferrous ammonium sulfate titrant used to titrate the sample (mL), V_s is the volume of sample (mL) and M is the exact concentration of ferrous ammonium sulfate titrant (mol L^{-1}) as determined in the standardisation titration above.

This procedure is suitable for samples with COD between 50 and 900 mg O_2 L^{-1}. The method can easily be modified for samples with higher or lower COD (see notes 1 and 2).

Notes

1. For samples with very high COD (>900 mg O_2 L^{-1}), use smaller aliquots of sample but dilute them to 50 mL. Thereafter carry out the refluxing and analysis as outlined above.
2. If the sample has very low COD, use the same experimental procedure as outlined above for regular samples except that more dilute standard solutions of potassium dichromate (0.00417 M) and ferrous ammonium sulfate (0.025 M) should be used. You can

prepare these by diluting the more concentrated standard solutions prepared above. Be careful to avoid any contamination by traces of organic matter.
3. Where preliminary tests on known samples have shown that shorter reflux times give the same result as after 2 h of refluxing, you may reduce the reflux time.
4. You may use different sample volumes (*e.g.* 20 mL), but you should then vary the volumes of reagents by a corresponding factor in order to keep the ratios of all reagents constant.
5. Chloride at high concentrations can interfere with the determination. This interference may be eliminated by adding 1 g of $HgSO_4$ to 50 mL of sample prior to analysis.
6. You may evaluate the method by determining COD on a standard potassium hydrogen phthalate solution. Prepare this standard solution as follows. Crush and dry some potassium hydrogen phthalate to constant weight at 120 °C. Dissolve 425 mg in water and dilute to 1 L. Store in a refrigerator. Unless biological growth is observed this solution is stable for up to 3 months. This solution has a theoretical COD of 500 mg O_2 L^{-1} (see Example 4.8).

Example 4.8

Calculate the COD of a solution containing 425 mg L^{-1} potassium hydrogen phthalate.

Potassium hydrogen phthalate is oxidised according to the following equation:

$$KC_8H_5O_4 + 7.5O_2 \rightarrow 8CO_2 + 2H_2O + K^+ + OH^-$$

Molecular weight of $KC_8H_5O_4$ = 204
Molecular weight of O_2 = 32

According to stoichiometry of the above reaction: 1 mol $KC_8H_5O_4$ = 7.5 mol O_2

$$
\begin{aligned}
425 \text{ mg L}^{-1} \, KC_8H_5O_4 &= 0.425/204 \text{ mol L}^{-1} \, KC_8H_5O_4 \\
&= 7.5 \times (0.425/204) \text{ mol } O_2 \text{ L}^{-1} \\
&= 7.5 \times (0.425/204) \times 32 \times 1000 \text{ mg } O_2 \text{ L}^{-1} \\
&= 500 \text{ mg } O_2 \text{ L}^{-1}
\end{aligned}
$$

COD of 425 mg L^{-1} potassium hydrogen phthalate solution = 500 mg O_2 L^{-1}

4.8.5 Questions and Problems

1. Explain why COD can be used as a measure of water pollution by organic compounds?
2. What are the main sources of organic compounds in natural water?
3. Describe the chemical principles of the dichromate and permanganate methods for COD determination.
4. In the past the permanganate method was commonly used in the water industry. Nowadays, the dichromate procedure is the most widely used method for COD determination and the permanganate method is becoming obsolete. Explain why this is so.
5. Compare and contrast the COD and BOD methods. What are their advantages and disadvantages?
6. Calculate the theoretical COD of industrial wastewater samples containing 400 mg L^{-1} of: (a) glycine ($C_2H_5O_2N$), (b) stearic acid ($C_{18}H_{36}O_2$) and (c) phenol (C_6H_6O).

4.8.6 Suggestions for Projects

1. Derive an empirical relationship between BOD and COD as follows. Determine both BOD and COD in various samples of natural waters and wastewaters and compare the results. Plot BOD against COD on a graph and calculate the correlation coefficient. If the correlation is significant, obtain a best fit line between the data points using the method of least squares and derive an equation relating COD to BOD.
2. Determine the COD values in a stream at different distances from a wastewater discharge point. Illustrate the data in graphical form with COD in mg O_2 L^{-1} as the y-axis and distance as the x-axis. This would portray the depletion of oxygen as a function of distance downstream from a point source discharge (*e.g.* a sewage outfall).
3. Carry out laboratory experiments with different solutions of organic compounds and measure their COD values. Compare the results with those calculated from theory.

4.8.7 Further Reading

APHA, "Standard Methods for the Examination of Water and Wastewater", 16th edn., American Public Health Association, Washington, 1985, pp. 532–538.

N. J. Bunce, "Environmental Chemistry", 2nd edn., Wuerz, Winnipeg, 1994, pp. 131–158.

C. N. Sawyer, P. L. McCarty and G. F. Parkin, "Chemistry for Environmental Engineering", 4th edn., McGraw-Hill, New York, 1994, pp. 545–551.

4.9 CHLOROPHYLL

4.9.1 Introduction

The content of chlorophyll is a common indicator of phytoplankton biomass and water quality, and it is determined in order to assess the ecological status of water bodies. Several types of pigments are present in green plants: chlorophyll *a, b* and *c*, carotenes and xanthophylls. Chlorophyll *a* is the most important; it is present at about 1–2% of the dry weight of planktonic algae. The absence or presence of various pigments can be used, together with other features, to separate the major algal groups and to characterise the ecological condition of surface waters (Table 4.15).

At present there are no national standards or international guidelines for the chlorophyll content in surface water. Data in Table 4.15 show the recommended guidelines which could be applied for rapid, preliminary determination of water quality as part of routine analysis. Depending on chlorophyll *a* levels, other chemical parameters, whose content is limited by national and international water quality standards (N, P, COD, BOD, O_2, *etc.*), could be selected for detailed analysis.

Chlorophyll *a* in phytoplankton can be determined by fluorimetry or spectrophotometry. Although fluorimetry is more sensitive, it is unreliable for fresh waters owing to interference by pheophytin *a*, a common

Table 4.15 *The content of chlorophyll* a *(μg L^{-1}) and eutrophication level of waters*

Eutrophication levels					
Very low	*Low*	*Medium*	*High*	*Very high*	
Ecological (trophic) condition					
Ultra-olygotrophic	*Olygotrophic*	*Mesotrophic*	*Eutrophic*	*Hyper-eutrophic*	*Region*
0.01–0.5	0.3–3.0	2–1	10–500	–	USA
$\leqslant 1.0$	$\leqslant 2.5$	2.0–8.0	8–25	$\geqslant 25$	Europe/OECD
0.01–0.1	0.1–1.0	1.0–10	>10	>100	Russia

Table 4.16 *The relationship between chlorophyll* a *content in plankton (Ch) and total phosphorus content (P_{tot}) in various water bodies of the world*

Regression equation	Region
$Ch = 0.46 \log P_{tot} - 3.87$	Lakes (world scale, International Biological programme)
$\log Ch = 1.46 \log P_{tot} - 1.09$	Lakes (world scale, $n = 143$)
$\log Ch = 1.45 \log P_{tot} - 1.14$	USA
$\log Ch = 1.583 \log P_{tot} - 1.134$	Japan
$\log Ch = 0.871 \log P_{tot} + 0.48$	United Kingdom
$Ch = 74.5 + 0.19 P_{tot}$	Netherlands/Europe
$\log Ch = 1.270 \log P_{tot} - 1.113$	Russia (Karelia)
$\log Ch = 1.49 \log P_{tot} - 1.32$	Poland/Europe
$\log Ch = 1.408 \log P_{tot} - 1.380$	Estonia/Europe
$\log Ch = 0.575 + 0.205 \log P_{tot}$	Latvia/Europe

Taken from RAS (1993). P_{tot} in mg P L^{-1}.

degradation product of chlorophyll *a*. Spectrophotometry is the most suitable technique for routine determination of chlorophyll *a*.

It is possible to relate chlorophyll *a* content to the concentration of total phosphorus in surface waters. Statistical regression equations shown in Table 4.16 could be used to calculate the approximate level of chlorophyll *a* content in surface waters from total phosphorus measurements. When using these equations, take into consideration the geographical area where you are living.

4.9.2 Methodology

Plankton are concentrated from the water sample by filtration. Chlorophyll *a*, *b* and *c* are extracted from the plankton concentrate into aqueous acetone solution and the absorbance of the extract is determined with a spectrophotometer. Pheophytin *a*, a degradation product of chlorophyll *a*, may interfere if it is present in the sample. A correction for this interference is possible with a simple modification of the analysis method.

4.9.3 Materials

- Spectrophotometer with a narrow band width (0.5–2.0 nm).
- Filtration assembly with filter holder and vacuum pump.
- Filter papers, glass fibre (GF/C or GF/A) 4 cm diameter, and solvent-resistant filters.

- Magnesium carbonate solution, 1%. Add 1.0 g finely powdered $MgCO_3$ to 100 mL water.
- Aqueous acetone solution, 90%. Mix 90 mL acetone with 10 mL magnesium carbonate solution to prepare 100 mL of solution. You may prepare larger volumes but keep the reagents in the same proportion.
- Centrifuge tubes, 15 mL graduated.
- Centrifuge.

4.9.4 Experimental Procedure

4.9.4.1 Sampling and Storage. The volume of water sample required for chlorophyll analysis varies from a few mL (waters with abundant "bloom") up to 1 L, depending on the population of algae cells. Collect sample in dark bottles. Preferably, the filtration of water for chlorophyll determination should be carried out immediately after sampling or within 24 h. If this is not possible, water samples may be stored in dark bottles at 4 °C for up to 2 weeks. Filter through a GF/C or GF/A filter paper. At least 1–2 g of plankton is required for extraction of measurable quantities of chlorophyll.

4.9.4.2 Analysis. Collect the plankton by filtration onto a glass fibre filter. Roll up the filter paper and insert it into a centrifuge tube. Add 10 mL of 90% aqueous acetone solution. Screw on the cap and leave overnight in a refrigerator at 4 °C. The next day, decant the extract if it is clear. If it is turbid, clarify either by centrifuging or by filtering through a solvent-resistant disposable filter. Decant clarified liquid into a 15-mL graduated centrifuge tube or a small graduated measuring cylinder and measure the total volume of extract.

Pour extract into a 1 cm cell and measure the absorbance at 630, 647, 664 and 750 nm. Subtract the absorbance measured at 750 nm from each of the absorbance values measured at the other three wavelengths. Use the corrected absorbance values in the following equations to calculate the concentrations of chlorophyll *a*, *b* and *c* in the extract:

$$Ca\ (\mu g\ mL^{-1}) = (11.85 \times A_{664}) - (1.54 \times A_{647}) - (0.08 \times A_{630})$$

$$Cb\ (\mu g\ mL^{-1}) = (21.03 \times A_{647}) - (5.43 \times A_{664}) - (2.66 \times A_{630})$$

$$Cc\ (\mu g\ mL^{-1}) = (24.52 \times A_{630}) - (7.60 \times A_{647}) - (1.67 \times A_{664})$$

where *A* is the corrected absorbance at the wavelength indicated as a subscript and Ca, Cb and Cc are the concentrations of chlorophyll a, b and c, respectively, in the extracts. Calculate the concentration of each substance in the original sample as shown for chlorophyll a below:

$$\text{Chlorophyll } a \text{ } (\mu\text{g L}^{-1}) = Ca \times V_e/V_s$$

where Ca is the concentration of chlorophyll a (or b or c) in the extract (μg mL^{-1}), V_e is the volume of extract (mL) and V_s is the sample volume (L).

Notes

1. The extraction efficiency of the above method may vary with different algae. You can improve the method by grinding the filter paper after collection of the phytoplankton in a tissue grinder if one is available. Place filter in grinder, cover with 3 mL of aqueous acetone solution and operate at 500 rpm for 1 min. After transferring sample to the centrifuge tube, rinse grinder with aliquots (2–3 mL) of aqueous acetone solution and transfer these to the centrifuge tube. Adjust volume to 10 mL and continue the analysis as outlined above.

2. Although glass fibre filters are recommended, you may use membrane filters (0.4 μm porosity) instead.

3. Filters with the phytoplankton sample can be stored frozen in plastic, air-tight bags for up to 3 weeks. This is only applicable to samples taken from neutral or alkaline waters. Samples from acid waters have to be analysed immediately to avoid chlorophyll degradation.

4. The above method may overestimate the concentration of chlorophyll a in samples containing pheophytin a, a product of chlorophyll a degradation. Pheophytin a absorbs at a similar wavelength to chlorophyll a. You can test for the presence of pheophytin a and obtain the corrected concentration of chlorophyll a with the following simple procedure. Place 3 mL of your final extract into a 1 cm cell and measure the absorbance at 664 and 750 nm. Then add exactly 0.1 mL of 0.1 M HCl, mix and measure the absorbance at 665 and 750 nm. If the ratio of absorbance at 664 nm before acidification and absorbance at 665 after acidification (A_{664}/A_{665}) is 1.7, the sample is considered to be free of pheophytin a. If A_{664}/A_{665} is in the range 1–1.7, some pheophytin a is present and a corrected concentration of chlorophyll a should be calculated from:

$$\text{Chlorophyll } a \text{ } (\mu\text{g L}^{-1}) = 26.7 \times (A_{664} - A_{665}) \times V_e/V_s$$

where A_{664} and A_{665} are the corrected absorbances before and after acidification, respectively, V_e is the extract volume (mL) and V_s is the sample volume (L). Absorbances are corrected in the same way as outlined earlier, by subtracting the absorbance at 750 nm from each reading:

A_{664} = absorbance at 664 nm before acidification − absorbance at 750 nm before acidification

A_{665} = absorbance at 665 nm after acidification − absorbance at 750 nm after acidification

The concentration of pheophytin a may be determined from:

$$\text{Pheophytin } a \ (\mu g \ L^{-1}) = 26.7 \times [1.7 \times A_{665} - A_{664}] \times V_e/V_s$$

where V_e and V_s are the same as above.

4.9.5 Questions and Problems

1. Explain why the content of chlorophyll can be used to characterise the ecological condition of surface waters.
2. What is the difference between various pigments of phytoplankton?
3. Characterise eutrophication processes in surface waters on the basis of the chlorophyll a content.
4. Explain the relationship between chlorophyll a and total phosphorus content in surface water.

4.9.6 Suggestions for Projects

1. Measure the chlorophyll a content in various surface waters. The level of eutrophication may be estimated using data shown in Table 4.15.
2. Determine the concentration of chlorophyll a and total phosphorus in surface waters of your locality. Is there a significant correlation between the two parameters? If there is, calculate a regression equation similar to those shown in Table 4.16 for your locality.
3. On the basis of a chemical survey of surface waters from different locations in your region, calculate correlation coefficients between chlorophyll a and other chemical indices like nitrogen, phosphorus, BOD, COD and DO and examine whether they are statistically significant or not. Explain the results from the point of view of eutrophication processes.
4. Determine the chlorophyll a content in wastewaters and effluents

from water treatment plants. Explain the results in terms of the degree of water purification.

4.9.7 Reference

RAS, "Methodological Problems of Research on Primary Production of Plankton of Inland Waters", Institute of Inland Biology RAS, St. Petersburg, 1993 (in Russian).

4.9.8 Further Reading

APHA, "Standard Methods for the Examination of Water and Wastewater", 16th edn., American Public Health Association, Washington, 1985, pp. 1067–1072.
OECD, "Eutrophication of Waters. Monitoring, Assessment and Control", OECD, Paris, 1982.
E. Roff, Spectrophotometric and chromatographic chlorophyll analysis: comparison of results and dissension of the trichromatic method, *Ergeb. Limnol.*, 1980, no. 14.
UNESCO, "Determination of Photosynthetic Pigments in Sea-Water", UNESCO, Paris, 1966.

4.10 NITROGEN

4.10.1 Introduction

Nitrogen compounds are of interest to environmental engineers because they are both essential nutrients, beneficial to living organisms, and pollutants, with potentially harmful consequences. Nitrogen can exist in seven different oxidation states: NH_3 ($-III$), N_2 (0), N_2O (I), NO (II), N_2O_3 (III), NO_2 (IV) and N_2O_5 (V), and its environmental chemistry is consequently quite complex. Bacteria can oxidise and reduce N under aerobic and anaerobic conditions, respectively. Nitrogen pollution can have potentially harmful effects in surface and ground waters, and these are causing considerable current concern. Major issues are:

- The nitrate content in drinking water.
- Eutrophication in many coastal and inland waters triggered by high ammonium and nitrate concentrations.
- Acidification of lakes, streams and ground waters caused by nitric acid in rain and by acidic runoff from fields treated with N fertilisers.

Levels of both NH_4^+-N and NO_3^--N are often below 1 mg L^{-1} in

---

Example 4.9

A wastewater sample was analysed and the results were reported as follows:

$$NH_3 = 52.8 \text{ mg L}^{-1}; \quad NO_3^- = 9.52 \text{ mg L}^{-1}; \quad NO_2^- = 0.57 \text{ mg L}^{-1}$$

Express the results of the analysis in the correct manner (*i.e.* in terms of mg N L^{-1}).

Calculate the conversion factors (CF) for each species by dividing the atomic mass of N by the molecular/ionic mass of the species:

for NH_3, CF $= 14/17 = 0.82$
for NO_3^-, CF $= 14/62 = 0.23$
for NO_2^-, CF $= 14/46 = 0.30$

Multiply the reported results by the corresponding conversion factors:

$0.82 \times 52.8 \text{ mg NH}_3 \text{ L}^{-1} = 43.3 \text{ mg NH}_3\text{-N L}^{-1}$
$0.23 \times 9.52 \text{ mg NO}_3^- \text{ L}^{-1} = 2.19 \text{ mg NO}_3^-\text{-N L}^{-1}$
$0.30 \times 0.57 \text{ mg NO}_2^- \text{ L}^{-1} = 0.17 \text{ mg NO}_2^-\text{-N L}^{-1}$

---

unpolluted fresh water and this can limit plankton growth. However, in some areas, nitrate and ammonium ions have risen to potentially harmful levels due mainly to discharges of wastewaters and agricultural runoff. Dissolved organic N is also important in waters. Nitrite concentrations are usually less than 0.1 mg L^{-1} but under reducing conditions, especially in wastewaters, this may increase significantly. Nitrogen species are also important in marine waters; however, less is known quantitatively about the processes of N transformation in these environments.

The analysis of N in various waters is related to exposure assessment, health consequences and the control of nitrate and ammonium contamination. The main reasons for N monitoring are the magnitude of the pollution, the possible adverse health effects and the high cost of controlling N, mainly nitrate, contamination. In wastewater treatment plants, information on N levels is required in order to evaluate the treatability of wastewater by biological processes. Biological processes require nutrients in order to effectively treat waste and in some cases it may be necessary to add N to the wastewater if the concentration is too low.

Given that the analysis methods for ammonium, nitrite, nitrate and organic N are quite different, these are described separately. Concentrations of N compounds are usually expressed as mg N L^{-1}, although some report the concentration as mg species L^{-1}. This can lead to some confusion, so caution is urged when reporting or examining results (see Example 4.9).

4.10.2 Ammonia

In any aqueous solution, ammonium ion is in equilibrium with dissolved ammonia:

$$NH_4^+ \rightleftharpoons NH_3 + H^+$$

All N that exists either as NH_4^+ or NH_3 is considered to be ammonia-N in this determination. At typical pH values of 6–8 and temperatures between 5 and 30 °C the relative part of NH_3 varies from 0.1% to 5.0% of the total sum of ammonium and ammonia. However, in wastewaters having pH values of 10.0–11.0 and higher, the relative fraction of NH_3 can increase to as much as 91–99%.

The EC guideline for NH_4^+-N is 2 mg L^{-1} for surface waters classified as A3 (*i.e.* water would require advanced treatment to make it usable for drinking). Enhanced levels of ammonia are indicative of recent pollution. Sewage is a major source of ammonia. Ammonia in sewage results from the breakdown of urea, $CO(NH_2)_2$, by urease bacteria. Free ammonia is more toxic than ammonium ion. Consequently, pH influences significantly the toxicity of NH_3/NH_4^+ in water. One can calculate the separate concentrations of free NH_3 and NH_4^+ at different pH values and temperatures using the data in Table 4.17.

There are several standard methods for the determination on ammonia in water: colorimetric, titrimetric and an instrumental method using an ammonia-sensitive membrane electrode. As with other compounds of environmental interest, there are two major factors that influence the selection of the method to determine ammonia: concentra-

Table 4.17 *The relative content of NH_3 in water as % of total $NH_3 + NH_4^+$*

T (°C)	pH 6	7	8	8.5	9	9.5	10	10.5	11
25	0.05	0.49	4.7	13.4	32.9	60.7	83.1	93.9	98.0
15	0.02	0.23	2.3	6.7	19.0	42.6	70.1	88.1	96.0
5	0.01	0.11	0.9	3.3	9.7	25.3	51.7	77.0	91.5

tion and presence of interferences. Direct determination of ammonia is generally carried out on drinking waters, unpolluted surface waters and high-quality nitrified wastewater effluents containing low concentrations of ammonia. In other samples, and where greater precision is required and interferences are present, a preliminary distillation step is performed. A distillation and titration technique is preferred for samples containing high ammonia concentrations. Nesslerisation is the classical colorimetric method for ammonia determination, but it is being gradually replaced by other methods less prone to interferences.

4.10.2.1 Methodology. Ammonium ion is first separated from interfering substances by preliminary distillation under alkaline conditions:

$$NH_4^+ + OH^- \rightarrow NH_3(g) + H_2O$$

Ammonia gas is collected, together with the condensed distillate, in a boric acid solution where it is converted back to ammonium:

$$NH_3 + H_3BO_3 \rightarrow NH_4^+ + H_2BO_3^-$$

Ammonium in the distillate can be determined by one of three methods: colorimetry by nesslerisation, colorimetry by the indophenol blue method or by titration with standard acid:

$$H^+ + H_2BO_3^- \rightarrow H_3BO_3$$

The nesslerisation method involves reacting ammonia with Nessler's reagent, a strongly alkaline solution of potassium mercury(II) iodide (K_2HgI_4) to give a yellowish-brown colloidal suspension:

$$2K_2HgI_4 + NH_3 + 3KOH \rightarrow \begin{matrix} & & ^I \\ Hg< & & \\ & >O + 7KI + 2H_2O \\ Hg< & & \\ & & _{NH_2} \end{matrix}$$

The colour intensity is proportional to the amount of NH_3 present in the sample and it can be measured using a spectrophotometer. The nesslerisation method can be used to determine ammonia concentrations as low as 20 μg NH_3-N L^{-1}, and as high as 5 mg NH_3-N L^{-1}. The method is prone to interference by some amines, chloramines, ketones, aldehydes and alcohols, which can also produce colours with Nessler's reagent. The colorimetric indophenol blue method is described in Section 2.5.2.

4.10.2.2 Materials. *(a) For distillation*

- Borate buffer solution. Dissolve 9.5 g of $Na_2B_4O_7 \cdot 10H_2O$ in water and dilute to 1 L. Place 500 mL of this solution in a 1 L volumetric flask. Add 88 mL of 0.1 N NaOH and make up to the mark with water.
- Round-bottom or Kjeldahl flask, 250 mL, with splash head and condenser
- Boric acid solution, prepared by dissolving 20 g of boric acid (H_3BO_3) in water and diluting to 1 L
- pH meter and electrodes

(b) For nesslerisation

- Spectrophotometer
- Nessler's reagent. Add 12 g of NaOH to 70 mL of 5% potassium iodide solution saturated with mercury(II) iodide and dilute to 100 mL. (*N.B.* This reagent is caustic and toxic. Avoid contact with skin. Wipe up any spillage immediately).
- Sodium hydroxide, 6 M. Dissolve 120 g of NaOH in water and dilute to 500 mL.
- Stock ammonium solution, 1 mg NH_3-N mL^{-1}. Dry some ammonium chloride in an oven at 105 °C. Dissolve 3.819 g of the dried ammonium chloride in water and dilute to 1 L.
- Working ammonium solution, 10 μg NH_3-N mL^{-1}. Pipette 1 mL of the stock solution into a 100 mL volumetric flask and make up to the mark with water.
- Ammonia-free water. High quality water from laboratory purification systems (Milli-Q, NANOpure, *etc.*) is adequate. Use freshly purified water and not water that has been standing around in the laboratory.

(c) For titration

- Mixed indicator solution. Dissolve 50 mg methyl red and 100 mg bromocresol green in 100 mL ethanol.
- Hydrochloric acid, 0.01 M, or sulfuric acid, 0.005 M (1 mL = 140 μg N)

4.10.2.3 Experimental Procedure. *(a) Storage of samples.* Ammonia should be determined in fresh samples whenever possible. If this is not possible, samples should be acidified to a pH of 1.5–2 by adding 0.8 mL of concentrated H_2SO_4 per litre of sample, and stored at 4 °C in tightly stoppered glass containers. Higher volumes of concentrated H_2SO_4 may be required to achieve this pH with some wastewater samples. Samples which have been acidified for storage purposes should be neutralised with NaOH or KOH immediately before analysis. It is important to

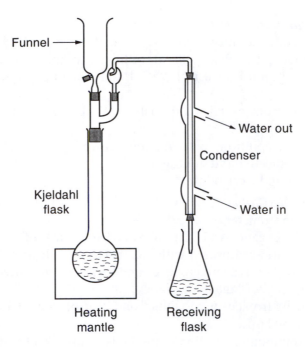

Figure 4.9 *Ammonia distillation apparatus*

avoid contamination of samples by NH_3 from the environment and laboratory. Preferably, "clean room" facilities should be used (see Appendix II). Natural waters, purified drinking waters and highly purified wastewaters of low colour may be analysed directly, without distillation, by direct nesslerisation as below in (c) or by the indophenol blue method as described in Section 2.5.2.

(b) Distillation. All glassware should be acid-washed, rinsed with pure water and stoppered. A portion of pure water should be distilled through the distillation apparatus to remove any ammonia. The apparatus for distilling ammonia is illustrated in Figure 4.9. Pipette 100 mL of the sample into a round-bottom or Kjeldahl flask. Add 5 mL of the borate buffer solution and adjust the pH to 9.5 with NaOH if necessary. Connect the flask to the condenser and turn on the cooling water. Place 10 mL of boric acid solution in a 250 mL conical flask and ensure that the tip of the condenser is immersed in the receiving solution. Distill slowly. Do not boil the flask to dryness as this may cause distillation of interfering substances. Collect about 50–60 mL of the distillate in the receiving flask. You may analyse the distillate by nesslerisation as in (c), titration as in (d) or by the indophenol blue colorimetric method as outlined in Section 2.5.2.

Notes

1. In some Kjeldahl distillation assemblies a dropping funnel with a valve is connected to the distillation flask. Sodium hydroxide is then poured through the funnel and the valve closed immediately so as to prevent any volatilisation of NH_3 after addition of NaOH. The amount of NaOH added is not critical as long as it is sufficient to bring the pH to values >9.5. The amount of NaOH required can be calculated from the pH of the sample, and then an amount of NaOH in excess of that calculated is added.
2. Semi-micro steam distillation units are used for NH_3 distillation in some methods for the determination of organic N. These use a steam generator to boil the sample solution. Semi-micro units are used for distilling smaller volumes of solution.
3. You may analyse the distillate with an ammonium ion selective electrode if one is available in your laboratory. Prepare a series of NH_4^+ calibration standards over the required range and prepare a calibration graph of potential (mV) against concentration as described for F^- in Section 4.13. Read off the concentration of NH_4^+ in the sample from the graph. Follow the instructions that come with the electrode and the pH meter/potentiometer, and use an appropriate reference electrode.

(c) Analysis by nesslerisation. Neutralise the distillate with NaOH and transfer it to a 100 mL volumetric flask. Make up to the mark with water and mix. Pipette 50 mL of this solution into a 50 mL volumetric flask and add 1 mL of Nessler's reagent. Mix well and allow to stand for 10 min for the colour to develop. Prepare a series of calibration standards by pipetting 0 (*i.e.* blank), 1, 2, 4, 6, 8 and 10 mL of the 10 μg N mL^{-1} ammonia working solution into 50 mL volumetric flasks. To each flask add 5 mL of boric acid previously neutralised with NaOH. Make up to the mark with pure water and then add 1 mL of Nessler's reagent to each flask. Mix well and allow to stand for 10 min for the colour to develop. Measure the absorbance of the samples and calibration standards on a spectrophotometer at 410 nm using a 1 cm cell with water as the reference. Subtract the absorbance of the blank from the absorbance readings of the samples and standards. Prepare a calibration graph of nett absorbance against μg N. Read off the amount of ammonia (expressed as N) in the sample from the calibration graph. Express the result as mg N L^{-1} in the sample from the following equation:

$$\text{mg NH}_3\text{-N L}^{-1} = \mu\text{g NH}_3\text{-N}/V$$

where V is the volume of solution anlaysed (50 mL).

If the absorbance of the sample is outside the calibration range, dilute an aliquot of the distillate to 50 mL with pure water and repeat the nesslerisation.

Notes

1. You may occasionally check the procedure by running a blank and standards through the distillation prior to nesslerisation. Distill under the same conditions as the samples (*i.e.* prepare 100 mL of each standard, add 5 mL of borate buffer and distill approximately 80 mL into 10 mL of boric acid). The same calibration curve should fit standards with or without distillation.
2. You may use direct nesslerisation for natural waters, purified drinking waters and highly purified wastewaters of low colour. In this case, add 1 mL of Nessler's reagent to 50 mL of the sample, mix and measure the absorbance as above after the appropriate time period. Prepare the calibration standards as above but without the addition of 5 mL of boric acid. Errors may arise if the sample contains high concentrations of calcium, magnesium, iron and sulfide, as these form turbidity when treated with Nessler's reagent.
3. For lower ammonia concentrations you may use a 5 cm cell rather than a 1 cm cell, as this increases the sensitivity.
4. If the sample contains residual chlorine you should add a dechlorinating agent at the time of collection. You may use freshly prepared sodium sulfite reagent (dissolve 0.9 g of Na_2SO_3 in water and dilute to 1 L). This reagent is unstable and should be prepared on a daily basis. As an alternative you may use a sodium thiosulfate reagent (dissolve 3.5 g $Na_2S_2O_3 \cdot 5H_2O$ in water and dilute to 1 L). This reagent is stable for up to one week. Add 2 mL of either reagent to remove 1 mg L^{-1} of residual chlorine in a 1 L sample.
5. If high levels of Ca and Mg are present, add 1 drop of EDTA solution (dissolve 50 g EDTA in water with 10 g NaOH and dilute to 100 mL) to undistilled samples and use 2 mL Nessler's reagent. Treat standards in the same way.

(d) Analysis by titration. Before starting the distillation, add 5 drops of the mixed indicator solution to 10 mL of the boric acid solution in the receiving flask. Carry out the distillation as described above in (b). Remove the receiving flask and titrate with 0.01 M HCl (or 0.005 M H_2SO_4) until the colour at the end point changes from green to pink. Calculate the concentration of NH_3-N from:

$$mg\ NH_3\text{-N}\ L^{-1} = 140 \times V_t/V_s$$

where V_t is the volume of titrant (0.01 M HCl or 0.005 M H_2SO_4) used up (mL) and V_s is the volume of sample before the distillation (mL).

Notes

1. If the concentration of NH_3-N in the sample is too high, you may titrate an aliquot of the distillate diluted to 100 mL.
2. If the concentration in the sample is too low, you may distill a larger volume (200 mL, 500 mL, *etc.*) but you may have to use larger round-bottom and receiving flasks.
3. You may standardise the HCl (or H_2SO_4) solutions by titrating against an accurately prepared Na_2CO_3 solution. This could be prepared in boric acid so as to reproduce the conditions of the analysis.

4.10.3 Nitrate

The presence of nitrate ions in unpolluted surface waters is due mainly to processes in the water body itself, such as nitrification. This is the oxidation of ammonium ions to nitrate by bacteria species under aerobic conditions. *Nitrosomonas* bacteria oxidise NH_4^+ to nitrite:

$$NH_4^+ + OH^- + 1.5O_2 \rightarrow H^+ + NO_2^- + 2H_2O$$

while *nitrobacter* bacteria oxidise nitrite to nitrate:

$$NO_2^- + 0.5O_2 \rightarrow NO_3^-$$

Atmospheric deposition is another important source of nitrate ions in surface waters. The background content of NO_3^--N in atmospheric deposition is about 0.9–1.0 mg L^{-1} but in many parts of the world the values increase up to 5–10 mg L^{-1} owing to various emission processes. Industrial, municipal and agricultural effluents containing levels as high as 50–100 mg NO_3^--N L^{-1} can introduce large amounts of nitrate into surface and ground waters, and this can end up in water supplies. Agriculture is a major source of nitrate pollution due to N fertilisers and runoff from animal feedlots. These sources are very difficult to control because of their diffuse character. Even if agricultural source controls are implemented, the response times in ground waters may be too long to make control effective. Denitrification and phytoplankton uptake are the main processes that reduce the nitrate content in surface waters and wastewaters. Nitrate ion in waters is a major environmental concern due to:

- High concentrations
- Eutrophication enhancement (together with phosphorus)
- Human health effects

The nitrate content of drinking water is rising at an alarming rate in both developed and developing countries owing largely to a lack of proper sewage treatment, and excessive fertiliser application. Nowadays, nitrate risk analysis (including exposure, consequences and control) is warranted in high-exposure areas of Europe, USA, Japan, South-East Asia and Latin America. Such analysis should consider total nitrate exposure, including drinking water. For risk assessment purposes, sufficiently accurate models are available to predict nitrate loads, and its fate in water consequent to changes in land use, wastewater treatment, *etc.* However, in many developing countries the monitoring of N in natural waters is inadequate, and often non-existent.

The current WHO drinking water guideline is 50 mg NO_3^- L^{-1} (or 11 mg N L^{-1}); however, the EU guideline is half this value. No adverse health effects have been observed with water concentrations <20–30 mg NO_3^--N L^{-1}, except for methaemoglobinemia in infants. Nitrate in drinking water is a major health concern because of its toxicity, especially to young children. Nitrate concentrations greater than 10 mg NO_3^--N L^{-1} in drinking water have been known to cause methaemoglobinemia in infants, a disease characterised by cyanosis, a bluish colouration of the skin, the so-called "blue-baby" syndrome. Infants up to three months old are especially prone to this disease. The actual toxin is not the nitrate ion itself, but rather the nitrite ion NO_2^-, which is formed from nitrate by the reducing action of intestinal bacteria, notably *Escherichia coli*. In adults, NO_3^- is absorbed higher up in the digestive tract before reduction can take place. In infants, whose stomachs are less acidic, *E. coli* can colonise higher up the digestive tract and reduce the nitrate before it is absorbed. Nitrite ion can combine with haemoglobin to form a *methaemoglobin* complex. The association constant for methaemoglobin formation is larger than that for oxyhaemoglobin formation, and so the nitrite ion ties up the haemoglobin, depriving the tissues of oxygen. At low pH values characteristic of the stomach, nitrite ion is converted to nitrous acid, which can react with secondary amines in the digestive tract to produce *N*-nitrosamines. *N*-nitrosamines are known carcinogens and mutagens. However, many studies have not confirmed an association between nitrates and stomach cancer in humans. Nitrate removal from drinking water can be achieved by microbial denitrification, selective ion exchange and electrodialysis, but all are expensive.

A large number of direct and indirect methods are available for nitrate analysis in water, but they all have their limitations. One of the simplest is the chromotropic acid method given below. Indirect methods include the preliminary reduction of nitrate to ammonia (or nitrite) (see Section 5.9.4). Other methods that can be used include ion chromatography (see Section 2.4.6) and potentiometry using an ion selective electrode.

4.10.3.1 Methodology. Two moles of NO_3^- react with one mole of chromotropic acid to form a yellow reaction product, the absorbance of which is measured at 410 nm. The method can be used to determine nitrate concentrations in the range 0.1–5 mg NO_3^--N L^{-1}. It is necessary to eliminate interference by nitrite, residual chlorine and certain oxidants which yield yellow colours when they react with chromotropic acid. Interference from residual chlorine and oxidizing agents can be eliminated by addition of sulfite. Urea eliminates nitrite interference by converting it to N_2 gas. Addition of antimony can mask up to 2000 mg $Cl^- L^{-1}$.

4.10.3.2 Materials
- Spectrophotometer
- Cooling bath
- Stock nitrate solution, 100 μg NO_3^--N mL^{-1}. Prepare by diluting a commercially available 1000 mg L^{-1} solution. Otherwise prepare as follows. Dry sodium nitrate ($NaNO_3$) in an oven at 105 °C for 24 h. Dissolve 0.607 g of the dried salt in water and dilute to 100 mL.
- Working nitrate solution, 10 μg NO_3^--N mL^{-1}. Pipette 50 mL of the stock solution into a 500 mL volumetric flask and make up to the mark with water.
- Sulfite–urea reagent. Dissolve 5 g urea and 4 g anhydrous Na_2SO_3 in water and dilute to 100 mL.
- Antimony reagent. Heat 0.5 g of antimony metal in 80 mL of concentrated H_2SO_4 until all the metal has dissolved. Cool the solution and cautiously add to 20 mL iced water. If crystals form after standing overnight, redissolve by heating.
- Purified chromotropic acid solution (0.1%). Boil 125 mL of water in a beaker and gradually add 15 g of 4,5-dihydroxy-2,7-naphthalene-disulfonic acid disodium salt, while stirring constantly. Add 5 g of decolourising activated charcoal and boil the mixture for 10 min. Add water to make up for loss due to evaporation. Filter the hot solution through cotton wool. Add 5 g of activated charcoal to the filtrate and boil for 10 min. Remove the charcoal completely from

the solution by filtering, first through cotton wool and then through filter paper. Cool and add slowly 10 mL of concentrated H_2SO_4. Boil the solution down to 100 mL in a beaker and stand overnight. Transfer crystals of chromotropic acid to a Buchner funnel and wash thoroughly with 95% ethyl alcohol until crystals are white. Dry the crystals in an oven at 80 °C. Prepare a 0.1% solution by dissolving 100 mg of the purified chromotropic acid in 100 mL of concentrated H_2SO_4 and store in a brown bottle. This solution is stable for two weeks. If the sulfuric acid is free from nitrate impurities the solution should be colourless.

- Sulfuric acid, concentrated. High purity.

4.10.3.3 Experimental Procedure. (a) Storage of samples. Results are most reliable when nitrate ion is determined in fresh samples. For short-term preservation of up to 1 day, samples can be stored in a refrigerator at 4 °C. If it is not possible to carry out the analysis promptly, samples can be preserved by adding 0.5–1.0 mL of concentrated H_2SO_4 per litre of sample and stored at 4 °C.

(b) Analysis. Prepare nitrate standards in the range 0.1–5 mg NO_3^--N L^{-1} by pipetting 1, 5, 10, 20, 40 and 50 mL of the working nitrate solution into a series of 100 mL volumetric flasks and making up to the mark with water. Filter the samples if significant amounts of suspended matter are present. Pipette 2 mL aliquots of samples, standards and a water blank into 10 mL volumetric flasks and add 1 drop of the sulfite–urea reagent to each flask. Place flasks in a tray of cool water with a temperature between 10 and 20 °C and add 2 mL of the antimony reagent. Swirl the flasks when adding each reagent. After the flasks have stood in the bath for about 4 min, add 1 mL of chromotropic acid reagent. Swirl the flasks again and allow to stand in the cooling bath for another 3 min. Make up to the mark with concentrated H_2SO_4. Stopper the flasks and mix the contents by inverting them four times. Allow the flasks to stand at room temperature for 45 min and again adjust the volume to 10 mL with concentrated H_2SO_4. Finally, mix very gently to avoid introducing gas bubbles. Allow the flasks to stand for at least 15 min before measuring the absorbance at 410 nm using a 1 cm cell with water in the reference cell. Subtract the absorbance reading of the water blank from the absorbances of samples and standards. Prepare a calibration graph of nett absorbance against mg NO_3^--N L^{-1} based on the standard measurements and read off directly the concentration of NO_3^- (expressed as mg N L^{-1}) in the samples.

Notes

1. Nitrate may be determined by ion chromatography according to the procedure described in Section 2.4.6.
2. You may determine nitrate with an ion selective electrode if one is available in your laboratory. Prepare a series of nitrate calibration standards over the required range and prepare a calibration graph of potential (mV) against concentration as described for F^- in Section 4.13. Then simply read off the concentration of nitrate in the sample from the graph. Follow the instructions that come with the electrode and the pH meter/potentiometer, and use an appropriate reference electrode.

4.10.4 Nitrite

The main sources of nitrite ion (NO_2^-) in unpolluted surface waters are the processes of organic matter mineralisation and nitrification by *Nitrosomonas* bacteria. Another process of nitrite formation in unpolluted water bodies is denitrification:

$$C_6H_{12}O_6 + 12NO_3^- \rightarrow 12NO_2^- + 6CO_2 + 6H_2O$$

The content of nitrite in unpolluted surface and ground waters is very low, generally $<0.01–0.02$ mg NO_2^- L^{-1}. Similar concentrations are typically found in unpolluted seawater. Higher nitrite concentrations may be encountered near effluent outlets from various plants applying nitrite salts (food, metallurgical industries, *etc.*). Nitrite is significantly more toxic than ammonia or nitrate. Nitrite is of environmental concern due to its toxicity to water plants and biota as well as because of human health effects mentioned earlier (Section 4.10.3). The EU limit for drinking water is 0.1 mg NO_2^--N L^{-1}.

4.10.4.1 Methodology. The method is based on the diazotisation of sulfanilic acid by nitrites in acid solution. The resulting diazo compound couples with 1-naphthylamine-7-sulfonic acid (Cleve's acid) to form a purplish-pink dye. The absorbance is measured using a spectrophotometer at 525 nm.

4.10.4.2 Materials
- Spectrophotometer
- Sulfanilic acid solution. Dissolve 0.5 g of sulfanilic acid in a mixture of 120 mL of water and 30 mL of acetic acid. Filter solution and store in the dark.

- 1-Naphthylamine-7-sulfonic acid (Cleve's acid) solution. Dissolve 0.2 g of the acid in a mixture of 120 mL of water and 12 mL of acetic acid. Heat, if necessary, to dissolve. Filter solution and store in the dark.
- Acetic acid, 20%
- Stock nitrite solution, 100 μg NO_2^--N mL^{-1}. Dry $NaNO_2$ in an oven at 110 °C. Dissolve 0.4926 of the dried $NaNO_2$ in water and make up to 1 L.
- Working nitrite solution, 1 μg NO_2^--N mL^{-1}. Pipette 1 mL of the stock solution into a 100 mL volumetric flask and make up to the mark with water. This should be prepared fresh when required.
- Concentrated hydrochloric acid
- pH meter with electrodes

4.10.4.3 Experimental Procedure. (a) Sample storage. Analysis should be performed as soon as possible after sampling to prevent bacterial conversion of NO_2^- to NO_3^- or NH_3. Acid preservation should not be used for samples to be analysed for NO_2^-. Samples can be stored for 1–2 days by freezing at -20 °C or by adding 40 mg $HgCl_2$ L^{-1} sample and storing in a refrigerator at 4 °C.

(b) Analysis. If the sample contains suspended solids, filter through a 0.45 μm membrane filter. Measure the pH of the sample. If it is alkaline, add concentrated HCl dropwise until it is slightly acidic. Pipette 40 mL of the sample into a 50 mL volumetric flask. Add 2 mL of the sulfanilic acid solution, mix and allow to stand for 20 min. Add 5 mL of Cleve's acid solution, make up to the mark with distilled water and allow to stand for a further 20 min. Measure the absorbance of the solution at 525 nm using a 1 cm cell and water in the reference cell. Prepare calibration standards by pipettting 0 (blank), 1, 2, 3, 4, 6, 8 and 10 mL of the working nitrite solution into a series of 50 mL volumetric flasks. Dilute each to approximately 40 mL and proceed as outlined above, adding the respective reagents and allowing to stand. Subtract the absorbance of the blank from the absorbance values measured for samples and standards. Draw a graph of nett absorbance against μg NO_2^--N. Read off the number of μg NO_2^--N present in the sample solution from the calibration graph, and calculate the concentration in the sample from:

$$\text{mg } NO_2^-\text{-N L}^{-1} = \mu\text{g } NO_2^-\text{-N}/V$$

where V is the sample volume (ml).

The sample volume analysed in this procedure is 40 mL, but it can be varied. If the sample contains very high nitrite concentrations, above the range of the standard curve, use a smaller volume of sample and dilute to about 40 mL before carrying out the analysis.

Notes

1. Coloured or turbid samples may lead to errors. Correct for this as follows. Pipette 40 mL of the sample into a 50 mL volumetric flask and add 4 mL of 20% acetic acid and make up to the mark with water. Measure the absorbance at 525 nm and subtract the result from the sample absorbance obtained following the normal procedure described above, before reading the mass of nitrite-N from the calibration graph. If you have used less than 40 mL of the sample in the original analysis, then use the same volume in the correction procedure.
2. Interference may occur if amines or strong oxidising agents are present in the sample.
3. Cells with longer path lengths (5 or 10 cm) may be used to determine lower nitrite concentrations.

4.10.5 Organic Nitrogen

Organic N consists of N present in organic compounds such as urea, amines, amino acids, amides, nitro compounds, *etc.* The main inner source of total organic N in water bodies is the decomposition of aquatic biota, generally phytoplankton. Another inner source is the excretion of N by water organisms; for example, algae can excrete up to 50% of assimilated N. Outside sources of organic N in surface waters and groundwaters are wastewaters (industrial, domestic and agricultural) and atmospheric precipitation. Nitrogen in domestic wastewater is mainly in the form of proteins and their degradation products (amino acids and polypeptides). Nitrogen in organic compounds is in the $(-III)$ oxidation state. The content of organic N in unpolluted surface waters can vary from 0.3 to 2 mg N L^{-1}. In eutrophic waters, organic N can reach 7–10 mg N L^{-1} and much higher values (>20 mg N L^{-1}) are found in polluted waters and wastewaters.

4.10.5.1 Methodology. The standard procedure for determining organic N is the Kjeldahl method. The sample is first boiled to expel ammonia and then digested. This involves oxidation of organic N to an ammonium salt by sulfuric acid at high temperature:

$$\text{Organic N} + \text{H}_2\text{SO}_4 \xrightarrow{\text{Cu, Hg, Se}} \text{NH}_4\text{HSO}_4 + \text{CO}_2(g) + \text{SO}_2(g) + \text{H}_2\text{O}$$

Mercury, Cu, or Se catalyst is employed to speed up the reaction. Excess sulfuric acid is neutralised and the pH raised so that NH_3 may be easily distilled:

$$NH_4HSO_4 + 2NaOH \rightarrow NH_3 + Na_2SO_4 + 2H_2O$$

The analysis is completed by any method suitable for ammonia determination, such as the nesslerisation procedure described earlier (Section 4.10.2.3). The procedure below uses a Cu catalyst.

4.10.5.2 Materials
- Kjeldahl flasks, 350 mL
- Kjeldahl heating unit
- Distillation apparatus. Same as described in Section 4.10.2.3
- Spectrophotometer. Same as described in Section 4.10.2.3
- All of the reagents required for the determination of ammonia as listed in Section 4.10.2.3
- Concentrated sulfuric acid
- Potassium sulfate
- Copper sulfate
- Sodium hydroxide–sodium thiosulfate reagent. Dissolve 500 g NaOH and 2 g $Na_2S_2O_3 \cdot 5H_2O$ in water and dilute to 1 L.
- Sodium hydroxide, 6 M. Dissolve 240 g of NaOH in water and dilute to 1 L.
- pH meter and electrodes

4.10.5.3 Experimental Procedure. (a) Storage of samples.
Analyse samples immediately after collection to obtain reliable results. If this is not possible, preserve samples by acidifying to pH 1.5–2.0 with concentrated H_2SO_4 and store at 4 °C.

(b) Analysis. The apparatus for digesting the samples is illustrated in Figure 4.10. Digest 100 mL of sample in a Kjeldahl flask. If analysing low sample volumes, dilute to 100 mL. Adjust the sample pH to 9.5 with NaOH and place in a Kjeldahl flask. Add some glass beads or boiling chips and boil off half of the solution. This removes any inorganic ammonia which may interfere with the analysis. If ammonia has been determined by the distillation method, the residue in the distilling flask may be used for organic nitrogen determination. Cool the solution and add carefully 10 mL of concentrated H_2SO_4, 5 g K_2SO_4 and 0.1g $CuSO_4$. Heat in a fume cupboard to remove acid fumes. Boil until the volume is reduced to about 20–50 mL and copious white fumes form in the flask. The white fumes indicate that digestion is taking place as the boiling

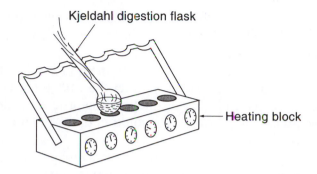

Figure 4.10 *Kjeldahl digestion heating unit*

point of sulfuric acid has been reached. Continue digesting. The mixture should turn black due to the dehydrating action of sulfuric acid on the organic matter. When all of the organic matter has been destroyed the solution should appear clear or straw-yellow in colour. Continue digesting for a further 20 min after this to ensure complete destruction of all organic matter in the sample. After digestion, allow the solution to cool. Add 50 mL of water and mix. Carefully add 20 mL of sodium hydroxide–thiosulfate reagent. Connect to the distillation apparatus described above (Section 4.10.2.3). Distill exactly as described in the section on ammonia analysis, collecting about 50–60 mL of the distillate in a 250 mL conical flask containing 10 mL of boric acid solution. Transfer the distillate to a 100 mL volumetric flask and make up to the mark with water. Pipette 50 mL of this solution into a 50 mL volumetric flask and add 1 mL of Nessler's reagent. Mix well and allow to stand for 10 min for the colour to develop. Measure the absorbance in the same way as for ammonia (Section 4.10.2.3). Prepare a calibration graph using the same standards as for ammonia determination and calculate the concentration in the same way. Digest 100 mL of water as outlined above, adding the same reagents as in the analysis of samples, distill, nesslerise and determine the absorbance. Use this blank to correct the results if necessary. As an alternative you may analyse the distillate using the titrimetric method described in Section 4.10.2.3, the indophenol blue colorimetric method outlined in Section 2.5.2 or an ammonium ion selective electrode if one is available.

Notes

1. Mercury is the most effective catalyst for the digestion; however, its use is restricted in many laboratories owing to its toxicity and problems of waste disposal. An alternative catalyst is Se but this,

too, is very toxic. The above procedure uses Cu instead of Hg. You may omit the catalyst altogether as many samples can be effectively digested without the use of a catalyst.

2. You may increase the sensitivity of the method by digesting larger volumes of sample (up to 500 mL) rather than 100 mL recommended in the above procedure. After distillation you proceed as described in the above procedure, *i.e.* make the solution up to 100 mL, effectively concentrating the sample.

3. Nitrate at concentrations > 10 mg L^{-1} can cause positive or negative interference, depending on the sample composition.

4.10.6 Questions and Problems

1. What are the main N compounds in natural waters and wastewaters and what are their typical concentrations?

2. Describe N transformations in water and the role of chemical, biological, biochemical and biogeochemical processes.

3. Discuss the environmental concerns over N pollution, both in terms of effects on ecosystems and human health.

4. Describe the chemical reactions used in the colorimetric methods of ammonium, nitrite and nitrate analysis.

5. Describe the possible interferences in the determination of various N forms and suggest how these could be eliminated.

6. What measures can you suggest for decreasing the nitrate content of polluted water?

7. The equilibrium constant for the reaction: $NH_4^+ \rightleftharpoons NH_3 + H^+$ is 5.6×10^{-10} mol L^{-1} at 25 °C. What is the ratio of NH_3 to NH_4^+ in water samples having the following pH values: (a) 4.5, (b) 6.3, (c) 8.2 and (d) 9.5.

4.10.7 Suggestions for Projects

1. Carry out a survey of nitrate content in drinking water in your locality. Collect tapwater samples from various houses in your area and compare with national drinking water standards. Calculate the mean and standard deviation of nitrate in drinking water for your locality on the basis of all your measurements.

2. Determine the content of various nitrogen forms (NH_3, NO_3^-, NO_2^-, organic-N) in different surface water samples (rivers, lakes, ponds) in your locality and compare with national standards for surface waters. Perform a simple statistical analysis of your results, *i.e.* calculate the mean and standard deviation for each N species, and calculate the correlation coefficients for various pairs of species (NH_3 *versus* NO_3^-,

NO_3^- *versus* NO_2^-, *etc.*). Refer to statistical tables (Appendix V) to see whether any of the relationships are statistically significant.

3. Determine the content of various N forms (NH_3, NO_3^-, NO_2^-, organic-N) in wastewater at different sewage treatment plants. You could analyse both the untreated wastewater and the treated effluent. Explain your results in terms of the control technologies employed at the plants. Compare the concentrations in the effluent with existing standards.

4. Determine the concentration of different forms of N (NH_3, NO_3^-, NO_2^-, organic-N) at a point where sewage or wastewater effluent is discharged into a river, and at several points downstream from the discharge (*e.g.* 2, 5, 10 km). Also determine dissolved oxygen (DO) and any other relevant parameters (*e.g.* BOD, pH) at the same points. Prepare a table of results and plot them on a graph with distance from source as the *x*-axis and concentration of respective species as the *y*-axis. Explain any changes in N forms, and in the other parameters.

4.10.8 Further Reading

APHA, "Standard Methods for the Examination of Water and Wastewater", 16th edn., American Public Health Association, Washington, 1985, pp. 373–412.

D. F. Boltz (ed.), "Colorimetric Determination of Nonmetals", Interscience, New York, 1958.

S. Saull, Nitrates in soil and water, Inside Science, *New Sci.*, 15 September 1990.

C. N. Sawyer, P. L. McCarty and G. F. Parkin, "Chemistry for Environmental Engineering", 4th edn., McGraw-Hill, New York, 1994, pp. 552–566.

E. A. Steward (ed.), "Chemical Analysis of Ecological Materials", 2nd edn., Blackwell, Oxford, 1989, 119–135.

P. W. West and T. P. Ramachandran, Spectrophotometric determination of nitrate using chromotropic acid, *Anal. Chim. Acta*, 1966, **35**, 317–324.

4.11 PHOSPHORUS

4.11.1 Introduction

Phosphorus, like N, is a nutrient required by living organisms, however, it can also be considered a pollutant if present at high concentrations under specific environmental conditions. The addition of P, as phosphate ion, to natural waters is one of the most serious environmental problems because of its contribution to the eutrophication process. Nitrate pollution also contributes to eutrophication (see Section 4.10.3), but phosphate appears to be the main culprit in freshwaters.

In natural waters, phosphorus is often the *limiting nutrient*, *i.e.* algal growth is limited by the supply of phosphorus but not, in general, by the supply of carbon and nitrogen. The uptake of these nutrients by biomass occurs in the approximate ratio C:N:P of 100:16:1. Typically, phosphorus concentrations in natural waters are much lower than those of nitrogen and carbon, and phosphorus can limit the primary productivity of a water body. Throughout the world there are numerous examples of increasing eutrophication, or *algal blooming*, of lakes and water reservoirs, *e.g.* Great Lakes in USA, water reservoirs along the Volga river in Russia, many artificial water reservoirs in Asia, *etc*. In general, the eutrophication process is greatly accelerated in subtropical and tropical climates, leading to enhanced algal blooming and consequently to increasing costs of water cleanup.

In natural waters and wastewaters, phosphorus is present almost exclusively in the form of phosphates, which can be present in solution, in particles or detritus, or in aquatic organisms. Phosphorous compounds in the environment are classified into three categories:

- *Orthophosphates*. Orthophosphates are water soluble salts of phosphoric acid, H_3PO_4, which dissociates in solution into several ionic species. The relationship between inorganic orthophosphates depends on the pH of the water (Figure 4.11) (see also Example 4.10). At pH values of environmental interest, phosphates are mainly present as $H_2PO_4^-$ and HPO_4^{2-}. Orthophosphates are applied as fertilisers on soils and carried by runoff into surface

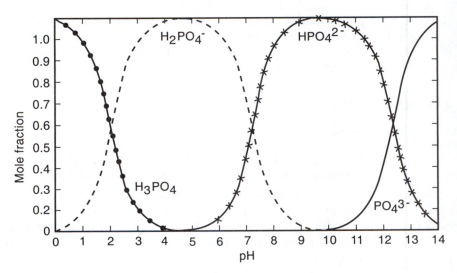

Figure 4.11 *Mole fraction of P(V) species in equilibrium as a function of pH*

Example 4.10

Calculate the percentage of each species of phosphoric acid present at pH 7.

Phosphoric acid is a polyprotic acid which dissociates in several steps, each of which is defined by an equilibrium constant:

$$H_3PO_4 \rightleftharpoons H_2PO_4^- + H^+ \qquad K_1 = [H_2PO_4^-][H^+]/[H_3PO_4]$$

$$H_2PO_4^- \rightleftharpoons HPO_4^{2-} + H^+ \qquad K_2 = [HPO_4^{2-}][H^+]/[H_2PO_4^-]$$

$$HPO_4^{2-} \rightleftharpoons PO_4^{3-} + H^+ \qquad K_3 = [PO_4^{3-}][H^+]/[HPO_4^{2-}]$$

When the concentrations are specified in mol L^{-1} the equilibrium constants at 25 °C have the following values: $K_1 = 7.52 \times 10^{-3}$, $K_2 = 6.23 \times 10^{-8}$ and $K_3 = 2.2 \times 10^{-13}$.

The total phosphate concentration is the sum of the above species:

$$P_T = [H_3PO_4] + [H_2PO_4^-] + [HPO_4^{2-}] + [PO_4^{3-}]$$

From the above equilibrium equations the concentration of each ionic species can be defined as follows:

$$[H_2PO_4^-] = K_1[H_3PO_4]/[H^+]$$

$$[HPO_4^{2-}] = K_2[H_2PO_4^-]/[H^+] = K_1K_2[H_3PO_4]/[H^+]^2$$

$$[PO_4^{3-}] = K_3[HPO_4^{2-}]/[H^+] = K_1K_2K_3[H_3PO_4]/[H^+]^3$$

The total phosphate concentration can then be expressed as:

$$P_T = [H_3PO_4] \left(1 + K_1/[H^+] + K_1K_2/[H^+]^2 + K_1K_2K_3/[H^+]^3 \right)$$

Therefore:

$$[H_3PO_4] = P_T/(1 + K_1/[H^+] + K_1K_2/[H^+]^2 + K_1K_2K_3/[H^+]^3)$$

Substituting this in the expression for $[H_2PO_4^-]$ gives:

$$[H_2PO_4^-] = P_TK_1/(1 + K_1/[H^+] + K_1K_2/[H^+]^2 + K_1K_2K_3/[H^+]^3) [H^+]$$

and continuing to substitute for each species into the subsequent equation gives:

$$[HPO_4^{2-}] = P_TK_1K_2/(1 + K_1/[H^+] + K_1K_2/[H^+]^2 + K_1K_2K_3/[H^+]^3)[H^+]^2$$

$$[PO_4^{3-}] = P_TK_1K_2K_3/(1 + K_1/[H^+] + K_1K_2/[H^+]^2 + K_1K_2K_3/[H^+]^3)[H^+]^3$$

Substituting the values of K_1, K_2 and K_3, and 10^{-7} mol L^{-1} for the concentration of H$^+$ (pH 7) into these equations, and taking $P_T = 100\%$ we can calculate:

$$[H_3PO_4] = 8.2 \times 10^{-4} \%$$

$$[H_2PO_4^-] = 61.6 \%$$

$$[HPO_4^{2-}] = 38.4 \%$$

$$[PO_4^{3-}] = 8.4 \times 10^{-5} \%$$

This may be repeated at different pH values to obtain a distribution diagram such as that shown in Figure 4.11.

waters. Superphosphate is a common plant fertiliser prepared as a mixture of calcium dihydrogen phosphate, Ca(H$_2$PO$_4$)$_2$, and gypsum, CaSO$_4$·2H$_2$O.

- *Condensed phosphates* (pyro-, meta- and other polyphosphates). Condensed phosphates are salts of pyrophosphoric acid, metaphosphoric acid and polyphosphoric acids. These are major constituents of many commercial cleaning products and laundry detergents. They are also used extensively in boilers to control scaling. Another use is in water supplies, where small quantities of phosphates are added during treatment to inhibit corrosion. In aqueous solution, polyphosphates slowly convert to orthophosphates. Sodium tripolyphosphate (Na$_5$P$_3$O$_{10}$) is the major phosphate in laundry detergents.
- *Organic phosphorus.* Organic phosphorus compounds are formed mainly by biological processes. Body wastes and food residues containing organic phosphates end up in sewage. Organic phosphorus may also be formed in biological wastewater treatment processes or in natural water bodies.

Phosphorus is cycled through the environment, as are other nutrients such as carbon and nitrogen, and it may enter water bodies as a result of natural processes such as the weathering of rocks, release from bottom sediments and excretion by animals. The main anthropogenic sources of P in water are the discharge of raw or treated sewage, agricultural drainage and certain industrial wastewaters. The disposal of detergents is a major contributor to water pollution. In the past, detergents caused widespread foaming on water surfaces owing to the slow biodegradation of surfactants. This problem was solved in the mid-1960s by manufacturing more biodegradable detergents. However, even these newer chemical preparations can make surface waters unfit for aquatic life and human recreation (*e.g.* fishing) by triggering eutrophication. The problem with modern detergents is not the surfactant but rather the polyphosphate builders. The main compound used, $Na_5P_3O_{10}$, produces the $P_3O_{10}^{5-}$ ion in water and this slowly hydrolyses to orthophosphates:

$$P_3O_{10}^{5-} + 2H_2O \rightarrow 2HPO_4^{2-} + H_2PO_4^-$$

The sale of phosphate-containing detergents has been banned in many countries as a result of these concerns.

Titrimetry can be used to analyse samples with high concentrations of orthophosphates (>50 mg P L^{-1}). This method is not suitable for general environmental analysis and it is only applicable to boiler waters. Spectrophotometry is the preferred method for determining orthophosphates. Polyphosphates and organic phosphorus are converted to orthophosphates and analysed in the same way as orthophosphates. Procedures have been developed to determine total phosphorus, total inorganic phosphate and orthophosphates. The concentrations of polyphosphates and organic phosphorus can then be calculated by difference from these measurements. Phosphorus determination is used to assess the potential biological productivity of surface waters. Phosphorus is routinely determined in wastewater treatment plants because it is an essential nutrient in biological methods of wastewater treatment. It is also determined to ascertain that the effluent discharged into surface waters complies with national standards.

The concentration of phosphates is generally reported in mg P L^{-1}. Some prefer to report it in mg PO_4 L^{-1}. When comparing results reported differently it is necessary to apply a conversion factor. Unpolluted waters generally contain total phosphate at concentrations <0.1 mg P L^{-1}, while municipal wastewater contains between 4 and 15 mg P L^{-1}.

4.11.2 Methodology

Phosphates react with ammonium molybdate under acidic conditions to form an ammonium phosphomolybdate complex:

$$PO_4^{3-} + 12(NH_4)_2MoO_4 + 24H^+ \rightarrow (NH_4)_3PO_4 \cdot 12MoO_3 + 21NH_4^+ + 12H_2O$$

The molybdenum in the ammonium phosphomolybdate complex is

reduced by ascorbic acid to produce a blue-coloured compound known as *molybdenum blue*. The absorbance of molybdenum blue is measured using a spectrophotometer at 880 nm.

Orthophosphates are determined by analysing samples without preliminary hydrolysis. Some condensed phosphates are also analysed at the experimental conditions. Owing to this slight interference, the term *reactive phosphorus* is sometimes used to define the phosphates that are determined without preliminary hydrolysis. Condensed phosphates are converted into orthophosphates by hydrolysis and the total inorganic phosphates (orthophosphate + condensed phosphate) determined. Some organic phosphorus may also be released during the treatment and interfere in the analysis. Condensed phosphates are determined by difference. The term "acid-hydrolysable phosphorus" is sometimes used instead of "condensed phosphate" to describe this fraction. Organic phosphorus is oxidised to orthophosphate and the total phosphate determined. Organic phosphorus is determined by difference.

Samples are filtered through a 0.45 μm filter prior to analysis if only dissolved phosphorus compounds are to be analysed. Unfiltered samples can also be analysed to give total phosphorus (dissolved + suspended) and the phosphorus in suspended matter calculated by difference.

4.11.3 Orthophosphates

4.11.3.1 Materials
- Spectrophotometer
- Potassium antimony tartrate solution prepared by dissolving 2.7 g $K(SbO)C_4H_4O_6 \cdot 0.5H_2O$ in water and diluting to 1 L
- Ammonium molybdate solution prepared by dissolving 40 g $(NH_4)_6Mo_7O_{24} \cdot 4H_2O$ in water and diluting to 1 L
- Ascorbic acid, 0.01 M, prepared by dissolving 1.76 g ascorbic acid in water and diluting to 100 mL. Store in a refrigerator at 4 °C. This solution can be used for up to one week.
- Sulfuric acid. Dilute 70 mL concentrated H_2SO_4 to 500 mL with water.
- Reaction mixture. Add 100 mL of above sulfuric acid solution, 10 mL potassium antimony tartrate solution, 30 mL ammonium molybdate solution and 60 mL ascorbic acid solution to a 250 mL bottle. Mix after addition of each reagent and allow to cool to room temperature before adding the next reagent. If the final solution is turbid, shake and stand for a few minutes until the solution becomes clear. This reagent is stable for only 4 h so prepare when required.
- Stock phosphate solution, 100 mg P L^{-1}. Dissolve 439.0 mg

anhydrous potassium dihydrogen phosphate (KH_2PO_4) in water and dilute to 1 L; 1 mL = 100 μg.
- Working phosphate solution, 10 mg P L^{-1}. Dilute 10.0 mL stock phosphate solution to 100 mL with water; 1 mL = 10 μg P.

4.11.3.2 Experimental Procedure. *(a) Sampling and Storage.* Analysis of P should be carried out as soon as possible after sample collection. If this is not possible, samples can be stored by deep-freezing at temperatures lower than $-10\,°C$ after adding 40 mg $HgCl_2$ per litre of sample. Neither acid nor $CHCl_3$ should be used as a preservative when phosphorus is to be determined. If only dissolved phosphorus compounds are to be analysed, filter samples immediately through pre-washed 0.45 μm membrane filters. Filters should be pre-washed by soaking in water for 24 h to remove any phosphorus from the filters, as they could contribute significantly to samples with low levels of phosphate. Preferably use glass bottles for sampling. Plastic bottles may be used for storing frozen samples. Glass sampling bottles should be washed with hot dilute HCl and rinsed several times with water. Commercial detergents containing phosphate should never be used for cleaning glassware used in phosphate analysis.

(b) Analysis. Prepare calibration standards by adding volumes of standard phosphate solution (10 μg P mL^{-1}) corresponding to between 5 and 60 μg P (*i.e.* 0.5–6 mL of standard solution) to a series of 50 mL volumetric flasks. Pipette 40 mL of sample into a 50 mL volumetric flask. To each flask, add 8 mL of the reaction mixture and make up to the mark with water. Prepare a water blank by pipetting 8 mL of the reaction mixture into a 50 mL flask and making up to the mark with water. Mix the solutions thoroughly and allow to stand for at least 10 min, but no longer than 30 min. Use the water blank prepared above in the reference cell of the spectrophotometer rather than the customary water. Measure the absorbance of each solution at 880 nm in a 1 cm cell. Plot a calibration graph of absorbance against μg P. This should be a straight line going through the origin. Read off the amount of phosphorus in the sample and calculate the concentration as:

$$\text{mg P L}^{-1} = \mu\text{g P}/V$$

where V is the volume of the sample (mL).

The volume of sample is 40 mL in the above procedure. If the sample contains more P than the highest calibration standard, dilute the sample to fit on the curve and correct for the dilution when calculating the result.

Notes

1. For more dilute samples you could use a cell with a longer path length (*e.g.* 5 cm). In this case prepare a more dilute working phosphate solution (1 μg P mL^{-1}) by diluting 1 mL of the stock phosphate solution to 100 mL. Prepare calibration standards containing between 0.5 and 15 μg P (*i.e.* 0.5 and 15 mL of the 1 μg P L^{-1} working solution).

2. Arsenates can interfere if present at concentrations above 0.1 mg L^{-1} by forming a blue complex with molybdate. This can be removed by reducing arsenate to arsenite by adding a few drops of 5% KI solution to the sample. Silicates can interfere at concentrations above 10 mg L^{-1}. Sulfides can interfere at concentrations exceeding 1 mg L^{-1}. Chromates and nitrates can cause negative interference at concentrations above 1 mg L^{-1}.

3. If the sample is turbid or highly coloured, prepare a sample blank as follows. Mix 100 mL of sulfuric acid solution with 30 mL ammonium molybdate solution. Do not add potassium antimony tartarate or ascorbic acid. Pipette 8 mL of this mixture into a 50 mL flask and make up to the mark with the sample. Determine the absorbance as outlined above and subtract the reading from the absorbance values of each sample, but not from the values determined with calibration standards.

4.11.4 Condensed Phosphates

4.11.4.1 Materials
- Same as in Section 4.11.3
- pH meter with electrodes
- Acid solution. Add slowly 300 mL concentrated H_2SO_4 to about 700 mL water in a 1 L volumetric flask. Allow to cool, add 4.0 mL concentrated nitric acid and dilute to the mark with water.
- Sodium hydroxide solution, 5 M
- Hot plate

4.11.4.2 Experimental Procedure. Hydrolyse the sample as follows to convert condensed phosphate to orthophosphate. Place 100 mL of sample into a conical flask and add 1 mL of the acid solution (sulfuric + nitric). Boil gently on a hot plate for 2 h inside a fume cupboard. If the volume drops to below 30 mL, add a little water. Cool and adjust the pH to between 7 and 8 with 5 M NaOH. Transfer to a 100 mL volumetric flask and make up to the mark with water to bring back to the original volume. Analyse 40 mL of this solution using exactly the same procedure as for orthophosphates (Section 4.11.3). Prepare a calibration graph using hydrolysed standards as follows. Pipet aliquots of the working phosphate solution (*e.g.* 3 mL to give 30 µg P) into a series of conical flasks containing 100 mL of water. Hydrolyse in exactly the same way as the sample by adding 1 mL of the acid solution. Boil down to about 30 mL, cool, neutralise and transfer into 50 mL volumetric flasks. Add 8 mL of the reaction mixture, make up to the mark with water and determine the absorbance after at least 10 min at 880 nm. Carry 100 mL of water through the same procedure and use this as the blank. Plot a calibration graph and determine the concentration of total inorganic phosphate in the sample. Calculate the concentration of condensed phosphate as follows:

condensed P = total inorganic P − orthophosphate P

4.11.5 Organic Phosphorus

4.11.5.1 Materials
- Same as in Section 4.11.3 and Section 4.11.4
- Potassium persulfate, $K_2S_2O_8$, or ammonium persulfate, $(NH_4)_2S_2O_8$

4.11.5.2 Experimental Procedure. Place 100 mL of sample into a conical flask and add 0.5 mL of concentrated H_2SO_4. Add either 0.8 g $(NH_4)_2S_2O_8$ or 1.0 g $K_2S_2O_8$. Boil gently for 1 h on a hot plate inside a fume cupboard. Add water if the volume drops below about 30 mL. Cool and adjust the pH to between 7 and 8 with 5 M NaOH. Transfer to a 100 mL volumetric flask and adjust to the mark with water. Analyse 40 mL of this solution by following exactly the procedure outlined for orthophosphates above (Section 4.11.3). Process standards and a water blank in the same way as samples and prepare a calibration graph. Determine the concentration of total P in the sample. Calculate the concentration of organic P as follows:

organic P = total P − total inorganic P

4.11.6 Questions and Problems

1. What are the main sources of P in natural waters and wastewaters?
2. Why is P considered as the trigger of eutrophication is surface water? Describe the eutropication process and give examples of development of "algal blooms".

3. What is the difference between *reactive phosphorus, acid-hydrolysable phosphorus* and *organic phosphorus*? Describe the methods used to determine these P fractions.

4. Describe the chemical principles on which the colorimetric method for phosphate determination is based.

5. The residence time of water in a lake is 2.7 years. If the input of phosphorus to this lake is halved, how long will it take for the concentration of phosphorus in the lake water to fall by 10%?

6. The concentration of inorganic phosphates in weak, strong and domestic wastewaters was reported as 9, 15 and 31 mg PO_4 L^{-1}, respectively. Express these concentrations in units of mg P L^{-1}.

7. A lake water sample has the following partial analysis: total carbonate $= 86$ mg L^{-1}; nitrate $= 0.12$ mg L^{-1}; ammonia $= 0.04$ mg L^{-1}; phosphate (as PO_4^{3-}) $= 0.08$ mg L^{-1}. Which is the limiting nutrient?

8. If the concentration of the $H_2PO_4^-$ ion is 1.0×10^{-4} mol L^{-1} in a water sample of pH 6.5, what are the concentrations of H_3PO_4, HPO_4^{2-} and PO_4^{3-}?

9. A water sample was analysed and found to have a pH of 8.2 and a total orthophosphate concentration of 9 mg P L^{-1}. Calculate the concentrations of $H_2PO_4^-$ and HPO_4^{2-} in mol L^{-1}.

10. What percentage of each species of H_3PO_4 exists at pH 5?

4.11.7 Suggestions for Projects

1. Carry out a survey of phosphates in different water bodies such as the lakes, ponds, streams and rivers. Compare the results with the content of total C and N and determine the limiting nutrient.

2. Determine the content of phosphates in wastewaters from different industrial, municipal and agricultural outlets. Attempt to relate the P content in the effluents to technologies employed at the wastewater treatment facilities.

3. On the basis of a visual survey you could choose a lake or water reservoir with detectable features of eutrophication. Use the development of algal blooms as an indicator. Measure the P content in the lake and in river/stream inlets and outlets, as well as the water flows. The input of phosphates from various streams and rivers could then be determined. If the residence time of the water is known, calculate by how much the P input has to be reduce in order to decrease the phosphate content in lake water by 20%.

4. Measure the phosphate content in different surface and ground waters in forested, cultivated agricultural and urban areas. Identify the

sources of P for different land use types and investigate the possible statistical relationships, for example between content of P in waters and percentage of forested area, agricultural land, fertiliser application, population density, cattle density, *etc.*

4.11.8 Further Reading

APHA, "Standard Methods for the Examination of Water and Wastewater", 16th edn., American Public Health Association, Washington, 1985, pp. 437–450.

N. J. Bunce, "Environmental Chemistry", 2nd edn., Wuerz, Winnipeg, 1994.

C. N. Sawyer, P. L. McCarty and G. F. Parkin, "Chemistry for Environmental Engineering", 4th edn., McGraw-Hill, New York, 1994, pp. 596–601.

4.12 CHLORIDE

4.12.1 Introduction

Chloride is a major ion in surface waters and wastewaters and its concentration in natural waters varies widely. High levels of chloride are found in seawater and water from saltwater lakes, while mountain streams generally have low levels. Estuarine water is strongly influenced by mixing with seawater and it has intermediate chloride concentrations. Domestic waste contains considerable quantities of chloride due to its presence in urine, and in the past, tests for chloride were used to detect contamination of groundwater by wastewater.

Chloride is not considered as being harmful to human health. Levels of chloride in water supplies are limited to 250 mg L^{-1} as at higher concentrations chloride imparts a salty taste which makes the water unpalatable, and the WHO guideline for drinking water is set at this value. High chloride concentrations are harmful to plants; some damage may occur at levels as low as 70–250 mg L^{-1}.

4.12.2 Methodology

A simple method of determining chloride in water is by means of precipitation titration with $AgNO_3$, also called *argentometric titration*. The most common argentometric method for chloride is the so-called *Mohr method*, which involves the direct titration of chloride with $AgNO_3$, using potassium chromate as an indicator. Reaction of $AgNO_3$ with chloride precipitates silver chloride:

$$Ag^+ + Cl^- \rightarrow AgCl(s)$$

At the end point, excess of silver ions react with chromate to precipitate silver chromate which has a distinctive reddish-brown colour:

$$2Ag^+ + CrO_4^{2-} \rightarrow Ag_2CrO_4(s)$$
$$\text{reddish-brown}$$

The solution must be neutral to slightly alkaline. At lower pH, CrO_4^{2-} is converted to $Cr_2O_7^{2-}$, while at higher pH, Ag^+ is precipitated as $AgOH(s)$. An excess of Ag^+ is needed to precipitate a visible amount of $Ag_2CrO_4(s)$ and it is necessary to carry out a blank correction. See Example 4.11.

4.12.3 Materials

- Standard $AgNO_3$ solution, 0.0141 M
- Primary standard NaCl solution, 0.0141 M
- Potassium chromate solution, K_2CrO_4, 5%
- Sodium hydroxide, 0.1 M
- pH meter and electrodes or pH paper

Example 4.11

Calculate the concentration of chromate required to precipitate silver chromate at the equivalence point in the Mohr titration, given that the solubility product is 1.8×10^{-10} for AgCl and 1.1×10^{-12} for Ag_2CrO_4.

The equilibrium constant for a precipitation reaction is called the solubility product, K_{sp}. For the titration reaction in Mohr's method:

$$K_{sp} = [Ag^+][Cl^-]$$

At the equivalence point:

$$[Ag^+] = [Cl^-] = K_{sp}^{1/2} = 1.3 \times 10^{-5} \, \text{mol L}^{-1}$$

This Ag^+ concentration can now be used in the equation of K_{sp} for Ag_2CrO_4:

$$K_{sp} = [Ag^+]^2 [CrO_4^{2-}]$$

The concentration of chromate ion required to precipitate silver chromate at the equivalence point of Mohr's titration is:

$$[CrO_4^{2-}] = K_{sp}/[Ag^+]^2 = 1.1 \times 10^{-12}/(1.3 \times 10^{-5})^2 = 6.5 \times 10^{-3} \, \text{mol L}^{-1}$$

4.12.4 Experimental Procedure

4.12.4.1 Standardisation of AgNO₃. Standardise the $AgNO_3$ solution by titrating with primary standard 0.0141 M NaCl solution. Pipette 20 mL aliquots of the standard NaCl solution in 100 mL conical flasks and adjust the pH to between 7 and 10. Add 1 mL of the chromate indicator solution and titrate with $AgNO_3$ to the first appearance of the red silver chromate. Discard the result of the first titration and repeat the titration three more times on separate aliquots of NaCl solution. The end point may not be easy to identify as the solution changes from a yellow to an orange-red colour. Carry out a blank titration by adding 1 mL of chromate indicator solution to 20 mL of laboratory water and titrating with $AgNO_3$. Subtract the volume of titrant required for the blank titration from the volume of titrant required for the titration of NaCl solution and calculate the molarity of the standard $AgNO_3$ solution.

4.12.4.2 Analysis. Pipette 100 mL of water sample into a 250 mL conical flask and adjust the pH to between 7 and 10 if necessary. Preferably use pH paper rather than a pH meter as the electrode may contaminate the sample. Add the indicator and titrate as above. Also repeat a blank titration using 100 mL of laboratory water and correct the volume of titrant used. Repeat the titration with three aliquots of each sample and take an average. Calculate the concentration of chloride in the sample from the concentration of the standardised $AgNO_3$ solution, the corrected volume of titrant used and the volume of sample as follows:

$$Cl^- \text{ (mg L}^{-1}) = 35\,453 \times M \times (V_1 - V_2)/V_s$$

where V_1 is the volume of titrant for the sample (mL), V_2 is the volume of titrant for the blank (mL), M is the molarity of $AgNO_3$ as determined by standardising with primary standard NaCl and V_s is the volume (mL) of sample used (100 mL in above case). If the sample contains very high concentrations of chloride, dilute the sample before analysis.

Note

You may determine the chloride ion concentration, together with the nitrate and sulfate concentrations, in your samples using the ion chromatographic method described in Section 2.4.6.

4.12.5 Questions and Problems

1. Discuss the advantages and limitations of using chloride concentrations as an indicator of pollution by domestic wastes.

2. What substances would you expect to interfere with the chloride determination and why?

4.12.6 Suggestions for Projects

1. Determine the chloride ion concentration in waters from different sources (*e.g.* drinking water, river water, rain water, seawater) and comment on your results.
2. Determine the chloride ion concentrations at various locations in an estuary during high and low tide. Plot the chloride ion concentration with distance from the mouth of the river. Comment on your results.
3. Correlate the chloride ion concentration in different samples with other measurements (*e.g.* sodium, conductivity) in the same samples.

4.12.7 Further Reading

APHA, "Standard Methods for the Examination of Water and Wastewater", 16th edn., American Public Health Association, Washington, 1985, pp. 286–294.

C. N. Sawyer, P. L. McCarty and G. F. Parkin, "Chemistry for Environmental Engineering", 4th edn., McGraw-Hill, New York, 1994, pp. 509–514.

4.13 FLUORIDE

4.13.1 Introduction

Fluoridation of water supplies was introduced in the 1940s after it was shown that fluoride reduced the incidence of tooth decay. The topic of fluoridation is hotly debated and the evidence regarding its benefits is inconclusive. Fluoride is a toxic substance and concern has been expressed about the health effects of consuming fluoridated water, especially among infants and children. It has even been suggested that fluoride is a carcinogen, although the evidence is again inconclusive. While fluoridation has been adopted by many countries throughout the world (USA, Russia, Canada, Australia, New Zealand), other countries are phasing out or banning fluoridation altogether (Netherlands, Sweden and other west European countries). Natural fluoride levels in water vary considerably and where concentrations are high these have to be reduced to acceptable levels during water treatment. Where natural levels are low, fluoride is artificially added in those countries where fluoridation is an accepted practice. The World Health Organisation has adopted a guideline value for fluoride in drinking water of 1.5 mg L^{-1}.

At fluoride levels greater than 1.5 mg L^{-1} there is an increased incidence in dental fluorosis (*i.e.* teeth become discoloured, brittle and mottled). On the other hand, decreasing fluoride concentrations are associated with a greater incidence of dental caries (*i.e.* tooth decay). A concentration of approximately 1 mg L^{-1} is considered optimum for dental health.

4.13.2 Methodology

Fluoride is determined by potentiometry using a fluoride *ion selective electrode* (ISE) and a reference electrode. The measured potential is proportional to the concentration of fluoride ion when the ionic strength of all standards and the sample is the same. This is achieved by adding a *total ionic strength adjustment buffer* (TISAB) to all solutions. The potential is measured against a reference electrode (*e.g.* calomel electrode) and plotted against the logarithm of the fluoride concentration $\log_{10}[F^-]$. The graph should be a straight line except at very low fluoride concentrations.

The electrode used for fluoride analysis is a solid-state electrode constructed from a single crystal of lanthanum fluoride. This electrode is suitable for fluoride concentrations down to about 10^{-6} mol L^{-1}; below this concentration the response levels off due to the solubility of LaF_3. The fluoride electrode can be used with a standard calomel electrode as reference. The pH limits the use of the electrode. At pH < 3.5, fluoride forms HF which is not measured by the electrode. The use of the buffer maintains the pH at the desirable value.

You may use this method not only to analyse drinking water but also other surface waters, mineral waters, waste effluents, *etc.*

4.13.3 Materials

- Fluoride ion selective electrode
- Calomel reference electrode
- pH meter with millivolt scale or millivoltmeter
- Magnetic stirrer and bar
- Stock solution of sodium fluoride, 100 mg F L^{-1}. Dry some NaF at 110 °C for 2 h. Weigh 221.0 mg NaF, dissolve in laboratory water and make up to the mark in 1 L flask.
- Total ionic strength adjustment buffer (TISAB), commercially available

4.13.4 Experimental Procedure

Prepare a series of 100 mL calibration standards each containing the following concentrations of fluoride in mg $F L^{-1}$: 0.1, 1, 2.5, 5, 10 and 20 by dilution from the 100 mg L^{-1} fluoride stock solution. Analyse all these solutions in the following manner.

Pipette 50 mL of the standard solutions into a beaker and add 10 mL of the TISAB solution. Place a stirring bar in the beaker and stir on a magnetic stirrer at a moderate rate. Immerse the tips of the fluoride electrode and the standard calomel electrode into the solution and measure the potential on the millivolt function of the pH meter. Measure the lowest concentration standard first and work your way up to the highest concentration standard. Between each measurement, rinse the electrodes with laboratory water and dab dry. Plot the potential (as the *y*-axis) *versus* $\log_{10}[F^-]$ (as the *x*-axis) and draw a calibration graph.

Measure the potential of the water sample in the same way (50 mL sample + 10 mL TISAB) and obtain the concentration of fluoride. Warm the sample to room temperature before taking the measurement. Reading off the calibration graph will give you $\log_{10}[F^-]$. The concentration of fluoride is then the antilogarithm of the value that you read off from the *x*-axis. Suppose that your water sample gave a potential that corresponded to 0.25 on the *x*-axis. The fluoride concentration is $[F^-] = 10^{0.25} = 1.78$ mg L^{-1}. Carry out several replicate measurements on each sample and take the mean.

4.13.5 Questions and Problems

1. Outline the electrochemical principles on which analysis using ion selective electrodes is based.
2. A water supply containing 200 mg L^{-1} of calcium is to be fluoridated by adding NaF to obtain a fluoride concentration of 1 mg L^{-1}. Can this concentration be achieved? What is the maximum concentration of soluble fluoride that can be achieved? (K_{sp} for $CaF_2 = 3 \times 10^{-11}$ mol^3 L^{-3}).
3. Is it necessary to fluoridate water, considering that most people use toothpaste containing fluoride? Is there a danger of overexposure to fluoride amongst those using fluoridated toothpaste and consuming fluoridated water?

4.13.6 Suggestion for Project

Carry out a detailed survey of fluoride in water in your locality. Analyse both tapwater and surface water samples. On the basis of your results decide whether: (a) it is necessary to fluoridate water in your locality; (b) addition, or removal, of fluoride is being carried out at the water treatment plants; and (c) the levels of fluoride are below or above the WHO guideline.

4.13.7 Further Reading

APHA, "Standard Methods for the Examination of Water and Wastewater", 16th edn., American Public Health Association, Washington, 1985, pp. 352–364.

C. N. Sawyer, P. L. McCarty and G. F. Parkin, "Chemistry for Environmental Engineering", 4th edn., McGraw-Hill, New York, 1994, pp. 583–588.

N. F. Gray, "Drinking Water Quality: Problems and Solutions", Wiley, Chichester, 1994, pp. 200–205.

4.14 SULFATE

4.14.1 Introduction

Sulfate (SO_4^{2-}) is a major ion occurring in natural waters and waste-waters. In fresh, unpolluted, waters the sulfate concentration is between 5 and 100 mg L^{-1}, while in salt lakes it can be as high as 4000–5000 mg L^{-1}. The main natural source of sulfate in surface and ground water is the processes of chemical weathering and dissolution of sulfur-containing minerals, predominantly gypsum ($CaSO_4 \cdot 2H_2O$). Other natural sources are the oxidation of sulfides and elemental sulfur, and the decomposition of animal and plant residues. Sulfur is a nutrient essential for the synthesis of proteins and it is released upon their degradation.

Direct anthropogenic sources of sulfate include industrial and municipal wastes, agricultural drainage and runoff. Sulfuric acid present in mine drainage may also contribute significant amounts of sulfate to surface waters into which it is discharged. Sulfide minerals such as pyrite (FeS_2) are oxidised by chemical and bacterial action:

$$2FeS_2 + 7O_2 + 2H_2O \rightarrow 4SO_4^{2-} + 2Fe^{2+} + 4H^+$$

Drainage from mines can also lead to significant acidification and high Fe concentrations may also degrade water quality. Deposition of sulfate in rain is another source in surface waters, the sulfate originating from anthropogenic SO_2 pollution and natural sources. This can also lead to serious acidification problems in weakly buffered rivers and lakes (see Section 2.1.2).

Sulfate is usually quite stable in natural waters and does not undergo major tranformations, unlike nitrogen compounds. However, sulfur reducing bacteria may convert sulfate to hydrogen sulfide (H_2S) and if sufficient iron is present under reducing conditions, iron sulfides may also be precipitated.

High concentrations of sodium sulfate ($Na_2SO_4 \cdot 10H_2O$) and magnesium sulfate ($MgSO_4.7H_2O$) can impart a bitter taste to water, which is detectable at concentrations > 500 mg L^{-1}. At concentrations > 600 mg L^{-1} these salts may act as laxatives and are to be avoided in drinking water supplies. The WHO recommended guideline value for sulfate in drinking waters is 250 mg SO_4^{2-} L^{-1}. Most groundwaters have sulfate concentrations < 100 mg L^{-1}, but higher levels can be found close to sedimentary rock deposits. Another problem with sulfate salts is scale formation in water boilers (see Section 4.4). Sulfates are also of concern in wastewater treatment plants. Bacterial reduction of sulfate under anaerobic conditions can produce sulfide ion:

$$SO_4^{2-} + \text{organic matter} \rightarrow S^{2-} + H_2O + CO_2$$

which, in equilibrium with hydrogen ion, can form hydrogen sulfide:

$$S^{2-} + H^+ \rightleftharpoons HS^-$$

$$HS^- + H^+ \rightleftharpoons H_2S$$

This can give rise to an unpleasant odour. Also, hydrogen sulfide released into the air from wastewater in sewers, and elsewhere in the plant, can accumulate on the walls of pipes. There it can be oxidised to sulfuric acid by bacteria such as *Thiobacillus* which are present in sewage:

$$H_2S + 2O_2 \rightarrow H_2SO_4$$

Sulfuric acid can corrode concrete pipes. Sulfate in untreated domestic wastewater is generally present at between 20 and 50 mg L^{-1}.

Sulfate in waters is commonly analysed by adding excess barium chloride to precipitate barium sulfate, which can then be determined either by turbidimetry or gravimetry. The turbidimetric method is very rapid and it is described below. Sulfate can also be determined by ion chromatography, together with chloride and nitrate.

4.14.2 Methodology

An excess of barium chloride ($BaCl_2$) is added to the sample. The barium ion reacts with the sulfate to precipitate barium sulfate crystals of uniform size:

$$Ba^{2+} + SO_4^{2-} \rightarrow BaSO_4(s)$$

The colloidal suspension is measured using a spectrophotometer and the sulfate concentration determined by comparison with standards. Suspended particles present in large amounts will interfere and these can be removed by filtration. Highly coloured samples may give erroneous results.

4.14.3 Materials

- Magnetic stirrer and bar
- Spectrophotometer and absorption cells (2–10 cm)
- Stopwatch
- Barium chloride ($BaCl_2 \cdot 2H_2O$), crystalline. Dry in an oven and pass through a sieve of 20 mesh into a sieve of 30 mesh. Use crystals retained in the 30 mesh sieve.
- Sodium chloride–hydrochloric acid reagent. Dissolve 60 g NaCl in water, add 5 mL concentrated HCl and dilute to 250 mL.
- Glycerol–ethanol solution. Mix 100 mL glycerol with 200 mL of ethanol in a bottle.
- Standard sulfate solution, 100 mg L^{-1}. Dissolve 147.9 mg anhydrous sodium sulfate and dilute to 1 L with water. 1 mL = 0.1 mg SO_4^{2-}.

4.14.4 Experimental Procedure

4.14.4.1 Sampling and Storage. Store samples at 4 °C to prevent biological reduction of SO_4^{2-} to S^{2-}. Unpolluted samples can be stored at room temperature for 2–3 days. Filter through a 0.45 μm membrane filter.

4.14.4.2 Analysis. Measure 100 mL of sample into a 250 mL conical flask and place on a magnetic stirrer. While stirring add 20 mL of the NaCl–HCl solution and 20 mL of the glycerol–alcohol solution. Add approximately 0.3 g barium chloride. Stir for 2 min exactly after adding the barium chloride. Immediately pour some solution into an absorption cell and measure the absorbance at 420 nm after exactly 3 min. Prepare a series of calibration standards by pipetting aliquots of the standard sulfate solution corresponding to between 0.5 and 5 mg SO_4^{2-} (*i.e.* 5–50 mL) into a 100 mL volumetric flask and making up to the mark with water. Analyse in the same way as samples. Prepare sample blanks by adding all the reagents except barium chloride to 100 mL of sample and measure the absorbance. Subtract from each sample reading the blank reading obtained using the same sample to compensate for sample colour and turbidity. Prepare a calibration graph of absorbance against mg SO_4^{2-}. Read off the amount of sulfate in the samples using the corrected absorbance reading and calculate the concentration in the sample as:

$$mg\ SO_4^{2-}\ L^{-1} = 1000 \times mg\ SO_4^{2-}/V$$

where V is the volume of the sample (mL).

The volume of sample is 100 mL in the above procedure. If the sample contains more sulfate than the highest calibration standard, dilute the sample to fit on the curve and correct for the dilution when calculating the result.

Notes

1. Keep all conditions constant when analysing samples and standards, *i.e.* constant stirring speed, amount of barium chloride, *etc.*
2. If a magnetic stirrer is not available, you can shake by hand for 5 min exactly in a constant and repeatable manner. Organic compounds at high concentrations can cause severe interference. Silica at high concentrations (> 500 mg L^{-1}) will interfere.
3. You could determine sulfate in surface water samples using ion chromatography as described in Section 2.4.6 for rainwater.

4.14.5 Questions and Problems

1. What are the environmental concerns with regard to sulfates?
2. What are the main sources of sulfates in natural waters and waste-waters?
3. Describe the chemical principles of the turbimetric method for sulfate determination. Comment on potential interferences.
4. Why is it very important to keep all the conditions constant when analysing sulfate by turbidimetry?
5. Why is it that sulfate analysis does not have to be carried out immediately after collection, unlike analysis of nitrogen and phosphorus compounds?
6. The concentration of sulfate in a water sample was reported as 15 mg S L^{-1}. Convert this to mg SO_4^{2-} L^{-1}.
7. What percentage of each species of sulfide (*i.e.* H$_2$S, HS$^-$ and S^{2-}) is present at pH values of (a) 6, (b) 8 and (c) 10. Given that H$_2$S is the

only species that can be volatilised, at what pH conditions would you expect odour problems to be encountered in wastewater treatment plants. The stepwise equilibrium constants for the dissociation of H_2S are $K_1 = 1.0 \times 10^{-7}$ and $K_2 = 1.3 \times 10^{-13}$ mol L^{-1}.

4.14.6 Suggestions for Projects

1. Measure sulfate concentrations in different water bodies and assess the influence of natural and anthropogenic factors.
2. Compare the sulfate content in various industrial and municipal wastewaters and attempt to identify sources of sulfate pollution.
3. Determine the content of sulfate in rainwater and surface waters in your locality. Calculate the deposition rate of sulfate as shown in Chapter 2 and comment on the possible contribution of rainfall.
4. Determine the sulfate content in snow by analysing snow melt-water. Discuss the potential consequences on the surface waters in your locality when large amounts of snow thaw in spring.
5. Carry out a survey of sulfate in tapwaters in your region. Perform a statistical analysis of the results (*i.e.* calculate the mean, standard deviation, *etc.*) and compare with national and international standards.

4.14.7 Further Reading

M. B. Cerny (ed.), "Biogeochemistry of Small Catchments", Wiley, New York, 1994, pp. 229–254.

APHA, "Standard Methods for the Examination of Water and Wastewater", 16th edn., American Public Health Association, Washington, 1985, pp. 464–470.

N. J. Bunce, "Environmental Chemistry", 2nd edn., Wuerz, Winnipeg, 1994, pp. 131–158.

C. N. Sawyer, P. L. McCarty and G. F. Parkin, "Chemistry for Environmental Engineering", 4th edn., McGraw Hill, New York, 1994, pp. 589–595.

4.15 IRON

4.15.1 Introduction

Despite being the second most abundant element in the earth's crust (after Al), iron (Fe) is present in relatively small amounts in natural waters. The Fe concentration in river and lake water is generally below 1 mg L^{-1}. However, in coloured, swamp, ground and especially in acid waters the Fe concentration may increase significantly. The main sources of Fe in natural water are the processes of chemical weathering of geological materials, which lead to their destruction and transformation of Fe from solid compounds to soluble and/or colloidal forms. Iron is present in soils and minerals as insoluble ferric oxide, iron sulfide (pyrite) and slightly soluble ferrous carbonate (siderite). The latter may dissolve in waters rich in CO_2:

$$FeCO_3 + CO_2 + H_2O \rightarrow Fe^{2+} + 2HCO_3^-$$

Pyrite may be oxidised to soluble iron sulfate:

$$2FeS_2 + 7O_2 + 2H_2O \rightarrow 2Fe^{2+} + 4SO_4^{2-} + 4H^+$$

Under acidic (pH < 3–4) and reducing conditions, Fe is present in the ferrous state (Fe^{2+}), which is relatively soluble. Exposure to air or addition of oxidants causes oxidation of ferrous iron to the ferric state (Fe^{3+}), which may hydrolyse to form insoluble hydrated ferric oxide:

$$Fe(II) \rightarrow Fe(III) \rightarrow Fe(OH)_3\downarrow$$

In natural and waste waters, Fe may occur in soluble form, in a colloidal state, in inorganic or organic Fe complexes or in relatively coarse suspended particles. Iron compounds may be either ferrous or ferric, suspended or dissolved. Fe(III) is the only stable valence state for iron in oxygen-containing waters; however, it is only soluble at pH < 5. Fe(III) can be reduced to Fe(II) under highly anaerobic reducing conditions. Clay and silt in suspension may contain acid-soluble Fe.

Iron is a micronutrient required by living organisms, and therefore an essential and desirable constituent of natural waters. Wastewaters from different industrial plants (iron smelting, textile, painting) and from agriculture are enriched with Fe. High concentrations of Fe are not known to have any adverse health effects; however, they may lead to other problems. In many developing countries, iron oxide particles are the result of flaking of rust from water pipes. This can impart an unpleasant "ironish" smell to drinking water. A bittersweet astringent taste can be detected at levels above 1–2 mg L^{-1}. Other problems of concern regarding Fe are connected with the staining of laundry and porcelain. Water for domestic and industrial use is generally required to contain less than 0.2–0.3 mg L^{-1} of Fe. Furthermore, Fe hydroxide can form deposits in pipes.

4.15.2 Methodology

The method is based on the reaction between ferrous Fe(II) ion and 1,10-phenanthroline to form an orange-red complex ion. As some iron in the sample may be present as precipitated ferric hydroxide, Fe is first brought into solution and reduced to the ferrous state by boiling with hydrochloric acid:

$$Fe(OH)_3 + 3H^+ \rightarrow Fe^{3+} + 3H_2O$$

Ferric Fe(III) is reduced to the ferrous state by reaction with hydroxylamine:

$$4Fe^{3+} + 2NH_2OH \rightarrow 4Fe^{2+} + N_2O + H_2O + 4H^+$$

Each ferrous ion reacts with three molecules of 1,10-phenanthroline to form a complex ion which is orange-red in colour. The pH is kept at pH 3.2–3.3 to ensure rapid colour development. The coloured solution obeys Beer's law and can be measured using a spectrophotometer. The test procedure may be modified to determine total, dissolved or suspended iron. Concentrations of Fe as low as 10 μg mL^{-1} can be determined using a 5 cm cell.

4.15.3 Materials

- Spectrophotometer
- Filtration apparatus
- Filter paper, 0.45 μm
- Hot plate
- Concentrated hydrochloric acid
- Hydroxylamine solution (10%) prepared by dissolving 10 g $NH_2OH \cdot HCl$ in 100 mL water
- Ammonium acetate buffer solution. Dissolve 250 g $NH_4C_2H_3O_2$ in 150 mL water and add 700 mL of concentrated glacial acetic acid.
- Phenanthroline solution prepared by dissolving 100 mg of 1,10-phenanthroline monohydrate, $C_{12}H_8N_2 \cdot H_2O$, in 100 mL water and adding two drops concentrated HCl. The solution should be discarded if it darkens.
- Stock iron solution, 1000 mg L^{-1}. Commercially available. This can also be prepared as follows. Dissolve 1.00 g iron wire in 50 mL 1:1 HNO_3 and dilute to 1 L.

- Working iron solution, 10 mg L^{-1}, prepared by dilution of the stock solution
- Standard iron solution, 1 mg L^{-1}, prepared by further dilution

4.15.4 Experimental Procedure

4.15.4.1 Sampling and Storage. Care should be taken not to contaminate the sample with iron. Use Pyrex or plastic bottles. Do not use metal caps as these may contaminate the sample with Fe. Clean bottles with concentrated HCl and rinse with water. Great care must be taken to obtain a representative sample. The form and concentration of Fe in waters collected from wells or taps may vary considerably with duration of flushing before sampling. Sometimes Fe oxide particles are collected with a water sample owing to flaking of rust from pipes. Colloidal Fe may adhere to the walls of sample bottles, especially those made of plastic. Shake the sample bottle often when determining Fe in suspension. Care must also be taken to prevent conversion of one form of Fe into another during transport and storage.

Addition of reagents and filtration may have to be done in the field in order to avoid reactions during storage. Take separate samples for total, dissolved and ferrous Fe(II) and proceed as specified in the procedures below.

4.15.4.2 Analysis. *(a) Total iron (dissolved + suspended Fe^{3+} + Fe^{2+}).* Shake sample bottle thoroughly and pipette a 50.0 mL aliquot into a 125 mL conical flask. Add 2 mL concentrated HCl, 1 mL hydroxylamine solution and some glass beads. Boil until the volume is reduced to about 20 mL. Cool the solution and transfer into a 100 mL volumetric flask. Add 10 mL of the sodium acetate buffer solution and 4 mL of the phenanthroline solution. Dilute to the mark with water and mix thoroughly. Stand for at least 10–15 min for maximum development of the orange-red colour. Carry out the spectrophotometric determination and calculation as detailed below in procedure (d).

(b) Dissolved iron. Filter 100 mL of sample through a 0.45 μm membrane filter into a vacuum flask containing 1 mL of concentrated HCl immediately after collection. To determine total dissolved Fe follow procedure (a) outlined above, and to determine dissolved ferrous Fe follow procedure (c) outlined below. Carry out the spectrophotometric determination and calculation as detailed in procedure (d) below. You may calculate the suspended iron concentration by difference as follows:

$$[\text{suspended Fe}] = [\text{total Fe}] - [\text{dissolved Fe}]$$

(c) Ferrous iron (Fe²⁺). Ferrous Fe(II) should be determined immediately upon collection because of the possibility of change in the Fe(II)/Fe(III) ratio with time in acid solutions. If the determination cannot be carried out at the site, the sample may be stored for up to several hours in a full (*i.e.* there must be no air pocket in the bottle), stoppered bottle, but the acid should not be added until the analysis can be performed. Acidify a 100 mL sample with 2 mL of concentrated HCl and immediately withdraw a 50 mL portion of acidified sample and place in a 100 mL volumetric flask. Add 20 mL phenanthroline solution and 10 mL ammonium acetate buffer solution and stir vigorously. Dilute to 100 mL with water and measure the colour intensity within 5–10 min, without exposing the flask to sunlight. This procedure is suitable for less than 50 μg total iron. Use a larger volume of phenanthroline or a more concentrated reagent if larger amounts of iron are present. Total Fe is determined according to procedure (a) above. The ferric Fe(III) concentration can be calculated by difference as follows:

$$[Fe^{3+}] = [\text{total Fe}] - [Fe^{2+}]$$

(d) Spectrophotometric determination. Pipette aliquots of working or standard Fe solutions, corresponding to between 1 and 200 μg Fe (depending on the expected concentration in your samples) into a series of 50 mL volumetric flasks and make up to the mark with water. Mix and transfer to a series of 125 mL conical flasks. Also pipette 50 mL of pure water into one conical flask to serve as a blank. Follow the procedure outlined in (a) above, adding the same volumes of reagents as when analysing samples to each of the conical flasks. Set the spectrophotometer at 510 nm and determine the absorbance of each solution in a 1 cm cell. Set the absorbance to zero with distilled water in a cell. Plot a calibration graph of absorbance against μg Fe. If you expect less than 100 μg Fe in the final 100 mL volume of solution [after following procedure (a)], prepare more dilute standards in the appropriate range, and use a 5 cm cell. Read off the mass of Fe in μg directly from the graph. Calculate the concentration of Fe in the sample from:

$$\mu\text{g Fe mL}^{-1} = \mu\text{g Fe}/V$$

where V is the original volume of the sample in mL (50 mL in this procedure).

Notes

1. If the absorbance is outside the range of the calibration standards you may use a smaller aliquot for analysis.
2. You may transfer the concentrated solution to a 50 mL flask rather than a 100 mL flask in order to increase the sensitivity.
3. If samples are turbid or coloured, take a second set of samples and treat them exactly as outlined in the procedure, but do not add the phenanthroline reagent. Use these samples as blanks to set the absorbance reading on the spectrophotometer to zero, instead of pure water. Read each developed sample with phenanthroline against the corresponding sample blank without phenanthroline.
4. Large amounts of organic matter can interfere in this method.
5. Some metal ions (Zn, Cu, Co and Ni) may interfere if present at high concentrations. Strong oxidising agents, phosphate, nitrite and cyanide may also interfere.
6. Iron may also be determined by atomic absorption spectrometric as described in Section 4.16.

4.15.5 Questions and Problems

1. Outline the sources of Fe in natural waters and wastewaters.
2. What is the difference in environmental behaviour of ferrous and ferric Fe?
3. Describe the processes leading to rust formation in water pipelines.
4. Outline the basic chemistry behind the colorimetric determination of Fe.
5. How can you analyse coloured or turbid samples?

4.15.6 Suggestions for Projects

1. Compare the content of total iron and its ferrous and ferric forms in natural stream water and swamp water. Explain the results.
2. Measure the content of ferrous and ferric iron, pH and Eh values in different water bodies (lakes, rivers, ponds, swamps, groundwater) and calculate statistical relationships between these chemical parameters.
3. Measure the content of total Fe in tapwater of your locality.
4. Compare the content of various Fe forms in wastewaters from different wastewater treatment plants. Explain the results on the basis of the treatment technology used.

4.15.7 Further Reading

APHA, "Standard Methods for the Examination of Water and Wastewater", 16th edn., American Public Health Association, Washington, 1985, pp. 214–219.

N. J. Bunce, "Environmental Chemistry", 2nd edn., Wuerz, Winnipeg, 1994, pp. 131–158.

C. N. Sawyer, P. L. McCarty and G. F. Parkin, "Chemistry for Environmental Engineering", 4th edn., McGraw-Hill, New York, 1994, pp. 577–582.

4.16 HEAVY METALS

4.16.1 Introduction

Heavy metals are defined as those metals with densities greater than 5 g cm^{-3}. Because they are present at much lower concentrations in waters compared to major ions (SO_4^{2-}, Cl^-, NO_3^-, Mg^{2+}, Ca^{2+}, *etc.*) discussed previously, heavy metals are commonly referred to as *trace metals*.

Major anthropogenic sources of heavy metals are industrial wastes from mining, manufacturing and metal finishing plants. Other anthropogenic sources of metals in surface waters include domestic wastewaters and runoff from roads. Metals are cycled through the environment and they may enter surface waters in precipitation. Therefore, metals emitted to the atmosphere by industrial processes (*e.g.* Hg from the combustion of coal, Pb from petrol) end up in surface or ground waters. They may also be leached from soils and rocks in contact with water. The disposal of massive quantities of metal wastes at landfills can lead to metal pollution of ground and surface waters. Acidification of surface and ground waters by acid rain could lead to increased leaching of metals and hence higher concentrations in water. Concern has been expressed that acid rain may lead to an increase in the concentrations of trace metals in drinking water owing to the acid-induced leaching of metals from pipes.

Many of the trace metals are highly toxic to humans (*e.g.* Hg, Pb, Cd, Ni, As, Sn) and other living organisms, and their presence in surface waters at above background concentrations is undesirable. Also, unlike many organic pollutants, metals cannot be degraded in the environment by chemical or biological processes. They can, however, react in the environment, but the resulting metal containing compounds may be even more toxic (*e.g.* methylation of Hg). Furthermore, toxic metals may be *bioconcentrated* in the food chain so that concentrations in the upper members of the chain can reach values many times higher than in water. The result is that some plants and animals may represent a serious hazard if consumed as food. This process (also termed *bioamplification*,

Table 4.18 *Concentrations of some heavy metals in water samples (μg L^{-1})*

Metal	Seawater	River water	Rainwater	Tapwater
Cd	–	0.1–20	–	<2
Cr	–	0.5–100	–	<4
Co	–	1–50	–	<5
Cu	5	1–300	0.1–150	15–25
Fe	50	1–5000	1–300	30–100
Pb	–	0.2–150	0.02–50	3–10
Mn	5	0.3–3000	0.3–30	5–20
Hg	0.03	0.1–0.2	–	–
Ni	0.1	1–150	0.1–100	<15
Ag	0.3	0.1–30	–	<3
V	0.3	2–300	–	5–15
Zn	5	2–1200	0.2–150	10–20

biomagnification or *bioaccumulation*) has been responsible for some major pollution incidents in the past. Toxic metal pollution has been identified in surface and coastal waters throughout the world and it is becoming a major environmental problem. Usually, treatment at source is the only practical way to control toxic metal pollution.

The concentrations of some heavy metals in different waters are shown in Table 4.18. Seawater has very low concentrations; however, coastal waters receiving sewage and other effluents can have considerably higher concentrations than those shown in the table. Concentrations in river-waters can vary considerably. Background concentrations in remote areas may be below the range shown in the table, while highly polluted waters in industrial zones may have even higher levels. Rainwater concentrations are generally lower than those observed in riverwater. Concentrations of some metals (Zn, Pb, Cu) in first-draw tapwater can be considerably higher than those shown in the table owing to the accumulation of metals leached from pipes in standing water.

The national and international guidelines and standards for heavy metals in waters intended for different uses are summarised in Appendix III. Some of the metals that are important in the hydrosphere are discussed in more detail below. For more information of sources and effects of these and other metals, see Section 5.13.

4.16.2 Aluminium

Aluminium is generally not considered to be a toxic metal. Elevated levels of aluminium have been found in patients suffering from Alzheimer's disease and this has led some to suspect a possible link with aluminium pollution. However, aluminium is believed to be responsible for the death of fish in acidified surface waters. Increasing amounts of aluminium are leached into solution as the pH decreases, and aluminium, together with low pH, can result in fish deaths, although the mechanism is still not fully resolved. Young fry are even more sensitive to high Al and low pH than are fish. High Al and low pH are also toxic to amphibians such as frogs. Loss of fish in acidified lakes and rivers can also lead to a decline in some birds that prey on fish. The WHO does not consider Al as being of health significance, rather a substance that "may give rise to complaints from consumers". Alum, $KAl(SO_4)_2 \cdot 12H_2O$, is used in the treatment of drinking water, but the quantity added, and other operating parameters (*e.g* pH), have to be carefully controlled. Very high levels of Al in drinking water have resulted from negligence or poor control of water treatment processes.

4.16.3 Copper

Copper is used extensively in alloys, plumbing, wires, paints, ceramics and pesticides. Copper is an essential element and is generally not considered harmful to human health. The main source of copper in drinking water is from the corrosion of pipes used in water distribution systems. At high concentrations, copper may impart an unpleasant taste to water and affect certain individuals suffering from a copper metabolism disorder called Wilson's disease.

4.16.4 Lead

Lead is present in various products (*e.g.* pipes, ammunition, solder, paint, petrol), but because of its high toxicity many of its uses have been discontinued. The major source of lead in the environment is the use of lead as a petrol additive. Atmospheric pollution by lead has caused considerable concern in the past and many countries have phased out the use of lead in petrol. However, lead already in the environment is cycled through the biogeochemical cycle, and lead originally released into the atmosphere has ended up in surface and ground waters. A major concern in the past has been the presence of lead in drinking water owing to the use of leaded plumbing in older houses. Lead can cause damage to the nervous system and the kidneys and it is a suspected carcinogen. Children exposed to high lead levels are particularly at risk. Lead in the environment is generally present as inorganic Pb^{2+}. Organometallic lead compounds, such as tetramethyllead, $(CH_3)_4Pb$, trimethyllead,

$(CH_3)_3Pb^+$, and dimethyllead, $(CH_3)_2Pb^{2+}$, are present at much lower concentrations than inorganic lead. These may result from the emission and degradation of some of the tetraalkyllead added to petrol or the biological methylation of inorganic lead in sediments. Lead is added to petrol as a mixture of tetramethyl- and tetraethyllead, but most of this is converted to inorganic lead salts during combustion. A small fraction may be emitted in organometallic form and this is eventually degraded to inorganic lead in the environment:

$$R_4Pb \rightarrow R_3Pb^+ \rightarrow R_2Pb^{2+} \rightarrow Pb^{2+}$$

where R is CH_3 or C_2H_5. In the past there have been cases of accidental spillages of organolead compounds at sea during transportation.

4.16.5 Mercury

Mercury is widely used in various products and processes (see Section 5.13.8) and it can enter the environment when wastes containing mercury are disposed. Other major sources are the mining and smelting of mercury itself, and the combustion of coal. The use of mercury in the mining of silver and gold is especially insidious since it can lead to air and water pollution and exposure of miners to potentially harmful levels. Mercury has the highest volatility of any metal and it is readily released to the atmosphere, from where it can enter surface waters in rain. All mercury compounds are toxic, although the toxicity varies with the form. Organic mercury compounds, such as methylmercury, CH_3Hg^+, and dimethylmercury, $(CH_3)_2Hg$, are especially toxic. These can be released into the environment either directly in wastewaters, or formed from inorganic mercury by bacteria in sediments. Mercury can be bioamplified in the food chain, so much so, that concentrations in fish and shellfish may be tens or even hundreds of thousands of times greater than in the water. A major incident of mercury pollution took place in Minamata Bay, Japan, in 1953. A chemical plant discharged wastewater containing methylmercury into a bay and this became concentrated in fish and shellfish which were subsequently eaten by the local fishermen and their families. Those affected suffered a severe form of organomercury poisoning, the so-called "Minamata disease". This resulted in 41 deaths, and more than 110 persons were afflicted. Some people suffered serious neurological disorders that left them paralysed for life. Furthermore, babies of afflicted mothers suffered congenital defects. A similar, but less severe, incident occurred in Niigata, Japan, in the early 1960s. Mercury bioaccumulated in fish can also cause death of birds and other animals when eaten. Mercury pollution has been found in surface waters at many other locations and this has led to the banning and control of many uses of mercury in developed countries. This cannot be said of developing countries (*e.g.* Brazil), where mercury is still widely used in the extraction of gold, leading to severe environmental pollution and affecting the health of miners. Because of low concentrations, mercury cannot be determined by conventional AAS and the procedure has to be modified. In the *cold vapour method*, mercury compounds are reduced to elemental mercury vapour and analysed by AAS in a special detection cell (see Section 5.13.19).

4.16.6 Tin

Tin is used in alloys, pigments, dyes and in the plating of metals. Several organotin compounds are used as sabilisers in poly(vinyl chloride) (PVC) products and as wood preservatives. Tributyltin was widely applied as an antifouling agent to the hulls of ships and boats in order to prevent growth of marine organisms. Tributyltin is highly toxic to all marine organisms; although it degrades in water, it tends to accumulate in sediments where degradation is much slower. Many countries have banned the use of tributyltin while others have imposed restrictions.

4.16.7 Analysis

Metals may be determined by colorimetric methods (see Section 4.15 for Fe) or by atomic absorption spectrometry (AAS) using flame or electrothermal atomisation. Flame AAS is more commonly available, but it is less sensitive. Flame atomic emission spectroscopy (AES) and inductively coupled plasma atomic emission spectroscopy (ICP-AES) can also be used for determining metal concentrations in water samples. Samples with low concentrations of metals should be reduced in volume by evaporation. Operating parameters for determining the more common heavy metals by AAS are given below. Those metals for which conventional flame AAS is not sensitive enough (*e.g.* As, Se, Hg) are not mentioned here. Arsenic and Se are usually analysed by AAS after hydride generation. Mercury is normally determined by cold vapour AAS (see Section 5.13.19).

Metals can be classified as:

- *Total metals* — metals determined in unfiltered samples (dissolved + suspended)

- *Dissolved metals* — metals in unacidified samples filtered through a 0.45 μm membrane filter
- *Suspended metals* — metals retained on a 0.45 μm membrane filter

4.16.8 Methodology

The concentration of total, dissolved and suspended heavy metals (Al, Ba, Cd, Cr, Co, Cu, Fe, Pb, Mo, Mn, Ni, Si, Sn, Zn and V) may be determined in natural waters, wastewaters and effluents. Samples are digested in concentrated nitric acid and analysed by AAS. Total metals are determined by digesting and analysing unfiltered samples. Dissolved metals are determined by digesting and analysing samples filtered through a 0.45 μm membrane filter. Suspended metals are determined by difference:

$$\text{Suspended metals} = \text{Total metals} - \text{Dissolved metals}$$

4.16.9 Materials

- Hot plate or steam bath
- Nitric acid, concentrated
- Atomic absorption spectrometer
- Lanthanum solution. Dissolve 67 g of lanthanum chloride ($LaCl_3 \cdot 7H_2O$) in 1 M HNO_3 by gently warming. Cool and dilute to 500 mL with pure water.
- Calcium solution. Dissolve 315 mg calcium carbonate ($CaCO_3$) in 25 mL 1:5 HCl. If the salt does not dissolve completely, heat and boil gently until it dissolves. Cool and dilute to 500 mL with water.
- Hydrogen peroxide, 30%
- Aluminium nitrate solution. Dissolve 139 g $Al(NO_3)_3 \cdot 9H_2O$ in 150 mL water and warm to dissolve completely. Cool and dilute to 200 mL.
- Potassium chloride solution. Dissolve 125 g KCl in water and dilute to 500 mL.
- Stock solutions of each metal to be analysed, 1000 mg L^{-1}. These are commercially available, otherwise they can be prepared as indicated in Table 4.19. 1 mL of stock solution = 1 mg.

4.16.10 Experimental Procedure

4.16.10.1 Sampling and Storage. Always use polypropylene or polyethylene bottles for sampling. If dissolved and suspended metals,

Table 4.19 *Preparation of 1000 mg L^{-1} stock solutions. Prepare by dissolving the amount stated in the primary solvent and then diluting to 1 L with water. Oven dry all salts at 110 °C for 2 h before weighing. When two solvents are specified (*e.g. water + acid*), dissolve in water first and then add the specified amount of acid before diluting with water to 1 L*

Metal	Reagent	Weight (g)	Primary solvent
Al	Al metal	1.000	20 mL conc. HCL
Ba	$BaCl_2 \cdot H_2O$	1.779	200 mL water + 1.5 mL conc. HNO_3
Cd	Cd metal	1.000	minimum volume of 1:1 HCl
Co	Co_2O_3	1.407	20 mL hot conc. HCl
Cr	$K_2Cr_2O_7$	2.828	200 mL water + 1.5 mL conc. HNO_3
Cu	Cu metal	1.000	15 mL 1:1 HNO_3
Fe	Fe wire	1.000	50 mL 1:1 HNO_3
Mo	MoO_3	1.500	10% HCl
Mn	$MnSO_4 \cdot H_2O$	3.076	200 mL water + 1.5 mL conc. HNO_3
Ni	NiO	1.273	minimum volume of 10% (v/v) HCl
Pb	$Pb(NO_3)_2$	1.598	200 mL water + 1.5 mL conc. HNO_3
Si	$Na_2SiO_3 \cdot 9H_2O$	10.12	water
Sn	Sn metal	1.000	100 mL conc. HCl
V	NH_4VO_3	2.296	800 mL water + 10 mL conc. HNO_3
Zn	Zn metal	1.000	20 mL 1:1 HCl

rather than total metals, are to be determined, filter immediately after collection and only acidify after filtration. Filter sample under vacuum through a 0.45 μm membrane filter (polycarbonate or cellulose acetate) in a filter holder. A typical filtering device is shown in Figure 4.2. Precondition the filter by rinsing with 50 mL of water and then with at least 100 mL of sample. If the filtration blanks are high, soak membrane filters in approximately 1:1 HNO_3 and rinse with water before use.

Preserve samples by adding 1.5 mL concentrated HNO_3 to 1 L of sample to give pH < 2. If the sample contains significant alkalinity or buffer capacity, add more HNO_3 and check the pH with a pH meter. Acidified samples may be stored at 4 °C for up to 6 months if they contain mg L^{-1} concentrations of metals. Samples containing metals at μg L^{-1} levels should be analysed as soon as possible. When handling samples, be extremely careful not to introduce contamination during sampling, handling, storage or treatment. Use the highest purity HNO_3 available in the laboratory and analyse water, filter and acid blanks regularly. If you are using plastic pipette tips to dispense the acid, soak in 2 M HNO_3 for several days and rinse with water before use to remove any contamination by Fe, Cu, Cd and Zn.

4.16.10.2 Digestion. Shake sample bottle thoroughly. Measure a volume (50–100 mL) of sample into a beaker and add 5 mL concentrated HNO_3. Boil slowly on a hot plate or a steam bath. Evaporate down to about 20 mL. Add a further 5 mL concentrated HNO_3, cover with a watch glass and heat. Continue adding concentrated HNO_3 and heating until the solution appears light coloured and clear. This indicates that digestion is complete. Do not allow to dry during digestion. Add 1–2 mL concentrated HNO_3 and heat slightly to dissolve any remaining residue. Wash down the beaker walls and watch glass with water. Transfer to a 50 mL volumetric flask, cool and make up to the mark with water.

4.16.10.3 Analysis. Prepare a series of calibration standards for each metal (if you will be determining several metals, you may prepare mixed calibration standards containing several metals) over the desired range by pipetting aliquots of the 1000 mg L^{-1} standard solution into a series of volumetric flasks (50 or 100 mL). Add 1 mL of concentrated nitric acid to each flask and dilute to the mark. Do not keep calibration standards for long periods and preferably prepare when required. Filter and digest laboratory water in the same way as the sample. Use this as a blank.

Switch on the AAS, select the appropriate hollow cathode lamp and adjust the lamp current to the value recommended in the instrument manual. Set the monochromator and slit settings to the the recommended value. Fine tune the wavelength setting and align the beam. Optimise the gain setting. Turn on the fume hood and switch on the flame according to the recommended procedure. You may get the laboratory instructor or demonstrator to do this for you.

Aspirate sample, standard and blank into flame of AAS equipped with the appropriate hollow cathode lamp and operating at the wavelength recommended in Table 4.20 for the metal being analysed. You should strictly operate the AAS under direction and supervision of a laboratory technician. Record the absorbance values for each standard and sample or, better still, use a chart recorder to record peak heights for each of the standards and samples. Aspirate laboratory water between measurements and allow the reading to return to a stable baseline before taking the next measurement. If using a chart recorder, measure the peak height from the baseline. Construct a calibration graph by plotting the absorbance, or peak height in cm, measured with standards against the metal concentration. Read off the concentration in the samples from the graph using the corrected reading obtained with samples. Calculate the concentration in the original sample as follows:

$$[\text{Metal}] \text{ mg } L^{-1} \text{ in sample} = C \times V_1/V_2$$

Table 4.20 *Recommended wavelengths, detection limits and optimum analysis ranges for the analysis of some metals by AAS*

Metal	Wavelength (nm)	Flame[a]	Detection limit (mg L^{-1})	Working range (mg L^{-1})
Al	309.3	N/A	0.1	5–100
Ba	553.6	N/A	0.03	1–20
Cd	228.8	A/A	0.002	0.05–2
Co	240.7	A/A	0.03	0.5–10
Cr	357.9	A/A	0.02	0.2–10
Cu	324.7	A/A	0.01	0.2–10
Fe	248.3	A/A	0.02	0.3–10
Mo	313.3	N/A	0.1	1–20
Mn	279.5	A/A	0.01	0.1–10
Ni	232.0	A/A	0.02	0.3–10
Pb	283.3 (217.0)	A/A	0.05	1–20
Si	251.6	N/A	0.3	5–150
Sn	224.6	A/A	0.8	10–200
V	318.5	N/A	0.2	2–100
Zn	213.9	A/A	0.005	0.05–2

[a] A/A = air/acetylene flame, N/A = nitrous oxide/acetylene flame.

where C is the concentration in mg L^{-1} of the metal in the final extract, V_1 is the volume of the final extract (50 mL in this procedure) and V_2 is the volume of original sample in mL.

If the concentration in the sample is higher than your most concentrated standard, dilute the sample to bring it within the range of the calibration standards and correct the result for dilution. If the blank, prepared as instructed above, gives a significantly different reading than the laboratory water, then subtract the reading of the blank from that of the samples before reading off from the chart. Do not subtract the blank reading from the standard reading as the standards were not filtered and digested.

Notes

1. Do not keep standard solutions containing less than 10 mg L^{-1} for more than 1 or 2 days as they will deteriorate due to absorption of metal onto the walls of the flask.
2. The detection limits given in Table 4.20 are only guidelines and may vary from instrument to instrument. Calculate your own detection limit as follows. Take repeated measurements of the absorbance while aspirating water into the flame, or record the output on a chart recorder. If you are using a chart recorder, measure the deviations of

the baseline noise at various intervals using a ruler. Calculate the standard deviation of either the absorbance or the baseline noise. Aspirate a standard with a low concentration of metal (*e.g.* 1 mg L^{-1}) and measure the absorbance. Calculate the sensitivity as follows:

$$\text{Sensitivity} = \text{Response/concentration of standard}$$

where the response is either in absorbance units or in units of length (mm or cm) if you are using a chart recorder. Calculate the detection limit from the standard deviation of the baseline noise and the sensitivity from:

$$\text{d.l.} = 3 \times \text{standard deviation of the noise/sensitivity}$$

3. You may have problems determining many of the metals in un-polluted samples owing to their low concentrations (compare the concentrations of metals observed in natural waters given in Table 4.18 with the detection limits of flame AAS in Table 4.20). Flame AAS analysis is suitable for polluted samples with higher levels of metals, so try to select appropriate sites when sampling. If you analyse a sample but are unable to obtain a response higher than with the blank, report the result as < d.l. (*i.e.* below the detection limit).
4. If the concentration of a metal is below the detection limit, you may consider concentrating a larger volume of sample. Measure a volume (500 mL or 1 L) of filtered sample and place 100 mL in a beaker together with 5 mL concentrated NHO_3. Boil down to 25–30 mL and keep adding more of the sample and evaporating. When all of the sample has been evaporated down to 25–30 mL, proceed as indicated above. Calculate a new detection limit, taking into account the concentration factor.
5. You may determine lower concentrations by extending the concentration ranges shown in Table 4.20 downward with scale expansion.
6. You may also determine lower concentrations of metals by using graphite furnace AAS if it is available in your laboratory.
7. You may determine higher concentrations of metals by choosing a less sensitive wavelength (see the operating manual for the AAS in your laboratory) or by rotating the burner.
8. When analysing for Fe and Mn, mix 50 mL of standard or sample with 10 mL of Ca solution before aspirating into flame.
9. When analysing for Cr, mix 50 mL of standard or sample with 0.5 mL of 30% H_2O_2.

10. When analysing for Al or Ba, add 1 mL KCl solution to 50 mL of standard or sample before aspirating in the flame.
11. When analysing for Pb, use the more sensitive line at 217.0 nm rather than the 283.3 nm line.
12. When analysing for V, add 1 mL aluminium nitrate solution to 50 mL of sample or standard before aspirating in the flame.
13. Alkali metals (Na, K) and alkaline earth metals (Mg, Ca) may be determined with AAS using the wavelengths listed in Table 2.8. Carry out the determination in the same way as outlined in the experimental procedure for heavy metals, but eliminate phosphate interference in the determination of Mg and Ca. When analysing for Mg and Ca, mix 50 mL of standard or sample with 5 mL of lanthanum solution. Prepare standards and use the wavelength settings given in Section 2.5.1 for rainwater analysis.

4.16.11 Questions and Problems

1. List the main sources of heavy metals in surface and ground waters. Give examples of ratios between natural and anthropogenic sources for different metals.
2. Describe the difference between total, dissolved and suspended forms of metals in waters. What types of pretreatment are necessary for the analysis of these forms?
3. What is the role of aquatic systems in the biogeochemical cycling of heavy metals? Give specific examples.
4. Some heavy metals are simultaneously essential micronutrients and harmful pollutants. Give examples of this dual role of metals in freshwater and marine ecosystems.
5. Give examples of some organometallic compounds of environmental concern.
6. Natural processes can sometimes convert a metal pollutant into a more harmful form. Give an example of this.

4.16.12 Suggestions for Projects

1. Determine several metals (*e.g.* Pb, Cu, Cd, Zn) in different water samples. You could investigate statistical relationships by computing correlation coefficients for pairs of metals and testing their significance. Attempt to identify the sources of heavy metals.
2. Determine various heavy metals in samples of surface water and corresponding bottom sediments (see Section 5.13 for sediment

analysis). Comment on the self-purification capacity of different water bodies on the basis of your results.

3. Carry out a survey of heavy metals in drinking water in your locality and compare the results with corresponding national and international guidelines and standards.

4. Collect a tapwater sample first thing in the morning without flushing the pipes. Then let the tap run for a while and take samples at several time intervals, recording the time since the tap was opened. Analyse the samples for heavy metals, including Fe, Pb, Zn and Cu. Does flushing influence the concentrations of any of the metals, and if so, which metals?

5. Determine heavy metal concentrations in samples collected in effluents from various industrial, municipal and agricultural sources. The input of these pollutants into surface waters can be calculated if the wastewater flow rates are known. You could ask operators of sewage and wastewater treatment plants whose effluent you analyse for this information. Discuss your results in terms of the source of metal pollution.

6. Analyse different samples for a specific metal by both the method of standard additions and the conventional calibration method outlined above. Plot the results of one method against those from the other and determine the best fit line through the data points. Also calculate the correlation coefficient. Comment on your results.

7. Determine iron in different samples using AAS and the spectrophotometric method outlined in Section 4.15. Plot the results of the spectrophotometric method against the results of AAS and determine the best fit line through the data points using the method of least squares. Also calculate the correlation coefficient. Comment on your results.

8. Determine iron several times in the same sample (*i.e.* replicate analysis) using AAS and the spectrophotometric method outlined in Section 4.15. Carry out all the measurements using one method in one day. If you have time you may use both methods and complete all the measurements in one day. Calculate the mean, standard deviation and coefficient of variation for the two methods. Comment on the repeatability of the two methods.

9. Do the same as in the previous project but perform only one analysis on one day using either or both methods. Complete the analysis of all the replicates over a longer period (weeks or months) by analysing on separate days, as you find convenient. Carry out the same kind of statistical analysis of your data as indicated above and comment on the reproducibility of the two methods.

10. Determine the concentrations of Pb and other heavy metals in road drainage waters at urban, suburban and rural sites. Correlate the concentration of each metal with traffic density and test for significance. If information on road traffic density is unavailable you may have to count the average number of cars passing by during a specified time period (*e.g.* 1 h) at each site. Also, investigate the statistical relationships between different metals by computing correlation coefficients for pairs of metals and testing their significance. Which metals other than Pb may be linked to traffic?

4.16.13 Further Reading

APHA, "Standard Methods for the Examination of Water and Wastewater", 16th edn., American Public Health Association, Washington, 1985, pp. 143–179.

D. C. Adriano, "Trace Elements in the Terrestrial Environment", Springer, New York, 1986, pp. 1–42.

A. L. Bandman, B. A. Filov and A. A. Potechin (eds.), "Dangerous Chemical Compounds, Vol. I, Inorganic Compounds of Elements of V–VIII Groups", Chemistry Publishing House, Leningrad, 1989 (in Russian).

US EPA, "1991 Water Quality Criteria Summary", Poster of the Office of Science and Technology, Washington, May 1, 1991.

CHAPTER 5

Soil, Sediment, Sludge and Dust Analysis

5.1 INTRODUCTION

The experiments outlined in this chapter can be used to analyse solid samples such as soils, sediments, sludges and dusts. The experimental procedures for these samples are all quite similar and they involve the extraction of the analyte species and their conversion into a form suitable for chemical analysis. This is in contrast to water samples which, in many cases, can be analysed directly, or with minimum treatment. Greater emphasis is placed on soil analysis than on the other solid samples owing to the important role soils play in biogeochemical cycling of nutrients and pollutants, and their vital role as the medium on which food is grown. The subject material of this chapter is closely linked with that of Chapter 6, which deals with plant analysis.

5.1.1 The Pedosphere

Pedos means soil in Greek and the term *pedosphere* is used to denote the soil cover of the terrestrial part of the earth. *Pedology* is the science of soils, and methods of studying and analysing soils and soil processes are an integral part of it. Soil is the main component of the biosphere, the vital layer of our planet populated by various organisms, from tiny bacteria to plants, animals and humans, and it provides a central link between different biospheric compartments. Soils are the most characteristic feature of the terrestial environment, providing a means of physical support for all terrestrial organisms: plants, animals and humans. They also supply nutrients required by living organisms. Since plants grow on soil, and animals graze on it, nutrients and toxic pollutants in the soil may be transported through the food chain.

Soils have numerous uses but the most vital is their use for growing crops, without which no human or animal could survive. Earliest human civilisations sprung up when man learned how to cultivate the soil and, to this day, agriculture is the most important of all human activities, since without it neither our society, not our race, would be able to exist. Even today, more than 50% of the world population lives on farms.

It is impossible to destroy the whole soil cover of our planet; however, it is possible to degrade the quality of the soil to such an extent that it becomes useless, harmful and even deadly. In fact, many early civilisations (*e.g.* Mesopotamia) died out when the soil cover on which they relied was degraded to a point where it was no longer capable of sustaining agriculture. Agriculture, as well as having many benefits, not least the maintenance of adequate food supplies to feed the world's population, can also have harmful effects on the environment.

The main characteristic of the soil is its fertility, or bioproductivity, and it is this property that is utilised by agriculture. The fertility depends on physical features such as soil depth and texture (*i.e.* proportion of sand, sandy-loam, clay-loam and clay in soil mineral mass), chemical composition (pH, buffering capacity, content of various nutrients) and physicochemical properties (*e.g.* aeration, water absorption capacity). It is possible to restore the fertility of degraded soils, but this involves considerable effort, time and expense. There are many examples where, owing to exhaustion of soil fertility, soils were transformed into deserts (*e.g.* Saharan North Africa, Central Asia). Natural processes of soil erosion can be significantly accelerated by human activity, and this is taking place today over large areas in different parts of the world, leading to a decrease in soil fertility. The mass of soil annually transported by wind and water erosion is estimated to be between 1.2 and 1.5 billion metric tons. Desertification of soils is a serious problem caused by poor agricultural practices. On a global scale, soil availability is not a problem. However, finding fertile soils of the required area and at the required location can be a problem, especially in many developing countries. These problems are due to water shortages and limited financial investments required to maintain soil fertility at the required level.

5.1.2 Soil

Soil is a mixture of minerals (*e.g.* clay, quartz), water, air and living organisms. Soils are formed by the weathering of parent rock and the decomposition of organic matter (the surface litter layer of dead leaves and twigs, fallen branches, *etc.*). Weathering can be either mechanical (*e.g.* abrasion, temperature changes) or chemical (*e.g.* hydrolysis, oxidation). The physical and chemical properties of soils can vary considerably with geographical loction, depending on the parent material, climate, type of weathering processes, *etc*. Soil forming processes are especially rapid in the tropics owing to the warm climate and high rainfall. Soils are classified according the size of the mineral particles:

- *clays* < 0.002 mm
- *silts* 0.002–0.02 mm
- *sands* 0.02–2 mm
- *gravel* > 2 mm

Different soils contain mixtures of the above particles in different proportions, and this is illustrated by means of a graphical representation using a *soil triangle* as shown in Figure 5.1 for the US Department of Agriculture (USDA) classification. A similar triangle is used in the UK with slight differences in the classification of the silt and clay loams.

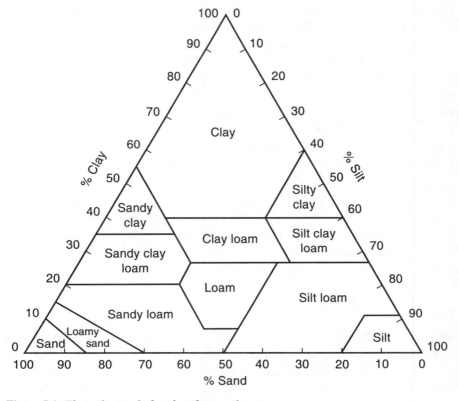

Figure 5.1 *The soil triangle for identifying soil texture*

The chemical composition of silt and sand closely reflects that of the parent rock. In temperate regions these soil fractions are generally dominated by quartz. Calcite, $CaCO_3$, and dolomite, $CaMg(CO_3)_2$, may predominate in areas where the soil was formed from limestone. In arid regions, the entire soil is composed of physically weathered parent material owing to the absence of water necessary for chemical and biological processes. In the tropics, Al and Fe oxides and hydroxides dominate in these fractions as a result of more intense weathering.

The clay fraction is composed mainly of clay minerals, humic substances and oxides of Al and Fe. Clay minerals are crystalline hydroxysilicates composed of sheets, or layers, of atoms. These may contain small amounts of Al, Fe, Mg and other cations. Clay minerals (kaolinite, montmorillonite, *etc.*) exhibit colloidal properties which are responsible for some of their characteristics, such as water retention and plasticity. Clay particles can absorb water to a remarkable degree, although the amount absorbed varies considerably between different clays. The surfaces of clay minerals are electrically charged and they can participate in ion exchange processes. This property is especially important as clays can hold vital plant nutrients (see Section 5.12). Humic substances result from the decomposition of organic matter.

The vertical cross-section through the soil is called a *soil profile*. Soil profiles are fundamental to understanding how the soil was formed in

Table 5.1 *Characteristics of horizons in a soil profile*

Horizon	General description	Specific characteristics
O	Litter zone	More or less decomposed leaf litter
A	Topsoil	Accumulated organic matter intimately mixed with mineral fraction. High content of nutrients and living organisms in the uppermost part, with lower nutrient levels in lower portions. Minerals leach into water and migrate downward through the soil.
B	Subsoil	Mineral layer containing silicate clay, iron, aluminium, humus, gypsum and silica. Precipitation of much of the material leached out of the A horizon. Lower content of organic matter and living organisms.
C	Weathered rock	Virtually lacking in organic materials. It may serve as parent material for the mineral fraction of the soil.
R	Bedrock	Unweathered rock (*e.g.* granite, basalt, quartzite, limestone or sandstone) from which soil may, or may not, have formed.

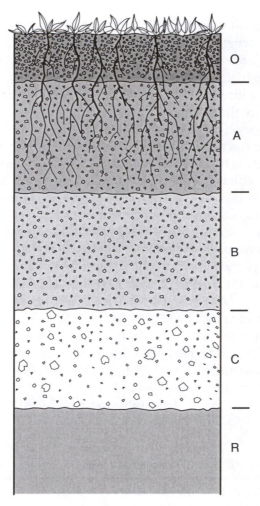

Figure 5.2 *Generalised soil profile*

relation to hydrological and geomorphological processes, and how the soil provides water, nutrients and anchorage for the plants that grow on it. The soil profile is widely used in soil, hydrological and geomorphological studies as a means of identifying and describing different soil types. A typical soil profile is illustrated in Figure 5.2. Different horizons can be distinguished within the soil profile and these have different physical and chemical properties. There are five basic horizons within a soil profile and these are denoted O, A, B, C and R. Characteristics of different horizons are summarised in Table 5.1. The relative thickness of the individual horizons can vary considerably from one soil to another. There are numerous soil types, but most fertile soils can be classified as either grassland soil (chernozem and chernozem-like soils) or forest soil (podzol and podsoluvisols). Chernozems have a deep A horizon and a shallower B horizon which is less abundant in nutrients and cannot support root growth. Podzols and podsoluvisols have a thin A horizon and a thicker B horizon rich in nutrients. Desert soils have poorly developed horizons because of low rainfall needed to promote leaching of minerals. The soil profile is influenced strongly by rainfall, vegetation and topography.

5.1.3 Soil Pollution

The human use of soil can lead to its deterioration by the introduction of various polluting substances, the degradation of soil organic matter and lowering of the fertility of the upper soil layer due to erosion. Degradation of soil organic matter and decreasing fertility due to erosion and over-use have been problems since the early days of agriculture; however, soil pollution has become a problem only since the advent of industrialisation. Physical factors and biological materials can cause soil pollution, but most soil pollutants are chemical substances. Soil quality determines its actual and potential use, *e.g.* the quality of soil intended for agricultural use depends on the content of both heavy metals and pesticide residues. Polluted soil cover may be of little use (*e.g.* construction).

Maintaining good soil quality and minimizing soil pollution and degradation is of fundamental importance and deserves the highest priority for a number of reasons:

- To protect our food supplies from toxic substances which can accumulate in the soil where they can have harmful effects on crops, as well as entering the food chain.
- To protect groundwater supplies, which are an important source of drinking water, from toxic pollutants which may seep through the soil into aquifers.
- To protect surface waters from being contaminated with agricultural chemicals such as fertilisers and pesticides, the former contributing to problems of eutrophication and the latter being bioaccumulated by aquatic organisms and passed down the food chain.

The quality of soil intended for agricultural crop production is of great concern as it must be free from substances harmful to human health. From a public health standpoint, it is most important to ensure that heavy metals and persistent organic compounds such as pesticides are not transported *via* food chains (domestic and wild animals) or directly as crops to humans. Biological pollution by pathogens can sometimes result in the transmission of disease carrying microorganisms from soils to waters and vegetables. However, soils have a remarkable self-purification capacity with regard to biological pollution and they can serve as filters, preventing the input of pathogens into groundwater sources of domestic water supplies.

In the developed countries of Europe and North America, soils are generally polluted by toxic chemicals such a pesticides, heavy metals, radioactive substances and hydrocarbons, and to a much lesser degree by pathogens. However, this cannot be said of the developing nations, where water-borne diseases such as typhoid fever and cholera are still prevalent, and often soil pollution acts as the source of these micro-organisms in surface waters. Soil acidification caused by deposition of atmospheric pollutants (see Section 2.1) and nitrogen fertlisers is a worldwide problem. Excessive inputs of acidifying compounds can overload the buffering capacity of soils and lead to their degradation.

Major categories of soil pollutants are listed in Table 5.2. Many of

Table 5.2 *Major chemical pollutants in the pedosphere*

Pollutant	Typical sources	Comments
Radioactive elements	Accidental discharges from nuclear power plants and industry. Transport of nuclear materials. Nuclear tests.	Very serious long-term soil pollution, especially in the case of the Chernobyl disaster. Standards are in force in many countries for radioactive soil pollution.
Organic chemicals	Agricultural use of herbicides, insecticides and fungicides. Industrial and domestic wastes and wastewaters. Oil pollution from leaking pipelines.	A wide variety of chemicals are discharged (petroleum hydrocarbons, pesticides, *etc.*). Actual and potential effects on animals, vegetation, and human health.
Heavy metals	Wastes from industry and agriculture. Urban drainage. Household wastes. Deposition of airborne pollution.	Many metal pollutants are hazardous to human health, and terrestrial ecosystems (Hg, Cd, Pb, Zn, Ni, *etc.*).
Acids	Acid deposition from the atmosphere. Drainage from mines. Nitrogen fertilisers. Industrial wastes. Land disposal sites.	Acids can harm terrestrial ecosystems, especially forests, owing to soil pH decrease and the release of free Al ions.
Nutrients	Agricultural use of fertilisers and sewage sludge.	May cause excessive accumulation in vegetables and forage. Nitrates may affect human health.

these pollutants are continuously discharged into soils through land waste disposal, inputs from the atmosphere and irrigation by municipal wastewaters on a daily basis. It is the occasional releases of large quantities of pollutants that attract the greatest public concern. A prominent example was the release of radioactive substances from the Chernobyl nuclear power plant in the former USSR in 1986, which caused massive contamination of soils throughout Europe with serious consequences for European agriculture; crops and many animals were declared unfit for human consumption and destroyed as far afield as the UK in the wake of the disaster. Accidental oil leakages from pipelines around the world can also cause enormous soil pollution in a relatively short period of time.

Sanitary and hazardous waste landfill sites are a potential source of soil and groundwater pollution. Huge amounts of domestic, commercial and industrial solid wastes of variable composition are disposed of at landfill sites. Modern secure landfills are supposed to control the leachate that drains from the waste and prevent, or at least detect, leakage. Impervious plastic or clay liners are employed to isolate the waste from the soil underneath and a system of drains collects the leachate in a draining basin, from which it can be transferred to a wastewater treatment plant. Usually several impermeable layers are employed to minimise any leakage, in case one of the layers ruptures. Burrowing animals may chew through the liners, causing leakage. Groundwater and leachate monitoring requirements for landfill sites are listed in Appendix IV. At some sites, landfill gas is collected and treated. Landfill gas consists of methane (30–60%), carbon dioxide (30–50%) and smaller amounts of hydrogen sulfide, carbon monoxide, *etc.* It can be recovered and used for generating heat or electricity.

A major and growing concern is the exposure of children playing in parks, gardens and other soil-covered locations to toxic substances present in the soil, especially toxic metals such as lead (see Sections 5.1.7 and 5.13.6).

Standards giving critical concentrations of various pollutants in soils have been adopted by many countries (see Appendix III). These generally specify threshold concentrations which should not be exceeded in order to avoid any potential health effects or other harmful consequences.

5.1.4 Soil Remediation

Once a soil has been contaminated to such an extent that it presents a hazard, several alternative measures can be taken: the soil may be left in place while limiting the use of the site, it may be excavated and disposed of at a hazardous waste landfill, it may be sealed with plastic liners and capped with a layer of clean topsoil, or it may be remediated. *Remediation* is essentially the process of cleaning up a contaminated soil and this can be done either *in situ* or off-site. Many soils in zones of intensive industrial activity are polluted with heavy metals, persistent organic pollutants, oil products, *etc.*, and remediation may be required for several reasons: to prevent the flow of toxic substances into ground and surface waters and plants, because a new use such as a housing development is intended for the site, *etc.* The type of pollution accumulated in the soil has to be considered when deciding which specific remediation method should be used. Furthermore, the choice of the remediation method may affect the subsequent use of the site. For example, some methods may not only eliminate the pollutants but may significantly change the chemical properties of the soil. Remediation is generally applied for heavy metals and organic pollutants, with the latter usually requiring bioremediation.

The more common methods for remediating contaminated soils are:

- *Washing*. This involves washing or flushing the contaminated soil on-site with a solvent to extract the polluting substance or substances. Solvents used include water, acids, bases, surfactants, complexing agents and organic compounds (petroleum ether, alcohols, *etc.*). For example, HCl, HNO_3 and H_2SO_4 can be used to extract heavy metals. Organic solvents are used to extract oils, tars and petroleum derivatives. Usually, mechanical treatment must first be applied to screen and reduce the soil particle size and enable more efficient extraction. After treatment the solvent is separated from the soil, and the solvent has to be further treated so as not to cause environmental pollution itself. This is done by using the same methods that are employed in wastewater treatment (see Section 4.1.9).
- *Vapour extraction*. Highly volatile substances can be desorbed out of the soil by pumping through horizontal or vertical wells drilled into the soil. Air sparging can also be employed. The area that is being remediated is usually covered with a plastic sheet, and the extracted vapours have to be treated (adsorption on activated carbon, scrubbing, *etc.*) in order to prevent pollution of the atmosphere. Vapour extraction is suitable for petroleum hydrocarbons and chlorinated hydrocarbons.
- *Thermal treatment*. This involves applying heat to the soil in order to volatilise or burn the polluting substances. The soil may have to

be taken to a thermal treatment facility or a plant may be set up on-site for the duration of the remediation project. In one method the soil is heated with steam to remove the pollutants, which are then condensed. This has been applied to insoluble hydrocarbons including benzene, toluene, xylene, kerosene, turpentine and chlorinated hydrocarbons, as well as soluble hydrocarbons such as phenol, methanol, alcohol and isopropanol. High temperature methods involve combustion of the pollutants; for example, chlorinated hydrocarbons are burned at temperatures of around 1300 °C. The extreme conditions of the treatment destroy the organic content of the soil and induce changes in the structure of clay minerals. Therefore, thermally treated soils have to be re-generated with relatively insoluble and bioavailable nutrients before they can be used for cultivation. Thermal treatment is the most efficient method of remediation but also the most expensive due to the fuel requirements.

- *Bioremediation.* This method is based on microbiological processes to reduce the pollutants. Degradation of pollutants by soil microorganisms takes place naturally; however, high contaminant levels may overload their capacity to break down the offending chemicals. In such cases the growth of indigenous bacteria is stimulated by providing optimum conditions (*e.g.* adequate supply of nutrients and oxygen) for their development, or strains of microorganisms specially developed to be pollutant-specific are applied to the soil in required amounts. Bioremediation can be effected on-site, or the soil may be moved to off-site *bioreactors* or *landfarms* where conditions can be carefully controlled to provide an optimum environment for effective bacterial action. Many organic pollutants, including benzene, toluene, xylene, naphthalene, phenol and other aromatic and aliphatic hydrocarbons, can be decomposed by microorganisms. Specially developed organisms have been used effectively to decompose phenol and polynuclear aromatic hydrocarbons (PAH).

- *Phytoremediation.* This is also biological remediation, but the term is used to distinguish it from bacterial methods mentioned above, to which the term "bioremediation" is generally applied. There are two approaches to phytoremediation: phytostabilisation and phytoextraction, and they are applied to heavy metal pollutants. *Phytostabilisation* is simply the application of agrotechnology and it involves growing pollutant-tolerant plants on the contaminated site. This lowers the mobility of pollutants by decreasing wind and water erosion and reduces leaching of heavy metals into ground waters. *Phytoextraction* is the permanent growing of special crops with a high ability to accumulate different metals by root uptake and to concentrate them in the above-ground tissue. The plant biomass undergoes special treatment later on. Phytoextraction is the most useful method for remediating polluted agricultural land in regions with overlapping industrial and agricultural activity. In many countries there are regions where the high content of heavy metals (Hg, Zn, Pb, Cd, Ni, *etc.*) in agricultural soils is due to heavy industrial pollution from one side, and ore deposits, such as sulfides, from the other. Metal availability to plants is one of the main restrictions of phytoextraction effectiveness. For example, Pb has a low solubility at normal pH values owing to adsorption on clay particles and oxides, chemical reactions with carbonates, phosphates and oxides, and complexation with organic matter. These all act to significantly reduce Pb uptake by plants. Synthetic agents such as EDTA salts may be used, either separately or in combination with some organic acids, to increase lead uptake by plants. Increasing metal mobility can also be stimulated by acidifying the soil with nitric acid or the perennial application of liquid manure. However, the application of chemical ligands has implications for biological activity in soils, especially fermentation, which can be either stimulated or inhibited.

Another option is to reduce the mobility and bioavailability of heavy metals, rather than increasing them as in the case of phytoremediation, so that they are not absorbed by plants. This is achieved by liming. Limestone is added to the soil to increase the pH to 7 or higher.

5.1.5 Sewage Sludge

Large quantities of sludge are generated in municipal sewage treatment plants. Approximately 1–2% of the wastewater ends up as a wet sludge and about 2–3 L of sludge are produced per person each day. The total dry solids content of sludge varies between 0.25% and 12%, and 60–70% of these solids consist of organic matter. Sludge is composed largely of highly polluting substances and it undergoes various treatments at sewage works in order to render it suitable for disposal or reuse. Among the more harmful components of sludge are pathogens (viruses, bacteria, protozoa, eggs of parasitic worms), toxic organic substances and toxic heavy metals. Concentrations of pollutants can be extremely high, especially for heavy metals, which can exceed $1000 \, \text{mg} \, \text{kg}^{-1}$, as shown in Table 5.3. The ultimate disposal of sewage sludge includes soil application, landfill, lagooning, incineration and disposal at sea. Owing

Table 5.3 *Typical chemical composition of wastewater sludges*[a]

Component	Concentration
pH	5.0–8.0
Alkalinity (mg L^{-1} as CaCO$_3$)	580–3500
Organic acids (mg L^{-1})	100–2000
Volatile solids (%)	0.8–12
Grease and fats (%)	5–30
Protein (%)	15–41
Cellulose (%)	8–15
Phosphorus (P$_2$O$_5$ %)	0.8–11.0
Nitrogen (N %)	1.5–6
Potassium (K$_2$O %)	0–1
Silica (SiO$_2$ %)	10–20
Al	<4000
As	1.1–230
Cd	1–3400
Cr	10–99 000
Co	84–17 000
Cu	10–2600
Fe	1000–154 000
Hg	0.6–56
Mn	32–9870
Mo	0.1–214
Ni	2–5300
Pb	13–26 000
Se	1.7–17.2
Sn	2.6–329
Zn	101–49 000

[a] All concentrations are with respect to dried sludges (*i.e.* % of total solids, or mg kg^{-1} of total solids) except for pH, alkalinity and organic acids which refer to wet sludge. Heavy metals are all in mg kg^{-1}.

to the high concentration of many harmful substances present in sludge, many countries have banned disposal at sea.

Application of sewage sludge to soils has been widely practiced for several decades. Sewage sludge has been applied to agricultural soils, forest soils and dedicated land disposal sites. Additionally, some sludge is marketed as a soil conditioner and applied to vegetable gardens, lawns, golf courses and parks. In the US, between 10% and 20% of sludge is composted and marketed for commercial use. The purpose of sludge application to soil is two-fold:

- To provide a partial replacement for expensive fertilisers, as sludge contains many of the nutrients required for plant growth.
- To further treat the sludge, as sunlight and microorganisms combine to destroy pathogens and many toxic organic substances.

Typical nutrient contents of stablised sewage sludge are: N 3.3%, P 2.3%, and K 0.3%. In most applications, sewage sludge can provide adequate supplies of nutrients for plant growth; however, P and K contents may be too low for some applications. The suitability of sludge for a particular soil application has to be carefully assessed and several factors have to be considered, including the chemical composition of sludge, national and local standards, the type of land use (agricultural, forest, *etc.*) and other site characteristics (topography, distance to underlying groundwater, *etc.*). Soil application of sludge is regulated by government authorities in order to avoid some of the potential problems listed below:

- *Pathogens* present in sludge could spread disease if there is human exposure.
- *Organic matter* in sludge could cause odour problems and act as a breeding ground for flies, mosquitoes and rodents, thus further increasing the potential risk of disease.
- *Nutrients* (N, P, K) present in sludge may present problems if they are transported to ground and surface waters. Nitrate contamination of groundwater is of special concern, as it may end up in drinking water. Nutrients in soil runoff that end up in surface waters may contribute to problems of eutrophication.
- *Toxic substances*, namely heavy metals and trace organic compounds, may pose a risk to plants, animals and humans. Cadmium is of particular concern, as it can accumulate in plants to levels that could be toxic to humans and animals, but are not harmful to plants themselves. Organic compounds, such as chlorinated hydrocarbons, can be toxic to grazing animals.

Standards for the metal content of sludges intended for agricultural use are given in Appendix III for various countries. Some industrial processes also produce sludge (*e.g.* paper mills) and its composition may differ significantly from that of sewage sludge, as well as varying from one industrial plant to another. In view of what was said above, the importance of monitoring sludge composition is evident.

5.1.6 Sediment

Sediments form in water bodies as a result of the gravitational settling of suspended matter. In fast flowing rivers, small particles may remain suspended, while in still waters, most suspended particles will settle to the bottom. Considerable sedimentation takes place in estuaries, where there is a reduction in the river flow velocity due to mixing with seawater and this can result in the formation of muddy tidal flats. The surface layer of an inter-tidal sediment has generally aerobic conditions, while lower layers are anaerobic (*i.e.* anoxic) due to the depletion of oxygen by microorganisms. These deeper layers are strongly reducing, and chemical species will be present in their reduced forms. As sedimentation is a continuous process, analysis of sediment cores can provide a historical record of the chemical composition of suspended particles. Since these particles originated at the surface of the water body, sediment cores can reveal chemical changes that have occurred in the environment in the past. The depth of the sediment core is proportional to time going back from the present. Toxic substances may accumulate in sediments. For example, organic compounds of low solubility and high molecular mass may be adsorbed onto sediments, where they are consumed by organisms which feed on the bottom of the water body (mussels, scallops, fish, *etc.*). This could represents a potential hazard to humans if these organisms are used as food. Sediments in harbours are especially contaminated with heavy metals, and the following ranges of concentrations have been reported (in $mg\,kg^{-1}$): As, 20–100; Pb, 80–500; Cd, 3–20; Cr, 90–250; Cu, 40–900; Ni, 20–100; Hg, 0.8–12; Zn, 250–2500.

5.1.7 Dust

Dusts in the urban environment are of greatest concern because of the presence of many toxic substances and the potential exposure of children who often play in the streets, school yards, parks, *etc.* The main source of road dust is the deposition of atmospheric aerosol particles. In the urban environment these particles originate mainly from road traffic, emissions from industries, from construction activities and from the flaking of paint. Particles larger than 10 μm in diameter are deposited quite rapidly to the earth's surface under the influence of gravity. While the major route of exposure to atmospheric aerosols is by inhalation, exposure to toxic substances in road dust is mainly by ingestion. Children playing in the street may tranfer dust particles onto their hands, and these could subsequently be ingested. Children often tend to pick up objects from the ground and chew them. This abnormal craving for unnatural foods in children is termed *pica*. Furthermore, in many developing countries, where urban pollution levels are especially high, a large number of children live on the roadside and spend most of the day on the streets in conditions of poor hygiene. Road dust may also be re-suspended into the air and then inhaled. Pollutants present in road dust can enter surface waters as road drainage water after rain showers. Furthermore, road dust can be carried into homes on items of clothing. The pollutant content of dust in homes, schools and places of work (factories, offices, *etc.*) can be as high as that of road dust, as shown in several studies. Greatest concern has been expressed about lead (Pb) in road dust, but other heavy metals are routinely observed in road dust: Cd, Cr, Cu, Ni, Zn, *etc.*

5.2 SAMPLING AND SAMPLE PREPARATION

The methods described in this chapter are intended mainly for soil analysis but they are equally applicable to dust, dried sediment and dried sludge. Soil tests are widely used in agriculture in order for farmers to determine the amount and type of fertiliser (N, P, K, Mg) and/or lime to apply in order to obtain high crop yields. The most common tests involve routine nutrient (N, P, K) measurements. Soil test kits and portable laboratories are commercially available (*e.g.* Hach) for on-site analysis. For farming purposes, field test are carried out once every 2 or 3 years.

5.2.1 Sampling

Sampling of inhomogeneous media such as soils, dusts, sludges and sediments presents some difficulties, and the sampling programme should be carefully planned from the outset if representative and meaningful results are to be obtained. The objectives of the monitoring programme should be clearly defined as these will determine when and where to sample, how many samples to collect and which extraction procedures and analytical methods to employ. The depth of soil sampling depends on the aims of monitoring. In studies of agricultural crop nutrition the top 20–25 cm is usually adequate as this is the zone where much of the root uptake of nutrients takes place. Analysis of the upper layers is also relevant in understanding soil interactions with other environmental compartments and the pathways of pollutants between them. For example, atmospheric dusts particles, fertilisers, pesticides and sewage sludge are all deposited onto the soil surface, while water and wind erosion of upper layers can transfer pollutants to surface waters and the atmosphere, respectively. However, samples from the lower horizons should also be analysed in order to assess the migration of substances within the soil profile. In addition to spatial and depth considerations, the time of sampling is an important factor in soil analysis. For example, the availability of many nutrients such as P increases during the spring and summer months and there is some evidence that inorganic N levels are higher in the spring. It is best to sample for nutrient analysis after harvest and before application of fertiliser, but you may also sample during the growing period.

Soils are variable both horizontally and vertically. A high degree of spatial variability in soil composition may exist even within quite a small area. Sampling should take into account the inherent variability of soil and other materials (dust, *etc.*). Very often the degree of spatial variability is not known beforehand and a number of samples have to be taken in order to obtain representative results. In practice, 3–5 soil or dust samples from an area of 10–20 m^2 should be regarded as a minimum. If site variability is to be evaluated with high precision, 15–30 samples may be required. This would enable statistical methods (*e.g.* factor and cluster analysis) to be applied. Soil variability is classified into three categories:

- Micro-variation (0–0.05 m)
- Meso-variation (0.05 m–2 m)
- Macro-variation (> 2 m)

Each field should be sampled separately. Parts of the field that differ in the appearance of soil or crop, or in the history of fertiliser or lime application, should also be sampled separately. Soil sampling can be either *random* or *systematic*, and several approaches may be adopted. Some of these are given below:

- Employ a random number of sampling points spatially distributed over the site area.
- Subdivide the area into grids and sample at regular intervals along definite grid cells.
- Traverse the area in a zig-zag pattern to provide a uniform distribution of sites.
- Select a test lot of small area within the larger field and apply random sampling to this test lot, rather than applying random sampling to the whole field.

In the last method the results are assumed to be representative of the entire field. The random sampling within the test lot gives an indication of the variability, which is assumed to be the same throughout the field. Still other variations exist. In *stratified random sampling* the entire area is divided up into smaller sections, and random sampling is applied within each section. This method gives greater precision than simple random sampling. In *composite sampling*, often employed in field surveys, individual samples are bulked into one sample. This method is intended to give an average value; however, it is not suitable for assessing spatial variability. The validity of the average value depends on the number of individual samples bulked and the adopted technical procedure, and this has to be standardised by taking samples of equal size from a specific depth.

If samples are collected from a heavily polluted site with disturbed or dead vegetation and visible staining due to chemical spills, then samples from an uncontamined adjacent area with same soil type and healthy vegetation should also be sampled for comparison. Samples should not be taken near the roots of large trees, near roads, foundations of buildings or other constructions. Samples should also not be taken near piles of manure, compost, lime or harvested crops.

Before taking samples you should remove leaves, grass and any large external objects. Surface samples can be collected with stainless steel spoons, scoops, spades or shovels, and stored either in plastic bags or glass or plastic containers. If wet samples are to be analysed, these should be stored in glass containers filled to the top (*i.e.* leave no airspace). You may obtain small samples of soil profiles from shallower depths using stainless steel cork borers available in the laboratory.

Soil profile samples are taken with soil corers. The simplest of these are metal or poly(vinyl chloride) (PVC) cylinders between 2 and 12 cm in diameter, the length of which depends on the depth of sampling. Corers can easily be improvised from PVC drainpipes by cutting a desired length of pipe (*e.g.* 30 cm) and sharpening the edges of one end of the tube. The corer tube is hammered into the soil. For the majority of analyses the gross disturbing of the soil during sampling is acceptable. The *auger* (Figure 5.3), simple in design and easy to use, is suitable for sampling hard soils. It consists of a sharpened spiral blade attached to a central metal rod which can be screwed into the soil. The sampler is screwed to the desired depth and the sample withdrawn. The sample should be transferred to a pan or tray for homogenising before storing in sample bags or bottles. Other, similar, soil samplers are also commercially available. The sampler should be cleaned with laboratory water between samples to avoid any cross-contamination. Corers and augers are practical for sampling many soil types but are less useful for water-logged or dry sandy soils. Special precautions should be taken when sampling peat and bottom sediments in order to minimise oxidation. For example, Fe is present mainly as reduced Fe^{2+} in anaerobic environments and exposure of the sample to the air will oxidise this to Fe^{3+}. If speciation is to be performed and different forms are to be analysed, it is essential that the conditions representative of the environment are

Figure 5.3 *Auger soil sampler*

preserved during transport, storage and analysis. In this case the sample should be immediately sealed within the core tube and handled in an inert atmosphere in a glove box upon arrival in the laboratory. It is impossible to sample stony soils with an auger and it is necessary to collect a large amount of soil to ensure that the sample is representative. The sampling of "concrete" horizons of many tropical and subtropical soils may require a shovel, or even a pick. Samples can be stored within the core tubes, or removed and stored in sealed plastic bags in a refrigerator at 4 °C so as to minimise bacterial activity. Samples should be handled with latex gloves.

Equipment used for sampling sediments varies depending on the location, water depth and texture of sediment. Scoops can be used near the shore, or they may be attached to poles and used from boats. Corers are used for softer sediment. The sample is extruded from the corer into a tray or pan and homogenised before storing in a sample container. The Eckman sampler (Figure 5.4) is the most widely used sampler and it is suitable for mud, sand and silt-type sediments. Other common collectors for sediment samples include the Peterson and Ponar samplers. Dust samples can be collected using a brush and dust pan, or a spatula, and stored in disposable polyethylene bags. You should collect at least 5 g of each dust sample.

It is important to avoid contamination of samples. When metal analysis is to be carried out, metal containers should not be used for

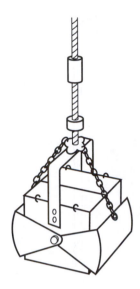

Figure 5.4 *Eckman sediment sampler*

collecting, mixing or storing samples. Plastic containers should be used instead.

You should record the time and place of sampling, and other relevant information concerning the soil and other sampled materials, as well as the site. This should include sampling depth or horizon, colour, consistency and texture of the soil, drainage conditions and details of the vegetation. When sampling sediments the water depth should be recorded. Any evidence of human influence on the terrestrial or aquatic ecosystem should be noted, for example grazing, fertilising, biomass burning, contamination or water eutrophication (*i.e.* algal blooming), *etc*. The exact location of each sampling site should be recorded in terms of a recognised grid reference system, so that information can be obtained regarding the site from maps (topographical, geological, hydrological features, *etc*.) or other sources. Precise latitude and longitude may be obtained using a relatively inexpensive portable GPS (Global Positioning System). When sampling sludges, a detailed description of the treatment technology used at the sewage works should be given. Dust samples should be characterised according to the site: pavement, road, car-park, indoor, *etc*. For street dusts, you should record the sampling area in m^2, traffic density, distance from the road, wind speed, the time of the last rainfall and its intensity, *etc*. Traffic density should be in terms of traffic counts, but if this is not available you may describe it qualitatively (*e.g.* low, medium, high). Such information can be recorded on standardised site data forms.

Table 5.4 *Recommended procedures*

Required analysis	Recommended procedure
pH, Eh, reduced ionic states	Immediate field analysis if possible. Otherwise, laboratory analysis of samples specially preserved to maintain integrity of field conditions.
Extractable nitrogen (NH_4^+, NO_2^- and NO_3^-), organic N and P fractions, humus fractions and all peat fractions	Laboratory analysis of fresh material or after short storage of 1–2 days at 4 °C
SO_4^{2-}, K, Na, Ca, Mg, Fe, Zn, Cd and other heavy metals, cation exchange capacity, nitrogen mineralisation capacity	Laboratory analysis after air drying, generally at 40 °C
Moisture content, loss-on-ignition, total elemental anlysis	Drying at 105 °C

5.2.2 Drying

With minimum transport and storage times, the changes in the concentrations of extractable nutrients or even of some organic constituents may not be significant. Samples may be *air-dried* by exposure to ambient air or *oven-dried* in an oven at 105 °C. Dried samples can be kept in a desiccator. Recommendations for different analyses are given in Table 5.4. Redox potential and pH should be determined as soon as possible, preferably in the field. It should be kept in mind that many of the treatment procedures may affect the sample composition, and consideration should be given to the physical and chemical properties of each analyte before deciding on the precise treatment. For example, some analytes, especially many organic compounds, are volatile or thermally unstable, and would be lost from the sample during oven drying. Also, contamination may be introduced at each stage of the analytical procedure.

5.2.3 Comminution

Soil may contain large aggregate grains and samples have to be ground and homogenised to break down the size of soil particles and achieve a homogeneous sample before they can be sieved. This is achieved by mechanical means such as grinding, crushing, milling, pulverising and other similar practices, often called *comminution*. High-speed grinding mills, or powered mortar and pestle, are commonly used. In the absence of these devices you may crush the soil manually using a mortar and pestle.

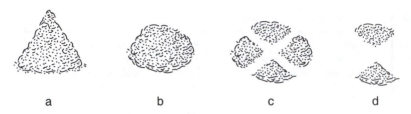

Figure 5.5 *Coning and quartering:* (a) *form sample into a cone,* (b) *flatten the cone,* (c) *quarter the sample,* (d) *discard two opposite quarters*

Further size reduction is achieved by passing through a sieve. Dried and ground samples (soil, dust, sediment, sludge) should be thoroughly mixed before sieving. This can be done by *coning and quartering* as follows. Shake the sample vigorously inside a sealed container and pour onto a clean glass surface in the form of a cone. Flatten the cone and form a circular layer of material. Divide the circular material into four even parts using a clean spatula and discard two opposite quarters (Figure 5.5). Combine the two retained quarters, and repeat the coning and quartering until the required sample size is obtained. Samples are then passed through a coarse 2 mm sieve, which has been adopted as an international standard in soil analysis.

5.2.4 Questions and Problems

1. Why do you think that it is important to prevent soil pollution?
2. List the main categories of soil pollutants and their sources.
3. Although unleaded petrol was introduced in developed countries of Western Europe and the US some time ago, why is urban lead pollution still a major issue in these countries?
4. How is soil texture characterised?
5. Discuss the advantages and disadvantages of applying sewage sludge to agricultural soils.

5.2.5 Suggestions for Project

Design standardised data forms for soil, sediment, sludge and dust sampling and analysis. These could consist of columns for entering the analytical results for the various components, as well as other relevant information such as time and place of sampling, sampling method (*e.g.* type of sampler), site description, visual observations, *etc*. There will be some differences, and similarities, between the forms for different sample types. These forms could be photocopied and used during monitoring programmes.

5.2.6 Further Reading

D. W. James and K. L. Wells, Soil sample collection and handling: techniques used on source and degree of field variability, in "Soil Testing and Plant Analysis", ed. R. L. Westerman, SSSA, Madison, 1990, pp. 25–44.

G. T. Patterson, Site description, in "Soil Sampling and Methods of Analysis", ed. M. R. Carter, Lewis, Boca Raton, 1993, pp. 1–3.

J. Crepin and R. L. Johnson, Soil sampling for environmental assessment, in "Soil Sampling and Methods of Analysis", ed. M. R. Carter, Lewis, Boca Raton, 1993, pp. 5–24.

P. R. O'Leary, P. W. Walsh and R. K. Ham, Managing solid waste, *Sci. Am.*, 1988, **259**, No. 6, 18–24.

R. G. Peterson and L. D. Calvin, Sampling, in "Methods of Soil Analysis: Part 1, Physical and Mineralogical Properties", 2nd edn., ed. A. L. Page *et al.*, ASA & SSSA, Madison, 1982, pp. 33–51.

K. H. Tan, "Soil Sampling, Preparation and Analysis", Dekker, New York, 1995.

5.3 DISSOLUTION AND EXTRACTION

Nutrients and pollutants can exist in different forms in soils, sediments and sludges and various methods are used to process the sample and convert the analyte into a form suitable for analysis. The adopted procedure depends on the nature of the analyte, the analytical technique that is to be used to measure the analyte and the kind of information that is being sought (*i.e.* total or available elemental concentration). The following categories of analyte can be determined:

- Total element
- Total organic
- Total inorganic
- Soluble ions
- Exchangable ions

Organic pollutants (dioxin, benzene, toluene, *etc.*) are generally present at very low levels, $\mu g \, kg^{-1}$ or below (*i.e. micropollutants*), and they are extracted into organic solvents (*e.g.* hexane). Extraction can be achieved by shaking the sample with the solvent and leaving the two phases in contact for several hours or by means of Soxhlet extraction. Headspace analysis and purge-and-trap techniques can also be used to analyse volatile organic compounds in soil and other solid samples.

Heavy metals are usually present in the $mg \, kg^{-1}$ range and the first step in their analysis involves dry ashing, wet ashing or fusion. *Dry ashing* involves heating the sample in a muffle furnace at 400–600 °C for 12–15 h. The resulting ash is then dissolved in a dilute acid solution. During dry ashing, volatile components are lost. *Wet ashing* consists of heating the sample in an oxidising acid solution (nitric, perchloric, *etc.*). Samples may also be fused with a fusion reagent (*i.e.* flux). *Fusion* involves mixing the sample with the flux (usually sodium carbonate) in a crucible and heating to high temperatures. After the fusion is complete, the crucible is cooled and the melt dissolved in acid. The crucible is heated to high temperature until solution is complete. Fusion is useful for substances that are only partially soluble in acids. Partially soluble substances may also be dissolved in acids but at high pressure inside stainless steel containers called "bombs". These are heated to 150–180 °C and can withstand pressures of 80–90 atm.

The analysis of individual components that could be incorporated into food chains is of more interest from the standpoint of environmental protection than the total elemental analysis of soils or other materials. Therefore, the determination of *available* fractions is more relevant. However, since an occasional need for total analysis may arise, procedures for bringing soil components into solution using hydrofluoric acid digestion are included in this chapter.

Most of the methods discussed here are intended for use with soils, but they can also be applied to other solid materials (sediment, sludge, *etc.*). Procedures for extracting soils and other solid materials are relatively simple; however, organic soils, sediments and almost all sludges present some difficulty. Samples with a large content of soil organic matter should first be ignited. Soils and other materials with high organic contents, such as peat and plant litterfall, as well as sludge, are in some respects more similar to plant materials and in many instances it may be more appropriate to use the methods outlined in Section 6.3.

Standard wet ashing methods for the determination of total element concentrations involve extremely dangerous hydrofluoric and perchloric acids. Skin contact and inhalation of vapours should be avoided at all costs, and some teaching laboratories may not be adequately equipped for handling these acids. Hydrofluoric acid must be handled with great care as it can cause serious and painful burns in contact with skin. The effects of HF may not be apparent until several hours after exposure. In case any HF does come into contact with skin, it should immediately be washed off with copious amounts of water and, if necessary, treated with a dilute calcium ion solution. Hydrofluoric acid dissolves silicates and it is this property that makes it especially useful in soil analysis; however, it also implies that HF will dissolve glass and therefore it should be handled only in Teflon (PTFE) or platinum equipment. It also attacks steel. Hydrofluoric acid should always be used in a fume cupboard. Specialised fume cupboards are required for both of these acids. Since HF attacks glass, fume cupboards for use with HF should not contain glass parts, or the glass windows have to be protected with transparent plastic sheets. Perchloric acid ($HClO_4$) is a very strong oxidising agent and it can give explosive reactions when hot acid comes into contact with organic and easily oxdised inorganic substances. The condensation of perchloric acid vapours on walls and shelves presents a grave danger. A fume cupboard lined with glass or stainless steel, and equipped with a shower system for washing down the walls with water, is required for using perchloric acid. Also, the fume cupboard should have its own exhaust system, isolated from other exhausts. When using $HClO_4$ the following advice should be taken:

- Work on a small scale.
- Never allow a digestion flask to dry out and beware of charring when accompanied by loss of fluidity.
- Do not digest materials with a high content of fat or oily substances, or any material containing hydroxy groups (carry out a preliminary oxidation with nitric acid initially at room temperature if in doubt).

The inclusion of sulfuric acid in the digestion mixture will prevent the flasks drying out. Most of the reported explosion incidents have been associated with perchlorates; metal perchlorates can detonate violently if heated to dryness, particularly in the presence of organic matter. Fume cupboards and wooden benches have been known to explode when subjected to friction.

Electrically heated block digestors are commercially available for heating the digestion tubes (Figure 4.10). These enable constant digestion conditions to be reproduced and several samples can be processed simultaneously.

Fusion requires the use of platinum crucibles, which can be quite expensive. Some teaching laboratories may not be adequately equipped for using the highly hazardous acids or they may not be able to afford the platinum crucibles. We have therefore given alternative procedures for individual experiments involving less hazardous materials (but all acids are dangerous and caution should be exercised!) and equipment readily available in most laboratories wherever possible. These methods, although not as effective as the fluxes or $HF-HClO_4$ digestions, are tried and tested, and have been widely used in environmental analysis. If the appropriate equipment is available you may use the more hazardous procedures, but only under strict supervision of trained personnel.

Soluble ions can easily be determined by analysing the aqueous phase of equilibrated soil/water mixtures. What is often of interest to soil scientists is the plant availability of nutrients. Simple empirical methods have been developed to simulate the action of water and plant roots on the uptake and movement of various ions. These availability tests involve chemical extraction to measure mobile forms of elements in soil. Ions (generally cations) are displaced from adsorption sites on the soil particles by other replacing ions added in great excess. The displaced ions are then determined. Monovalent ions, especially H^+, are held more strongly than larger divalent ions. However, divalent ions generally have higher displacing power. Monovalent ions are more widely used in practice. Ammonium acetate and acetic acid are common extractants. These methods generally measure exchangeable and soluble ions and can be used to derive an *availability index*. The availability index is used, together with other considerations (*e.g.* crop yield), to make fertiliser recommendations for different soils. The determination of water-soluble ions is described in Section 5.7.

5.3.1 Fusion

5.3.1.1 Sodium Carbonate Fusion. *(a) Methodology*. Silicates are fused with anhydrous sodium carbonate at high temperature (1000–1200 °C) inside platinum crucibles. A large excess of sodium carbonate is required to ensure complete dissolution, especially when aluminium and iron oxides are prominent in soil, dust or sediments. An oxidising flux must be maintained by including either sodium peroxide or potassium nitrate, because under reducing conditions Mn and Fe can react with platinum. Sodium peroxide also attacks platinum and it should be used in small quantities. The melt is dissolved in hot hydrochloric acid.

(b) Materials
- Platinum crucible
- Meker burner
- Water bath
- Sodium carbonate, anhydrous
- Hydrochloric acid, concentrated
- Sodium peroxide (Na_2O_2)

(c) Experimental procedure. Weigh approximately 0.6 g sodium carbonate and 0.05 g sodium peroxide into a clean platinum crucible. Add 0.10 g of finely ground, oven-dried, sample and mix with a glass rod. Spread a further 0.30 g sodium carbonate on top of the mixture and cover with the lid. Support the crucible upon a platinum or silica triangle.

Heat over a Meker burner with a low flame initially and with the lid slightly open. Increase the burner flame intensity gradually until the contents are liquefied; then continue heating for another 15–20 min. Remove the crucible using platinum-tipped tongs and gently swirl so that the molten material is distributed around the crucible walls. Cool and then immerse the crucible and contents in about 50 mL water inside a 250 mL polypropylene beaker. Add 3 mL concentrated HCl, cover with a watch glass and heat for 30 min on a boiling water bath. Transfer the solution into a 100 mL volumetric flask and make up to mark with water when cool. Prepare a blank by following the same procedure with the reagents but without the sample.

Notes

1. Avoid metal or pipe-clay triangles because they can damage the platinum crucible. Brass or iron tongs should not be used as they will stain the platinum.
2. Remember that is important to maintain oxidising conditions during the fusion. Only the top of the blue cone of the flame should be in contact with the crucible. The Meker burner is designed to provide a very hot oxidising flame and it is to be preferred to a muffle furnace.
3. If the melt is green, this means that manganate is present. In this case, add a few drops of ethanol and mix before adding HCl. This will prevent any attack on the platinum.
4. Platinum crucibles are quite expensive and they should be kept clean, polished and in proper shape. The rolling of platinum crucibles to crack away the fused mixture, although recommended by some, should be discouraged. Moulds can be used to re-shape dented platinum crucibles.

5.3.1.2 Sodium Hydroxide Fusion. (a) *Methodology.* Silicon is fused with sodium hydroxide at a lower temperature (600 °C) than with carbonate. Aluminium can also be determined by fusing with NaOH, but in general the carbonate fusion is preferred for Al and other elements in soil and other solid materials. Nickel crucibles are used since hydroxides can react with platinum. The melt is dissolved in water before transferring to an acidic solution.

(b) Materials
- Clean nickel crucibles
- Meker burner
- Sodium hydroxide, 25% w/v
- Hydrochloric acid, 1 : 1 (*i.e.* equal volumes of concentrated HCl and water)

(c) Experimental procedure. Pipette 15 mL of NaOH solution into a clean nickel crucible with a capacity of about 75 mL and carefully evaporate to dryness. Weigh 0.10 g of finely ground oven-dried sample and transfer into the crucible. Cover the crucible and heat over the burner to about 600 °C for 10 min. Remove the crucible and swirl the melt around the crucible walls to spread evenly. Allow to cool and add 50 mL of water. Leave overnight to dissolve the melt. Pour the crucible contents into a plastic beaker and add 30 mL 1 : 1 HCl. Allow to cool, transfer into a 100 mL volumetric flask and dilute to the mark with water. Prepare a blank by following exactly the above procedure with the reagents but without the sample.

Notes

1. Nickel crucibles should be cleaned with dilute HCl before use.
2. Potassium hydroxide (KOH) may be used instead of NaOH.

5.3.2 Acid Digestion

The most effective digestion involves hydrofluoric and perchloric acids. Other acid digestions (HNO_3, HCl, H_2SO_4, aqua regia) are also used, but these are not suitable for "total" element analysis. Instructions for some of these digestions are given in the text, where appropriate.

5.3.2.1 Hydrofluoric–Perchloric Digestion. *(a) Methodology.* Silicates are digested in a mixture of hydrofluoric and perchloric acids. The former dissolves the silicates while the latter breaks down the organic matter. These two acids are a standard mixture for treating soil and other solid materials. Platinum or PTFE crucibles can be used. A suitable fume cupboard must be used for this digestion (see Section 5.3 and Note 4 below). Heating on a sand bath is preferred.

(b) Materials
- Hot plate or sand bath
- Platinum or PTFE crucibles
- Perchloric acid, 60%
- Hydrofluoric acid, 40%
- Nitric acid, concentrated
- Sulfuric acid, concentrated

(c) Experimental procedure. Weigh 0.10 g of finely ground oven-dried sample and place in a platinum or PTFE crucible. Moisten with a little water. Add 1 mL 60% $HClO_4$ and 7 mL 40% HF. Cover the crucible with the lid but leave a little gap open and digest slowly for 2 h. Allow to cool

and wash the inside of the lid into the crucible with water from an aspirator bottle. Return to heat and evaporate the contents until fumes of $HClO_4$ appear. Cool, add 1 mL H_2SO_4 and heat again to drive off $HClO_4$. Cool and dilute with a little water. Filter into a 100 mL volumetric flask and dilute to the mark with water. Prepare a blank by following the same procedure with all the reagents but without the sample.

Notes

1. If a large residue remains after the first digestion, repeat the digestion or use a smaller sample weight.
2. Only PTFE or platinum vessels should be used when handing HF digests.
3. Since PTFE softens at about 300 °C, sand baths and hot plates should be kept at a low setting.
4. Since HF attacks glass, the glass panes of fume cupboards must be covered with a transparent plastic sheet or painted over with clear varnish. The fume cupboard should also be equipped with a shower to wash out $HClO_4$.
5. The recommended sample weights should not be exceeded if calcium carbonate is the main mineral component of the soil, since H_2SO_4 is used in the procedure. Carbonates are prominent non-silicate minerals, especially in calcareous soils.
6. If the amount of organic matter is high, add 3 mL HNO_3 and 1 mL $HClO_4$ to the crucible and heat until fumes of $HClO_4$ appear. Only then add HF and proceed as above.

5.3.3 Dissolution of Individual Elements

Recommended procedures for determining the total concentration of individual elements are specified in Table 5.5. The recommended weights are only guidelines, and these may be varied depending on the concentration in the sample. The reagent amounts should also be scaled up when using larger sample weights. The final acid strength must be the same in both standards and samples. Blank solutions should also be prepared and analysed, especially when determining low concentrations. These contain all the reagent but not the sample. If the organic matter is high it should be destroyed by ignition at 450 °C before proceeding with the fusion. This may give low recoveries of Cl, P and S. Prior treatment with a nitric–perchloric acid mixture is required before the hydrofluoric–perchloric digestion if the amount of organic matter in the sample is high. A reference soil of known chemical composition, if available, should be analysed with each batch of samples. Follow the dissolution

Table 5.5 *Recommended dissolution procedures for individual elements. Make up final solution to 100 mL. Complete analysis using methods specified in experiments for the individual elements in water*

Element	Weight of sample	Dissolution procedure	Analysis method	Comments
Al	0.1 g	Na_2CO_3 or NaOH fusion	AAS as in Section 4.16	–
Ca	0.1 g	Any fusion or HF–$HClO_4$ digestion	AAS as in Section 2.5.1	Take < 0.1 g of sample for calcareous soils if using digestion
Cl	0.1 g	Na_2CO_3 fusion	Titration as in Section 4.12 or IC as in Section 2.4.6	Adding a small amount of $NaNO_3$ improves flux. Dissolve melt in water (not in HCl) and adjust to pH 8 with 0.1 M H_2SO_4
Mg	0.1 g	Any fusion or HF–$HClO_4$ digestion	AAS as in Section 2.5.1	–
P	0.1 g	Any fusion	Spectrophotometry as in Section 4.11.3	Follow fusion by boiling with H_2SO_4 to convert to orthophosphate
P	0.05 g	Acid digestion with 2 mL HNO_3 + 1 mL $HClO_4$ + 0.5 mL H_2SO_4	Spectrophotometry as in Section 4.11.3	–
P	0.2 g	Acid digestion with H_2SO_4–H_2O_2 for 2 h	Spectrophotometry as in Section 4.11.3	–
K	0.1 g	Any fusion or HF–$HClO_4$ digestion	AAS as in Section 2.5.1	Fusion may give high blanks
Si	0.1 g	NaOH fusion	AAS as in Section 4.16	–
Na	0.1 g	HF–$HClO_4$ digestion	AAS as in Section 2.5.1	–
S	0.5 g	Any fusion	Turbidimetry as in Section 4.14 or IC as in Section 2.4.6	Mix soil with 2.5 g Na_2CO_3 and preheat at 450 °C for 30 min before fusion. Addition of 0.2 g $NaNO_3$ or 0.1 g Na_2O_2 improves flux
Zn	0.1 g	Any fusion or HF–$HClO_4$ digestion	AAS as in Section 4.16	–
Heavy metals	1.0 g	HF–$HClO_4$ digestion	AAS as in Section 4.16	–

and analytical procedures specified in Table 5.5 to determine elements of interest.

5.3.4 Extractions

5.3.4.1 Ammonium Acetate (pH 7) Extraction. This extractant is suitable for determining available Na^+, K^+, Mg^{2+} and Ca^{2+} in neutral and acid soils and may also be used for Mn^{2+}. It is also used in the determination of cation exchange capacity (CEC).

(a) Materials
- Ammonium acetate solution (1 M), pH 7, prepared by adding 575 mL glacial acetic acid and 600 mL aqueous ammonia solution (specific gravity $= 0.9$) to about 200–300 mL water in a large aspirator and mixing. Dilute to 10 L with water and mix thoroughly. Adjust the pH to 7.00 $\pm$ 0.05 if necessary by adding dropwise acetic acid or ammonia.
- Filtration apparatus (*e.g.* Buchner funnel) and filter paper, Whatman no. 44

(b) Experimental procedure. (i) Soils, dusts and sediments. Weigh 5 g air-dried and sieved (<2 mm) sample and place in a 250 mL bottle. Add 125 mL of the above ammonium acetate solution. Cap and shake on a rotary shaker for 1 h. If a shaker is not available, shake occasionally by hand. Filter the solution, discarding the first 5–10 mL. Filter the rest of the solution into a polyethylene bottle. Process two blanks containing extractant only. Determine the cations using AAS or a flame photometer as in Section 2.5.1.

(ii) Peat, organic bottom sediments and sludges. Weigh 25 g fresh peat, sediment or sludge and place in a pyrex beaker. Add approximately 100 mL of the above ammonium acetate solution. Stir and leave for 10 min. Filter the supernatant liquid into a bottle but do not discard any filtrate. Wash the peat onto the filter paper with a further 50 mL of ammonium acetate solution and allow to filter. Continue to filter by adding successively smaller aliquots of ammonium acetate solution until the filtered leachate volume reaches 250 mL. Process two blanks with extractant only. Determine the cations using AAS or a flame photometer as in Section 2.5.1.

Notes
1. Determine the moisture content at the time of weighing, using a separate sample aliquot, in order to correct the results to a dry basis.

2. Each addition of the ammonium acetate solution should saturate the sample.

5.3.4.2 Ammonium Acetate (pH 9) Extraction. This is a possible alternative for highly calcareous soils since relatively little calcium carbonate is dissolved at this pH. It may be used for the same elements as the pH 7 solution. Carry out the extraction in a fume cupboard to remove ammonia fumes.

(a) Materials
- Ammonium acetate solution, pH 9. Prepare 10 L using the same procedure as outlined above in Section 5.3.4.1 but use 740 mL NH_3 solution. The volume of glacial acetic acid is the same as above. Adjust to pH 9.00 ± 0.1 if necessary.
- Filtration apparatus and filter paper, Whatman no. 44

(b) Experimental procedure. Extract by following precisely the procedure outlined above in Section 5.3.4.1. Analyse in the same way.

5.3.4.3 Acetic Acid (2.5% v/v) Extraction. Acetic acid extraction should be used only for non-calcareous soils. This extraction procedure is suitable for most cations. The extraction conditions are less favourable for P extraction except on sandy soils. This extraction is not based on an ion exchange mechanism like the ammonium acetate method, but the two methods give comparable values in acid soils. Acetic acid essentially recovers only acid soluble ions.

(a) Materials
- Glacial acetic acid solution prepared by diluting 250 mL glacial acetic acid to 10 L with water and mixing thoroughly
- Filtration apparatus and filter paper, Whatman no. 44

(b) Experimental procedure. Follow exactly the procedure as given above for ammonium acetate (Section 5.3.4.1) but use the dilute glacial acetic acid solution instead.

5.3.5 Questions and Problems

1. Why are fusion, or very strong acid digestion, required for the complete dissolution of soil components?
2. What are the different categories of analyte that can be determined in a soil?
3. Why is it usually not necessary to know the total concentration of an element in a soil?

4. What is the purpose of the various extraction procedures (*e.g.* acetic acid) and what could the results of such tests be used for?
5. What safety precautions would you take if you wanted to use HF or $HClO_4$?

5.3.6 Suggestions for Projects

1. If you have the appropriate facilities in your laboratory, determine the concentrations of one or more heavy metals (*e.g.* Fe, Mn) using both the HF–$HClO_4$ digestion and the nitric acid digestion (as described in Section 5.13.16) in several soils and comment on any differences. You may do the same with dust, sediment and sludge samples.
2. Determine the available K^+, Na^+, Ca^{2+} and Mg^{2+} in different soils using the extraction procedures given above (*i.e.* ammonium acetate, acetic acid) and compare to the water soluble concentrations as determined in Section 5.7.

5.3.7 Further Reading

S. E. Allen (ed.), "Chemical Analysis of Ecological Materials", 2nd edn., Blackwell, Oxford, 1989, pp. 24–45.

C. H. Lim and M. L. Jackson, Dissolution for total elemental analysis, in "Methods of Soil Analysis: Part 1, Physical and Mineralogical Properties", 2nd edn., ed. A. Klute *et al.*, ASA & SSSA, Madison, 1982, pp. 1–12.

5.4 PHYSICAL PARAMETERS

5.4.1 Introduction

Physical parameters are important in defining the soil texture, structure, friability and type. Soil structure refers to the way soil particles clump together, while *friability* is the ability of a soil to crumble. Good soils tend to form clumps that crumble easily, whereas poor soils (*e.g.* sandy soils) do not clump and therefore lack structure. On the other hand, sandy soils are friable while clays are not. Physical properties of soils are closely related to the chemical composition, and they can influence the mobility and pathways of nutrients and pollutants within the soil as well as many physicochemical and biological processes, including plant growth. The way particles are arranged within a soil can influence the uptake of air and water. Air and water can occupy spaces between soil particles and participate in essential soil processes. The air provides oxygen required by soil organisms and plant root cells, and it can influence chemical reactions (oxidation–reduction). Water can supply nutrients to soil organisms and plant roots, it can leach various substances from the soil, provide a medium for aqueous chemistry to take place (hydrolysis, complexation, precipitation, *etc.*) and transport pollutants through the soil system and into ground and surface waters. Also, the transport of colloids (*e.g.* particles consisting of macro-molecules of humic substances) takes place through cracks in the soil structure. Well granulated soils have high permeability, enabling greater water and air infiltration. Soils with a high clay content tend to have poor structure with reduced permeability and consequently low water, air and root penetration.

Commonly determined physical parameters include:

- Bulk density
- Specific gravity
- Moisture content
- Loss on ignition
- Particle size

These parameters are relatively easy to determine in soil, dust, sediment or sludge using the procedures described here. All the methods involve gravimetry after some initial manipulation of the sample as described in the experimental procedures. Most of the procedures involve drying the sample at 105 °C in an oven to constant weight. This is achieved by repeating the drying and weighing cycle several times until successive weighings differ by no more than 1–2 mg. Drying to constant weight may take from 10 to 24 h, or even longer, depending on the type of soil and oven used. The mass of sample that you should use is not specified. You can use anywhere between 10 and 50 g, depending on the amount of sample available. Larger samples will give a higher accuracy when weighed. For soil moisture determination you may use as little as 1 g of sample if sample availability is a problem, but you will need a sensitive balance.

5.4.1.1 Materials
- Analytical balance
- Oven
- Desiccator
- Muffle furnace and crucible for loss-on-ignition

5.4.2 Bulk Density

Bulk density is the dry weight of a unit volume of soil expressed in

$g\,cm^{-3}$. Bulk density is inversely related to pore space and it has an important influence on root penetration and soil permeability, which in turn can affect the flow of material (air, water, nutrients, pollutants) within a soil. Bulk densities of soils generally vary between 0.8 and $1.7\,g\,cm^{-3}$. Peats have very low densities (0.1–$0.3\,g\,cm^{-3}$). Bulk density can be used to determine if a soil has the physical characteristics required for plant growth, building foundations and other uses. Soils having high bulk densities have low pore space and therefore low permeability and infiltration. High density soils are also inhibitive to root penetration. Several different methods are available for determining soil density. One method involves sampling a core of soil using a cylindrical metal sampler. The volume of the core is calculated from the diameter and length of the core, and the core is weighed. The density can then be calculated from the weight/volume relationship. Another simple method is given below.

5.4.2.1 Experimental Procedure. Dry a soil sample (*e.g.* 50 g) in an oven at 105 °C to constant weight. Weigh the soil sample. Pour the sample into a measuring cylinder a little at a time while gently tapping the cylinder to compact it. Measure the volume ($1\,mL = 1\,cm^3$). Determine the bulk density as:

$$\text{Bulk density } (g\,cm^{-3}) = \text{weight (g)}/\text{volume } (cm^3)$$

5.4.3 Specific Gravity

Specific gravity is defined as the ratio of mass of a given volume of soil to the mass of an equal volume of water. Specific gravity is directly related to the bulk density and it can be used in much the same way to characterise soil quality.

5.4.3.1 Experimental Procedure. Dry a soil sample (*e.g.* 50 g) in an oven at 105 °C to constant weight. Pour the soil into a pre-weighed glass bottle of suitable volume to the brim. Weigh the bottle containing the soil. Empty the bottle, rinse it with water and then fill with water to the brim. Weigh the bottle containing the water. Calculate the specific gravity as:

$$\text{Specific gravity} = \frac{M_2 - M_1}{M_3 - M_1}$$

where M_1 is the weight of the empty bottle, M_2 is the weight of the bottle containing the soil and M_3 is the weight of the bottle containing the water.

5.4.4 Water Content

The measurement of water or moisture content is important in the analysis of solid environmental samples but especially of soils because of the many essential roles that water plays in soil systems. Water is essential to plant growth. It also acts as a solvent and transporting agent for numeorus substances including essential nutrients and harmful pollutants. It provides a medium in which numerous chemical reactions and microbial activity can proceed. It is largely responsible for the appearance of soil profiles because percolating water washes substances out of the topsoil and deposits them lower down the profile. It also maintains the texture and compactness of the soil in a condition suitable for plant roots and soil animals. Too much, or too little, water can have undesirable consequences on plant growth. Rainfall and irrigation are the main sources of water in soils. Soils lose water through evaporation, uptake by plants and percolation through the soil. Soil moisture can be determined either on freshly collected samples or on samples that have been air dried and sieved. Air-dried and sieved samples are used for most chemical analyses and knowledge of their water content can be useful.

5.4.4.1 Experimental Procedure.   Accurately weigh a freshly collected and homogenised sample (*e.g.* 10–20 g). The sample should not be sieved but large stones, twigs and roots should be removed. Dry in an oven at 105 °C to constant weight. Cool in a desiccator. Weigh the sample once more and record the final weight. Calculate the water content in percent as:

$$\text{Water content (\%)} = 100 \times (M_1 - M_2)/M_1$$

where M_1 is the initial weight (g) and M_2 is the final weight (g). This is *fresh water content*. In the same way, determine the water content of a sample which has been air dried and sieved through a 2 mm sieve. This will give you the *air-dry water content*. Since the air-dry moisture is frequently used for later correction of results to dry weight basis, it is more convenient to calculate % dry matter as follows:

$$\text{Dry matter (\%)} = 100 \times M_2/M_1$$

5.4.5 Loss-on-ignition

Loss of sample mass on ignition can serve as a rough indicator of the amount of organic matter present in the sample. Results of quantitative chemical analysis based on dry ashing procedures are usually reported as "loss-on-ignition".

5.4.5.1 Experimental Procedure. Dry a sample in an oven at 105 °C to constant weight. Accurately weigh 1 g of this dried sample and pour into a pre-weighed dry crucible. Optionally, a few drops of H_2O_2 may be added at this stage to promote oxidation. Place in a muffle furnace and gradually increase the temperature to 500 °C. Leave inside the oven at this temperature for at least 4 h, or overnight if convenient. Cool, transfer to a desiccator and allow to cool to room temperature. Weigh and calculate loss-on-ignition in % as:

$$\text{Loss-on-ignition (\%)} = 100 \times (M_1 - M_2)/M_1$$

where M_1 is the initial weight (g) and M_2 is the weight after ignition (g).

5.4.6 Particle Size

Soil texture is determined by particle size (see Section 5.1.2). Clay particles are smaller than $2\,\mu m$ and they differ significantly from the larger soil particles (silt, sand, gravel, stones). The larger particles are all formed exclusively by physical breakdown of rocks and minerals and their chemical composition is not significantly different from the parent rock. On the other hand, clay is a combination of silicon, aluminium and oxygen which is structurally different from the other materials. Soil texture is important because it has an influence on soil structure, water holding capacity, nutrient storage, water movement and aeration. Particle size analysis of a soil sample can be carried out using the following procedure. This is based on separation of coarser sizes using a sieve set and smaller fractions, including clay, on the basis of sedimentation inside a suspension. The settling method is based on Stoke's law, which states that the velocity with which a particle settles out of a liquid medium is dependent on the radius of the particle:

$$v = 2gr^2\,(\rho_s - \rho_l)/9\eta$$

where v is the settling velocity ($m\,s^{-1}$), g is the gravitational force per unit mass ($9.81\,N\,kg^{-1}$), r is the radius of the solid particles (m), ρ_s and ρ_l are the densities of solid particles and the liquid respectively ($kg\,m^{-3}$), and η is the viscosity of the liquid ($N\,s\,m^{-2}$). The more that soil particles deviate from ideal spheres, the slower will be their sedimentation rate.

This can cause an error in the result since non-spherical particles will be assigned to a size fraction larger than they really are.

5.4.6.1 Materials
- Sodium diphosphate ($Na_2P_2O_7$). Dissolve 100 g in water and make up to 1 L with water.
- Weighing bottles, glass
- Sieve set with pore sizes: 2, 0.63, 0.2 and 0.063 mm

5.4.6.2 Experimental Procedure. Accurately weigh a sample of air-dried soil and pass through the 2 mm sieve. Stones and roots larger than 2 mm will be separated out. Weigh the mass of material > 2 mm.

Weigh accurately 20 g of the soil that has passed through the 2 mm sieve and place in a bottle together with 25 mL of the sodium diphosphate solution. Leave for about 8 h. Then add 200 mL of water and shake. Pass the suspension through the sieve set arranged in order of decreasing pore size and placed directly above a 1 L measuring cylinder. Clamp a large funnel with a surface area greater than that of the sieves above the cylinder before you start sieving the suspension. Wash each sieve with water until particles smaller than the nominal size of the sieve pores have passed through. Dry the sieves in an oven at 105 °C. Weigh each sieve before and after sieving the suspension.

Make up the cylinder containing the sieved suspension to 1 L with water, stopper and shake for 1 min. An alternative is to stir thoroughly with a hand stirrer using an up-and-down motion. Leave to stand and note the time. After 9.5 min, remove 10 mL of the solution from 20 cm below the surface using a pipette and place in a pre-weighed glass weighing bottle. Rinse the pipette with water and add the rinse water to the same weighing bottle. Particles < 20 μm are contained in this suspension. After 18.5 min, remove a further 10 mL of solution from 10 cm below the surface and transfer to a pre-weighed weighing bottle. Rinse the pipette into the bottle. This suspension contains particles < 10 μm. After 3 h and 5 min, take another 10 mL aliquot from 4 cm below the surface and transfer to a pre-weighed weighing bottle. Rinse the pipette into the bottle. This suspension contains clay particles smaller than 2 μm. Dry the weighing bottles, before and after adding the suspension, in an oven at 105 °C to complete dryness. Cool in a desiccator and weigh. Determine the mass of each particle fraction in the weighing bottles by difference. Subtract from each measurement 25 mg in order to correct for the presence of sodium diphosphate reagent and multiply by 5 to give each fraction as % of the original weight.

Tabulate the percentage (%) of the total mass that is present in each size fraction: <0.002, <0.01, <0.02, 0.063–0.2, 0.2–0.63, 0.63–2 and >2 mm.

5.4.7 Questions and Problems

1. Discuss the role of water in soils.
2. Differentiate between the following pairs of concepts:
   (a) density and specific gravity
   (b) water content and loss-on-ignition
   (c) fresh water content and air-dry water content
3. The weight of a soil core 5 cm in diameter and 10 cm long was found to be 225.6 g. What is the density of the soil?
4. A fresh soil sample was collected and its volume determined to be 70 cm^3. The soil weighed 110 g. After air drying it weighed 95 g and after oven drying it weighed 84 g. Calculate the water content in % by weight and by volume (density of water is 1.0 g cm^{-3}). Also calculate the bulk density of the soil. In view of your results, which weight would you recommend as a reference for calculations: fresh, air-dried or oven-dried? Explain why.
5. Explain the principle on which the experimental method for determining size fractions $< 20\,\mu$m is based and give an example of possible sources of error.
6. Calculate the settling velocity and the time taken by soil particles of 0.002, 0.01 and 0.02 μm to settle through 10 cm of water, given that the density of water is 1.0 g cm^{-3} and the viscosity of water is 1.0×10^{-3} N s m^{-3}. The soil density was determined to be 1.5 g cm^{-3}.

5.4.8 Suggestions for Projects

1. Determine the particle size distribution in different soils. Classify the soil texture (sand, clay, *etc.*) according to their particle size using the soil triangle (Figure 5.1). Measure the water content, density, and specific gravity of these soils and explain the results.
2. Sample a soil profile and determine the particle size distribution, density, specific gravity, moisture content and loss-on-ignition in samples collected from the different horizons. Explain your results and comment on the structure of the soil profile.
3. Measure loss-on-ignition in different sludge samples. Explain the results in terms of the technologies used at the sewage treatment works.
4. Determine the soil moisture and other parameters during different seasons (*i.e.* wet and dry seasons in the tropics; four seasons in temperate climates) in the same soil. Obtain typical rainfall amounts for the seasons and comment on your results.

5.4.9 Further Reading

S. E. Allen (ed.), "Chemical Analysis of Ecological Materials", 2nd edn., Blackwell, Oxford, 1989, pp. 14–16.

G. C. Topp, Soil water content, in "Soil Sampling and Methods of Analysis", ed. M. R. Carter, Lewis, Boca Raton, 1993, pp. 541–557.

J. L. B. Culley, Density and compressibility, in "Soil Sampling and Methods of Analysis", ed. M. R. Carter, Lewis, Boca Raton, 1993, pp. 529–539.

B. H. Sheldrick and C. Wang, Particle size distribution, in "Soil Sampling and Methods of Analysis", ed. M. R. Carter, Lewis, Boca Raton, 1993, pp. 499–511.

W. H. Gardner, Water content, in "Methods of Soil Analysis: Part 1, Physical and Mineralogical Properties", 2nd edn., ed. A. Klute *et al.*, ASA & SSSA, Madison, 1982, pp. 493–544.

G. R. Blake and K. H. Hartge, Bulk density, in "Methods of Soil Analysis: Part 1, Physical and Mineralogical Properties", 2nd edn., ed. A. Klute *et al.*, ASA & SSSA, Madison, 1982, pp. 363–382.

G. W. Gee and J. W. Bauder, Particle-size analysis, in "Methods of Soil Analysis: Part 1, Physical and Mineralogical Properties", 2nd edn., ed. A. Klute *et al.*, ASA & SSSA, Madison, 1982, pp. 383–411.

5.5 ELECTROCHEMICAL MEASUREMENTS

5.5.1 Introduction

Although these measurements in soils, sediments and sludges are very important, there is no universally accepted standard method for their determination and each analyst seems to have his or her own preferences. Measurements carried out in the field give the most representative results; however, if this is not possible, samples can be analysed in water suspensions in the laboratory. Usually the suspension is mixed vigorously and left to stand for a period of time before the measurement is taken. Different weight by volume ratios (w/v) of soil to water have been employed (between 1:1 and 1:10), as have different equilibration periods (from no delay to 1 h) before analysis. Also, some recommend analysing fresh samples while others recommend air-dried or oven-dried samples. Drying the sample will affect the measured parameters, as will increasing the dilution of the suspension. A soil/water ratio of 1:1 w/v is more commonly used; however, slurries or pastes can be formed at this ratio when soils with a high content of clay or organic matter are used. We have selected a 1:2 w/v ratio for soil to water and an equilibration time of 30 min. This should be sufficient for the soil particles to settle to the bottom with most soils. You should insert the electrodes in the clear supernatant. Some recommend inserting the electrodes into the soil or a thick paste; however, this could lead to errors. You can use the methods outlined here for soil, sediment or sludge samples; they are of little relevance to dusts and are not applied to them.

5.5.2 Conductivity

Soil conductivity can be used as a measure of total soluble ions in place of the laborious determination of individual ions. Soil conductivity can be used to assess the viability of saltwater flooded soils. *In situ* field determination of conductivity using portable probes is easy, convenient and quick. Soil conductivity is used in monitoring surveys and in assessing irrigation and drainage needs. It is also a useful determination in heavily fertilised soils such as soils in greenhouses and vegetable gardens. Conductivity of soil has also been related to plant growth. Effects of conductivity on plants are summarised in Table 5.6. Conductivity measurements are not as widely applied in soil analysis as are pH measurements.

5.5.2.1 Experimental Procedure. Weigh 20 g of fresh soil or other sample (*e.g.* sediment) from which stones, twigs and larger materials have been removed and place in a beaker or wide-neck bottle. Add 40 mL of laboratory water and stir vigorously on a magnetic stirrer if one is available, or manually. Allow to stand for 30 min and determine the conductivity as described in Section 2.4.4, making sure that the electrode cell is in the supernatant above the settled particles.

5.5.3 pH

The pH of a soil, as well as sludge and bottom sediment, is one of the most frequently measured parameters due to the importance of pH in

Table 5.6 *Plant responses to soil conductivity*

Soil conductivity (mmho cm^{-1})	Plant response
> 1.00	Expected to cause severe damage to most plants
0.71–1.00	May cause slight damage to most plants
0.46–0.70	May cause slight to severe damage to salt-sensitive plants
0.26–0.45	Suitable for most plants if recommended amounts of fertiliser are used
0–0.25	Suitable for most plants if recommended amounts of fertiliser are used

regulating numerous processes. The pH gives an indication of the acidity or alkalinity and this makes it valuable for soil characterisation. Many chemical reactions are pH dependent and knowledge of the pH can enable us to predict the extent and speed of chemical reactions. Also, the availability of different nutrients depends strongly on the pH. For example, soils with a pH of around 7 have a higher availability of Mg, Ca, K, N and S, while Fe, Zn and Cu are less available at high pH. The soil pH can be used as a quick diagnostic tool to inform us about the state of the soil. The soil pH depends on a number of factors: parent material, climate, vegetation, fertiliser and lime application. The pH of most mineral soils is between 5.5 and 7.5. Soils with higher pH values, as high as 10.5, may be encountered in calcareous, dolomitic and sodic soils from arid areas. Soils may become acidic if rainwater leaches out basic cations (Mg^{2+}, Ca^{2+}, Na^+, K^+). Soils with pH values below 5.5 are likely to contain exchangeable Al leached from clay minerals at high levels which could be toxic to plants. Acid soils may also contain toxic levels of Mn. The main sources of H^+ in soils include:

- The dissociation of carbonic acid formed when atmospheric CO_2 is absorbed in the soil
- Organic acids formed from the decomposition of organic matter
- Strong acids such as H_2SO_4 and HNO_3 in rainwater, originating from industrial air pollution
- The oxidation of ammonium salts to nitrate

Basic cations of alkali metals (Na^+, K^+) and alkaline earth metals (Ca^{2+}, Mg^{2+}) can raise the pH of soils by producing OH^- ions in solution. These cations can originate from the weathering of rocks and minerals, wind blown dust, irrigation water, runoff water, rainwater and liming. Calcareous soils have pH values between 7.0 and 8.2 as a result of calcium carbonate hydrolysis. Saline soils have neutral to slightly alkaline pH values. These soils are generally found in arid regions and they contain high levels of accumulated soluble salts (*e.g.* NaCl, $CaCl_2$, KCl). These salts can cause injury to plants. In sodic soils, usually found in arid regions, where sodium is the major component of rocks and minerals, pH values can range from 8.5 to 10.5. Irrigation can also build up salts in the soil. Acid soils can be treated by liming with materials that contain Ca, the most common liming agent being limestone. Soil pH may vary from season to season and from year to year, generally increasing during periods of high rainfall and decreasing during periods of low rainfall.

Measurement of pH can be problematic in any sample (Section 2.4.5), but it is especially so in soils, sediments and sludges. As mentioned earlier, there is no standard accepted method and the result will be affected by the soil:water ratio and method of measurement. For example, forming a paste around the electrode, as some recommend for field analysis, can give significantly different results. Also, many recommend stirring while taking the pH reading. This can lead to stirring errors caused by so-called "streaming potentials". The solution should be stirred so as to equilibrate the mixture, but it should then be allowed to rest and readings should be taken on still solutions. Measurement of pH is also used in tests for lime requirement.

5.5.3.1 Experimental Procedure. Weigh 20 g of fresh soil or other sample (*e.g.* sediment) from which stones, twigs and larger materials have been removed and place in a beaker or wide-neck bottle. Add 40 mL of laboratory water and stir vigorously on a magnetic stirrer if one is available, or manually. Allow to stand for 30 min and determine the pH as described in Section 2.4.5, making sure that the electrode is in the supernatant and not in contact with the settled soil particles. Contact with the solid particle could lead to errors of as much as 1 pH unit.

5.5.4 Redox Potential

Redox potential determines the geochemical mobility of pollutants and nutrients (especially S, N, P, heavy metals) in various compartments of the environment and consequently their influence on ecosystems. Oxidation and reduction reactions are especially important in soils. Microbial respiration in soils provides electrons that drive most redox reactions and these act to reduce available O_2. However, if the rate of O_2 used up in respiration exceeds O_2 availability, reduction of other substances takes place and this can affect the speciation of nutrients either directly or indirectly. The relative degree of oxidation or reduction in a soil or bottom sediment has a marked effect on their nature, their chemical reactions, their microbial and faunal populations and associated terrestrial or aquatic vegetation. This is illustrated by the sharp contrast between aerobic conditions in well-aerated fertile brown earths and the water-logged anaerobic conditions of gleyic soils or bog peats. A parallel situation occurs in well-aerated natural waters and anoxic bottom sediments. Redox potential is expected to vary with depth in sediments and soils since they contain microorganisms which consume oxygen. Well-aerated surface layers may have high Eh values, indicating oxidising conditions due to the availability of oxygen. Deeper layers may be completely devoid of oxygen, giving rise to highly reducing conditions and consequently low Eh values.

Although redox potential is relatively easy to measure, it is not always easy to produce meaningful results. The electrode potential can change considerably from the field to the laboratory and for representative results to be obtained measurements should be taken in the field. In some studies, samples have been stored and analysed in an atmosphere of inert gases so as to minimise changes between the field and the laboratory. An estimation of total oxygen, particularly in water-bottom sediments systems, is usually of more value under these conditions. Furthermore, any assessment of redox values should always be carried out in conjunction with pH. The field procedure outlined below will produce redox potentials more representative of field conditions, although the manipulation necessary to take the field measurement may give rise to minor errors. The laboratory procedure recommended below will give the redox potential of the soil under the conditions of the experimental determination. As such, this may provide a useful reference value when comparing the relative redox potentials for different soils measured in the same way.

5.5.4.1 Experimental Procedure. If possible, carry out redox estimations in the field, taking care to exclude air and to ensure good electrode contacts. Make a puddle and insert the platinum and reference electrodes. If the soil is dry, wet with laboratory water to make a puddle.

If the sample is to be transported to the laboratory, perform the analysis as follows. Weigh 20 g of fresh soil or other sample (*e.g.* sediment) from which stones, twigs and larger materials have been removed and place in a beaker or wide-neck bottle. Add 40 mL of laboratory water and stir vigorously on a magnetic stirrer if one is available, or manually. Allow to stand for 30 min and determine the electrode potential as described in Section 4.3.3, making sure that the electrode is in the supernatant and not in contact with the settled soil particles. Clean the platinum electrode surface by overnight immersion in 50% H_2SO_4 or HCl.

Notes

1. For all three determinations, you may use different sample sizes as long as you keep the w/v ratio the same. If you use a different w/v ratio (*e.g.* 10 g sample + 100 mL water), report the w/v ratio with your result.
2. We recommend that you use fresh samples for all three determinations in order to get results representative of the field conditions. You may, however, determine these parameters on air-dried and sieved (2 mm mesh) samples if fresh samples are not available.
3. For field measurement, form a small puddle in the soil. This is easy if the soil is naturally wet, as after rain or if it is waterlogged, as are bogs

and marshes. If there is not sufficient soil water to do this, then use laboratory water to form a puddle. Mix the soil with the water in the puddle. Insert the various probes and record the reading. Fairly robust portable monitors are commercially available, specifically designed for soil analysis. Some recommend that you should form a paste around the electrode, but this is not advisable because the electrode must be in contact with the solution, rather than the paste, to respond effectively. Inserting an electrode into a paste can lead to serious errors in pH measurement as well as a greater likelihood of electrode breakage.

4. Although all these determinations pose some difficulties, the redox potential determination is the most problematic. Exposing a sample collected at depth from an anaerobic environment to air could lead to serious errors. As stated previously, laboratory determinations may not be representative of the field, yet field determinations also have their problems. The laboratory procedure recommended above is not ideal but it provides a measure of the redox potential of the soil itself, rather than the actual conditions of the soil in the field. Dried and ground samples may also be used, although the results may be even less representative.

Example 5.1

Calculate the ionic strength of a soil solution having the following composition (composition taken from Campbell *et al.*, 1989):

Ion	Na^+	K^+	Mg^{2+}	Ca^{2+}	NO_3^-	Cl^-	SO_4^{2-}	$H_2PO_4^-$	pH
$mg\,L^{-1}$	11	15	3	84	12 (N)	57	11 (S)	2 (P)	7.7

First convert to molar units by dividing by the respective molar masses:

Ion	Na^+	K^+	Mg^{2+}	Ca^{2+}	NO_3^-	Cl^-	SO_4^{2-}	$H_2PO_4^-$	H^+
$10^{-4}\,mol\,L^{-1}$	4.8	3.8	1.2	21	8.6	16	3.4	0.64	2×10^{-4}

Ionic strength is defined as:

$$I = \tfrac{1}{2}\Sigma\, c_i \times z_i^2$$

where c_i is the concentration of ion i in $mol\,L^{-1}$ and z_i is the charge of ion i. For the above solution (expressing concentrations in brackets in $10^{-4}\,mol\,L^{-1}$):

$$I = \tfrac{1}{2}[4.8 + 3.8 + (1.2 \times 4) + (21 \times 4) + 8.6 + 16 + (3.4 \times 4) + 0.64$$
$$+ 2 \times 10^{-4}] \times 10^{-4} = 6.8 \times 10^{-3}\,mol\,L^{-1}$$

5.5.5 Questions and Problems

1. A soil sample was analysed using a soil : water w/v ratio of 1 : 1 and a pH of 6.24 determined. If the same sample was to be analysed using a soil : water w/v ratio of 1 : 5, what would you expect the pH value to be?

2. Explain the influence of various soil–water ratios on pH values. Give special consideration to problems relating to activity (temperature, ionic size and strength, and solvent density).

3. What are the sources of soil acidity and alkalinity?

4. What are the potential effects of increasing acidity in soils and how can these be counteracted?

5. Would you expect to observe any relationship between pH and conductivity and if so, why?

6. Outline the differences and similarities between pH and Eh. Give specific examples of the relationship between pH and Eh in different natural systems.

7. What do you expect the Eh values of an aerated sandy soil and a peat sediment at the bottom of a stream to be? Explain in terms of the chemical composition of the soil and sediment.

8. Explain the role of polyvalent elements like S, N, Fe and Mn in the regulation of redox potential in various environmental systems.

9. The dissolution of toxic aluminium in acidified soils and waters is a major concern. The dissolution of aluminium hydoxide can be illustrated by the following equation:

$$Al(OH)_3(s) \rightleftharpoons Al^{3+} + 3OH^-$$

The solubility product for $Al(OH)_3$ is 2×10^{-32}. What is the relationship between soluble Al^{3+} and the pH? Plot the concentration of Al^{3+} as a function of pH [*Note.* You do not need to know the concentration of $Al(OH)_3(s)$ to solve this problem] and discuss the environmental implications.

10. Denitrification involves two stages, the reduction of nitrate to nitrous acid, and the subsequent reduction of nitrous acid. The first stage can be represented as:

$$NO_3^- + 3H^+ + 2e^- \rightleftharpoons HNO_2 + H_2O$$

The standard reduction potential for this reaction is 0.934 V. Write the Nernst equation for this reaction and derive an equation relating Eh to pH for the condition where the concentrations of NO_3^- and

HNO$_2$ are equal. Plot this equation on an Eh–pH diagram together with the upper and lower limits of Eh in nature (Figure 4.5) and discuss the implications.

11. Using the concentrations given in Example 5.1, calculate the ratio of anions to cations and state whether the soil solution analysis is satisfactory according to the criteria given in Section 2.3.

12. Calculate the theoretical conductivity of the soil solution given in Example 5.1.

13. Why should the redox potential be determined immediately after sampling, and preferably in the field?

5.5.6 Suggestions for Projects

1. Measure the conductivity, Eh and pH values of different soils, and investigate any relationships. Determine the correlation coefficients for pairs of parameters and determine if they are significant or not. Plot pairs of parameters on an *x*–*y* graph if they are significant and determine the best-fit lines by the method of least squares.

2. Measure the conductivity, Eh and pH values in soil at several sites from the foot to the top of a hill and if possible up to a nearby stream, river or lake. Explain why the differences in Eh values vary more widely than the corresponding differences in pH values.

3. Compare the values of conductivity, pH and Eh of sludges from different sewage treatment plants.

4. Measure the conductivity, pH and Eh of soil before and after a rain shower. Explain the results.

5. Determine the conductivity, pH and Eh of soil at a specific site(s) during different seasons (*i.e.* wet/dry seasons in the tropics, four seasons in temperate climates) and discuss your results.

6. Determine the variability in conductivity, pH and Eh in soil at a particular site by analysing many samples at close proximity to each other and calculating the mean and standard deviation. Choose different sites in forests, agricultural fields, pastures, paddy soils, *etc.*, if possible, and compare the variability at the different sites.

7. Measure the conductivity, Eh and pH values in fresh and dried samples of soil and bottom sediment. Explain the results.

8. Measure the pH of fresh soil in water and in 1 N KCl at the same w/v ratio of 1 : 2. Typically, the pH values in 1 N KCl are less than in distilled water. Explain the results.

9. Investigate the effect of the experimental conditions on the conductivity, pH and Eh of soil, sediment and sludge samples. For example, select a particular sample and investigate the effect of:

- Varying the w/v ratio of sample : water in the suspension between 1 : 1 to 1 : 10
- Varying the equilibration time from 5 min to 2 h
- Drying the sample (analyse fresh and dried samples)

Present your results in tabular and graphic form. On the basis of your results, make recommendations for what you would consider to be the best analytical procedure.

5.5.7 Reference

D. J. Campbell, D. G. Kinniburgh and P. H. T. Beckett, The soil solution chemistry of some Oxfordshire soils: temporal and spatial variability, *J. Soil Sci.*, 1989, **40**, 321–340.

5.5.8 Further Reading

S. E. Allen (ed.), "Chemical Analysis of Ecological Materials", 2nd edn., Blackwell, Oxford, 1989, pp. 16–18.

A. L. Page, R. H. Miller and D. R. Keeney, "Methods of Soil Analysis: Part 2, Chemical and Microbiological Properties", 2nd edn., ASA/SSSA, Madison, 1982, pp. 199–224.

5.6 ALKALINITY

5.6.1 Introduction

Alkalinity in soils arises mainly because of the dissolution of calcite ($CaCO_3$):

$$CaCO_3 + H_2O \rightleftharpoons Ca^{2+} + HCO_3^- + OH^-$$

This results in an increase in pH. Calcareous, dolomitic and sodic soils are alkaline due to their content of $CaCO_3$, $CaCO_3 \cdot MgCO_3$ and Na_2CO_3. Some soils from arid and semi-arid regions may have pH values as high as 9.9, while sodic soils may have pH values up to 10.5. The bicarbonate, carbonate and hydroxide levels increase as the soils become more alkaline. Alkalinity in soils can have important effects on soil processes, with consequently undesirable effects on plants. Some examples include:

- Chlorosis of leaves of sensitive plants. High bicarbonate concentrations inhibit Fe and Mn uptake and utilisation by plants and this in turn interferes with the production of chlorophyll.

- Reduced yields of trees and bush fruit crops. Bicarbonate interferes with Fe and Mn metabolism. The low solubility of Fe and Mn in alkaline soils and strong binding of these metals on humus also contribute.
- Phosphate deficiency. Bicarbonate can reduce the uptake of phosphorus. The low solubility of phosphate in alkaline soils also contributes.

Alkalinity in soils may also arise from high bicarbonate levels in irrigation waters. If irrigation water has high concentrations of calcium, bicarbonate and carbonate ions, $CaCO_3$ may precipitate, blocking pores in the soil and reducing water permeation. This could influence other physicochemical processes in the soil which in turn could lead to harmful effects (*e.g.* increasing sodium toxicity, deterioration of soil structure). Soil alkalinity is determined on soil in water suspensions prepared in the same way as for pH and conductivity determinations.

5.6.2 Experimental Procedure

Weigh 20 g of fresh soil (or other solid sample) from which stones, twigs and larger materials have been removed and place in a beaker or wide-neck bottle. Add 40 mL of laboratory water and stir vigorously on a magnetic stirrer if one is available, or manually. Allow to stand for 30 min and filter through a Whatman no. 42 or 44 filter paper. Determine the alkalinity of the filtrate by titration with H_2SO_4 or HCl as described in Section 4.5. Calculate the alkalinity in the filtrate in mg $CaCO_3 \, L^{-1}$. Convert to the alkalinity of the soil from:

$$\text{Alkalinity (mg } CaCO_3 \text{ g}^{-1}) = A \times V/M$$

where A is the alkalinity in the extract (mg $CaCO_3 \, L^{-1}$), V is the volume of water in L (*e.g.* 40 mL = 0.04 L) and M is the weight of the soil sample (g). Calculate the concentrations of carbonate (if present) and bicarbonate as shown in Example 4.3 from the alkalinity measurements and convert these to mg $CO_3^{2-} \text{ g}^{-1}$ and mg $HCO_3^- \text{ g}^{-1}$.

Notes
1. You may use the extract prepared in the pH measurement (see Section 5.5.3) to determine the alkalinity after filtration.
2. If the alkalinity of the extract is too high you may dilute it before

analysis but take into account the dilution factor when calculating the alkalinity and concentrations of bicarbonate and carbonate in the soil sample.

5.6.3 Questions and Problems

1. Explain the origin of soil alkalinity.
2. In soil water an equilibrium will be established involving both calcite and CO_2 from the atmosphere. Write all the equilibrium equations necessary to describe the calcite/water/air equilibrium.
3. Calculate the pH of soil water simultaneously in equilibrium with atmospheric CO_2 ($CO_2 = 0.036\%$) and with calcite (for calcite: $K_{sp} = 8.7 \times 10^{-9} \, mol^2 \, L^{-2}$; for CO_2: $K_H = 0.031 \, mol \, L^{-1} \, atm^{-1}$, $K_1 = 4.3 \times 10^{-7} \, mol \, L^{-1}$, $K_2 = 5 \times 10^{-11} \, mol \, L^{-1}$) and calculate the solubility of calcite in this solution.
4. The partial pressure of CO_2 in soil air can be considerably higher than in the atmosphere. Calculate the solubility of calcite in soil water in contact with a soil atmosphere containing CO_2 at a partial pressure of $0.05 \, atm$, and estimate the equilibrium pH of the solution.

5.6.4 Suggestions for Projects

1. Determine the alkalinity, pH, conductivity and concentrations of water-soluble ions in different types of soils. Investigate any possible relationships between different parameters. Try to correlate the conductivity with individual water-soluble ions. Discuss the relationship between soil composition and alkalinity on the basis of your results.
2. Determine the alkalinity of soil irrigation water and of the soil.

5.7 SOLUBLE IONS

5.7.1 Introduction

Soluble ions, sometimes also referred to as soluble salts, are the major dissolved inorganic solutes. The concentration of soluble ions is related to soil conductivity. Ideally, one would want to know the concentrations of individual soluble ions in the soil water under different field conditions. Although in principle this is possible, it is extremely laborious and therefore impractical. Field measurements of soil conductivity are often used instead; however, these can be related only to total soluble ion concentrations. The concentration of soluble ions is often employed in characterising saline soils, and it is of great use in the management and reclamation of these soils.

Soluble ions can be determined either on soil water samples collected in the field or on water extracts of soil samples, the latter being more common. The ions in the extract can then be determined using methods suitable for water analysis. Soil/water ratios that have been used vary considerably, from saturation to $1:10$ w/v. In the procedure given here a $1:2$ w/v ratio is used rather than saturation since this was the ratio used for conductivity, pH and redox potential. This is also a good compromise value between the reported ranges of soil/water ratios used in different laboratories.

5.7.2 Experimental Procedure

Weigh 20 g of air-dried and sieved (<2 mm) soil and place in a flask or bottle. Add 40 mL of laboratory water, stopper, and shake on a mechanical shaker for 1 h. Filter through a Whatman no. 42 or 44 filter paper and analyse.

5.7.2.1 Analysis. Soluble ions that are commonly analysed are: Mg^{2+}, Ca^{2+}, Na^+, K^+, Cl^-, NO_3^-, NH_4^+ and SO_4^{2-}. Complete the analysis of the filtered water extract using methods given in Table 5.7. Refer to the relevant experimental sections indicated for instructions on analytical procedures. If an ion chromatograph (IC) is available it can be used for the simultaneous determination of Cl^-, NO_3^- and SO_4^{2-}. Express your results in $mg\,L^{-1}$ and $meq\,L^{-1}$ of the water extract and also in $mg\,kg^{-1}$ and $meq\,(100\,g)^{-1}$ of the soil sample.

After completing the determination of soluble Na, Ca and Mg, you may calculate the *sodium adsorption ratio* (SAR) from:

$$SAR = [Na^+]/([Ca^{2+}] + [Mg^{2+}])^{1/2}$$

Table 5.7 *Methods and procedures for determining soluble ions in soil*

Ion	Method	Experimental procedure
Mg^{2+}, Ca^{2+}, Na^+, K^+	AAS	as in Section 2.5.1
Cl^-, NO_3^-, SO_4^{2-}	IC	as in Section 2.4.6
Cl^-	Titrimetry	as in Section 4.12
NH_4^+	Colorimetry or titrimetry	as in Sections 4.10.2, 2.5.2
NO_3^-	Spectrophotometry	as in Section 4.10.3
NO_2^-	Spectrophotometry	as in Section 4.10.4
SO_4^{2-}	Turbidimetry	as in Section 4.14

where the concentrations are expressed in $mmol\,L^{-1}$. SAR is a useful index of soil sodicity. Soils with SAR > 13 are assumed to be sodic. SAR is also frequently determined in irrigation water.

Notes

1. Samples for soluble ions should not be oven dried. Use air-dried samples only. Oven drying can convert gypsum ($CaSO_4 \cdot 2H_2O$) to the more soluble plaster of paris ($CaSO_4 \cdot 0.5H_2O$).
2. Determine Ca^{2+} immediately on fresh extract. Prolonged standing may precipitate $CaCO_3$.
3. You may also analyse soluble ions in saturated soil. Saturation is achieved by adding water to the soil until a paste is formed. Leave to stand overnight. Free water should not collect on the soil surface. This is then filtered using a Buchner funnel and highly retentive filter paper while applying a vacuum, and the filtrate analysed.
4. Ammonium is determined following distillation as for water samples. Dilute to 100 mL before distilling according to the procedure outlined in Section 4.10.2. Calculate the concentration of NH_4^+ in the 40 mL original solution.

5.7.3 Questions and Problems

1. What is the difference between soluble and extractable ions? Which would you expect to give a higher result and why?
2. Explain how soils may become saline.
3. Saturation extracts of non-saline non-sodic (A), saline (B), sodic (C) and sodic saline (D) soils were analysed and the following results obtained:

Soil type	Cond.	pH	Ca^{2+}	Mg^{2+}	Na^+	HCO_3^-	SO_4^{2-}	Cl^-
A	0.84	7.9	1.4	0.9	5.2	6.6	1.4	0.4
B	12.0	8.0	18.5	17.0	79.0	7.2	31.1	47.0
C	3.16	9.6	0.6	0.2	29.2	27.1	2.3	7.5
D	16.7	7.8	16.2	19.2	145.0	3.3	53.0	105.0

Conductivity is in $mmho\,cm^{-1}$ and the ionic concentrations are in $mmol\,L^{-1}$. For each soil type calculate: (a) the ionic strength, (b) the SAR, (c) the theoretical conductivity and (d) the ratio of cations to anions. Using the criteria given in Section 2.3, assess the quality of the analysis (data from Richards, 1954).

5.7.4 Suggestions for Projects

1. Assess the usefulness of conductivity measurement as a substitute for soluble ion determination. Collect soil samples from various sites under different environmental conditions (*e.g.* dry, wet) and determine the soluble ion concentrations of the major anions and cations. Also measure the pH and conductivity. For each sample, calculate the theoretical conductivity from the ionic concentrations (see Example 2.2). Plot the theoretical against the measured conductivity. Calculate the correlation coefficient and test its significance. Plot the regression line and discuss your results.
2. Determine the concentrations of soluble Na, K and Ca in various soils and irrigation waters and calculate the SAR for each sample.

5.7.5 Reference

L. A. Richards, Diagnosis and improvement of saline and alkali soils, in "Agriculture Handbook No. 60", US Department of Agriculture, Washington, 1954.

5.7.6 Further Reading

J. D. Rhoades, Soluble salts, in "Methods of Soil Analysis: Part 2, Chemical and Microbiological Properties", 2nd edn., ed. A. L. Page *et al.*, ASA & SSSA, Madison, 1982, pp. 167–179.
H. H. Janzen, Soluble salts, in "Soil Sampling and Methods of Analysis", ed. M. R. Carter, Lewis, Boca Raton, 1993, pp. 161–166.

5.8 ORGANIC MATTER

5.8.1 Introduction

Organic matter in soil and sediments is the organic fraction derived from living organisms, decomposed and partly decomposed plant and animal residue. As a result of decomposition, inorganic nutrients in plant tissue (N, P, K, Ca, Mg, Fe, Cu, Zn and Mn) are released into the soil or bottom sediments and humus is formed. *Humus*, a by-product of organic matter decomposition, plays a very important role in both terrestrial and aquatic ecosystems because it increases the cation exchange capacity of soils and serves as an important reservoir of nutrients (N, P and S). Humus is composed of humic and fluvic acids and it is resistant to further microbial decomposition. During decomposition, the C/N ratio generally decreases from about 80:1 in fresh plant material to 8–15:1 in humus. The C/N ratio is frequently used as a measure of humification and humus quality. Humic substances can act as complexing agents for heavy metals. Organic matter also plays an important role in soil structure, aggregation, infiltration and retention of water, and other physical characteristics.

The organic fraction undergoes constant physical and chemical change as a result of decomposition and mineralisation processes. The end product of these processes is the production of CO_2, H_2O, inorganic and organic acids, and nutrients. Organic matter is often used as an index of soil fertility. The amount of organic matter varies considerably from less than 1% in recent soils to as much as 90% in bog peat soils. Mineral soils generally contain between 1% and 20% organic matter. A fertile, loamy top soil has an average organic matter content of about 5%. Low organic matter contents are generally found in warm arid climates, while higher contents are encountered in cool moist climates. In general, soils with relatively higher organic matter content are considered more fertile than soils with low organic matter content. Depending on climatic conditions and management practices, the level of organic matter in soils generally stabilizes at a certain value. Organic matter serves as the main source of energy and food for microorganisms in soil and sediments. These organisms are vital to many biochemical reactions involved in nutrient cycling. Decomposition of organic matter releases nitrogen which is used by growing plants, and the rate of nitrogen release is used to estimate the nitrogen supply by the soil. This can then be used to decide on the amount of fertiliser required. The carbon content of organic matter is generally around 58%.

Various instruments are commercially available for the determination of total organic carbon content in soils. These are generally based on the combustion of the sample to produce CO_2, which is then determined by a non-dispersive infrared analyser. The interference by inorganic carbon has to be removed. Some of these instruments also analyse H and N. Oxidation by dichromate can also be used to determine organic carbon. The analysis is then completed either by titration or by colorimetry. The dichromate method, although very convenient and useful, recovers variable amounts of organic carbon and is prone to interference by variable amounts of inorganic carbon. The analyte determined in the dichromate method is sometimes referred to as *oxidisable organic carbon* or *oxidisable organic matter*.

In the quantitative determination of organic matter, it is customary to measure the organic-C concentration. The organic matter content is then obtained by multiplying the organic-C concentration by a factor of 1.72.

5.8.2 Methodology

Organic carbon is determined by means of a potassium dichromate back-titration. A known excess of $K_2Cr_2O_7$ is added to the sample together with H_2SO_4 and the organic carbon is oxidised to CO_2:

$$2Cr_2O_7^{2-} + 3C_{organic} + 16H^+ \rightarrow 4Cr^{3+} + 3CO_2 + 8H_2O$$

In this reaction, 2 moles of dichromate oxidise 3 moles of carbon. The unreacted excess of dichromate remaining after reaction is determined by back-titration with ferrous sulfate:

$$Cr_2O_7^{2-} + 6Fe^{2+} + 14H^+ \rightarrow 2Cr^{3+} + 6Fe^{3+} + 7H_2O$$

In this titration reaction, 1 mole of dichromate oxidises 6 moles of Fe^{2+}. Organic carbon is calculated by difference. Ferrous iron (Fe^{2+}), if present in the soil, will interfere by reacting with chromate according to the same equation as that shown for the titration reaction above.

5.8.3 Materials

- Reflux apparatus consisting of 500 mL conical flask, condenser, and hot plate
- Standard potassium dichromate solution ($K_2Cr_2O_7$), 0.083 M. Dry some $K_2Cr_2O_7$, primary standard grade, in an oven at 105 °C for 2 h. Dissolve 24.518 g in water and dilute to 1 L. This solution is stable indefinitely.
- Standard ferrous ammonium sulfate solution, 0.2 M. Dissolve 78.39 g ferrous ammonium sulfate, $Fe(NH_4)_2(SO_4)_2 \cdot 6H_2O$, in water. Add 20 mL concentrated H_2SO_4, cool and dilute to 1 L.
- Ferroin indicator solution. Purchase ready-made indicator solution or prepare by dissolving 0.7 g $FeSO_4 \cdot 7H_2O$ and 1.485 g 1,10-phenanthroline monohydrate ($C_{12}H_8N_2 \cdot H_2O$) in water and diluting to 100 mL.
- Sulfuric acid, concentrated

5.8.4 Experimental Procedure

Weigh 0.5 g of air-dried homogenised and sieved soil, sediment or sludge and place in a conical refluxing flask. Add 10 mL of the standard $K_2Cr_2O_7$ solution and swirl to mix. Carefully add 15 mL concentrated H_2SO_4. Dispense the acid a little at a time since it generates heat, and swirl gently to mix. Connect the flask to the condenser and turn on cooling water. Cover open end of condenser with a small beaker. Place on a hot plate and reflux for 1 h (Figure 4.8). Cool and rinse down the condenser with distilled water, collecting the water in the flask. Disconnect flask from condenser and add about 100 mL water. Swirl to mix and add 5 drops of ferroin indicator. Titrate with ferrous ammonium sulfate to the end point, at which the colour changes from blue-green to violet-red. In the same way, analyse a blank consisting of all the same reagents but without the soil. Calculate the organic carbon from:

$$\text{Organic carbon (mg g}^{-1}) = \frac{18 \times C \times V}{M} \times (1 - V_1/V_2)$$

where C is the concentration in $mol\,L^{-1}$ of the dichromate solution (0.166 M), V is the volume of dichromate solution used (10 mL), V_1 is the volume of titrant used up in the sample determination (mL), V_2 is the volume of titrant used up in the blank determination (mL) and M is the weight of sample used (g). Calculate organic carbon in % as:

$$\text{Organic carbon (\%)} = \text{Organic carbon (mg g}^{-1})/10$$

Organic matter can be calculated as:

$$\text{Organic matter (\%)} = 1.72 \times \text{Organic carbon (\%)}$$

The factor of 1.72 is due to the fact that soil organic matter has a carbon content of about 58% (*i.e.* $1/0.58 = 1.72$). See also Example 5.2.

Notes

1. If refluxing apparatus is not available, follow the above procedure up to the point where the flask is connected to the condenser. Instead, leave flask to stand for 30 min and then titrate in the same way as indicated above. Also run a blank through the procedure. Use the same equation as above to calculate the organic carbon content in $mg\,g^{-1}$; however, multiply the result by a factor of 1.3. This method provides a somewhat lower recovery of organic carbon (*ca.* 77%) and for that reason a factor of 1.3 is used to correct the result.
2. When analysing calcareous soils or sediments, inorganic carbon has to be removed. This can be done by treating 0.5 g of soil with an excess of 5% H_2SO_4 solution for several hours. Dry the sample by leaving overnight in an evacuated desiccator containing NaOH pellets. The treatment is repeated until CO_2 is no longer evolved on addition of H_2SO_4.

Example 5.2

Results of soil analysis are usually converted to units of $kg\,ha^{-1}$ or $t\,ha^{-1}$ by assuming a soil depth of 15 or 20 cm. A soil sample with bulk density of $1.3\,g\,cm^{-3}$ was analysed and found to have an organic-C content of 1%. What is the organic-C content of 1 hectare of soil?

$$1\ ha = 10^4\,m^2$$

Assuming a soil depth of 15 cm, the volume of 1 ha of soil is $= 0.15 \times 10^4 = 1500\,m^3$

The density of the soil is $1.3\,g\,cm^{-3} = 1.3\,t\,m^{-3}$

The mass of 1 ha soil $= 1.3 \times 1500 = 1950\,t$

The organic-C content is 1% or 0.01. The organic-C content in 1 ha of soil is therefore:

$$\text{Organic-C} = 0.01 \times 1950 = 19.5\,t\,ha^{-1}$$

---

3. Chloride may cause an interference if it is present at high levels. It may be removed by washing the soil prior to analysis. Otherwise, if the Cl concentration is known, the results may be corrected:

$$\text{Organic C }(\%) = \text{Measured organic C }(\%) - [\text{Cl }(\%)/12]$$

where measured organic C is that determined without washing the sample.

5.8.5 Questions and Problems

1. What are the similarities and differences in sources of organic matter in soil, sediments and sludge?
2. Discuss the biogeochemical role of organic matter in terrestrial and aquatic ecosystems.
3. Why is it necessary to use a mixture of potassium dichromate and sulfuric acid when determining the oxidisable organic matter?
4. Explain why the pretreatment procedure is used before the TOC determination in calcareous soils and bottom sediment?
5. A representative oven-dried soil sample was analysed and found to have 2.0% organic carbon and a bulk density of $1.2\,g\,cm^{-3}$. Calculate

the organic carbon, and organic matter, present in one hectare of land (1 ha = $10\,000\,\text{m}^2$). Assume the soil extends down to a depth of 20 cm.

5.8.6 Suggestions for Projects

1. Compare the organic matter content in agricultural and neighbouring virgin soils.
2. Compare the organic content in soil and sediments from a nearby pond or lake.
3. Analyse different samples using an instrumental method for determining organic carbon (carbon analyser) if it is available in your laboratory. Also determine organic carbon in the same samples using the dichromate method and compare the results. Plot the results of one method against those of the other, determine the correlation coefficient and calculate the regression line by the method of least squares. Comment on your results.
4. Measure the content of organic matter in soil by sampling at sites along the slope of a hill, from the bottom to the top. Explain the role of water erosion in organic matter redistribution.
5. Compare the content of organic matter in soils of various textures: sand, sandy-loam, loam, clay. What is the reason for the differences? Also determine the nitrogen content and calculate the C : N ratios.
6. Determine the content of organic matter in sludges from various plants (municipal sewage works, paper mills, *etc*). Explain your results.
7. Determine the loss-on-ignition for various soils and also measure their organic content. Compare the results. Plot the results of one determination against the other and determine the correlation coefficient. Explain your results.

5.8.7 Further Reading

S. E. Allen (ed.), "Chemical Analysis of Ecological Materials", 2nd edn., Blackwell, Oxford, 1989, pp. 91–97.

K. H. Tan, "Soil Sampling, Preparation and Analysis", Dekker, New York, 1995, pp. 223–232.

A. L. Page, R. H. Miller and D. R. Keeney, "Methods of Soil Analysis: Part 2, Chemical and Microbiological Properties", 2nd edn., ASA/SSSA, Madison, 1982, pp. 539–579.

H. Tiessen and J. O. Moir, Total and organic carbon, in "Soil Sampling and Methods of Analysis", ed. M. R. Carter, Lewis, Boca Raton, 1993, pp. 187–199.

5.9 NITROGEN

5.9.1 Introduction

Nitrogen is an essential nutrient for all forms of life. It is a structural component of amino acids from which proteins are synthesized. Animal and human tissue (muscle, skin, hair, *etc.*), enzymes, and many hormones are composed mainly of proteins. Nitrogen chemistry and cycling in the environment are quite complex due to the great number of oxidation states. The biogeochemical cycling of nitrogen has been extensively studied in different ecosystems and the main processes are listed below.

- *Fixation* is the conversion of atmospheric N_2 to organic N
- *Mineralisation* is the conversion of organic N to inorganic N
- *Nitrification* is the oxidation of NH_4^+ to nitrite (NO_2^-) and nitrate (NO_3^-)
- *Denitrification* is the conversion of inorganic N to atmospheric N_2
- *Assimilation* is the conversion of inorganic N to organic N

These processes can be mediated by microorganisms and plants. Other important processes include *volatilisation* of NH_3 gas from the soil into the atmosphere and nitrate *leaching* from the soil by drainage water. Crop demands for N can quite often exceed the natural supply and fertilisation is then required to sustain high yields. However, excess N from fertiliser can have toxic effects. The concentration ranges for various forms of N generally encountered in the environment are given in Table 5.8.

Nitrogen deficiency is often a problem when N concentrations in the soil are low; however, high N levels in soils can contribute to several environmental problems:

- Acidification of soil, surface water and groundwater by nitric acid in rainwater and N fertilizers
- Eutrophication of aquatic ecosystems
- Groundwater pollution by nitrate leached from soils
- Decrease in biodiversity of terrestrial and aquatic ecosystems

Nitrogen pollution can have numerous other indirect effects on ecosystems: increased leaching of toxic Al in soils, surface waters and groundwater, formation of atmospheric ozone, imbalances in other nutrients (*e.g.* K, P, Ca), *etc.*

The determination of N presents specific difficulties due to the great

Table 5.8 *Concentration ranges of N in some environmental samples (dry samples are reported on a dry weight basis)*

Sample type	N content
Mineral soils	0.1–0.5%
Organic soils (peat)	0.5–1.5%
Soil extractions NH_4^+-N	2–30 mg kg^{-1}
$\quad\quad\quad\quad\quad NO_3^-$-N	1–50 mg kg^{-1}
Sludges	2–20%
Sediments	0.5–2.0%
Sediment extractions NH_4^+-N	1–60 mg kg^{-1}
$\quad\quad\quad\quad\quad\quad NO_2^-$-N	1–20 mg kg^{-1}
$\quad\quad\quad\quad\quad\quad NO_3^-$-N	1–80 mg kg^{-1}
Plant tissue	1–3%
Animal tissue	4–10%
Rainwater NH_4^+-N	0.03–4 mg L^{-1}
$\quad\quad\quad NO_3^-$-N	0.1–20 mg L^{-1}
Freshwater NH_4^+-N	0.1–2 mg L^{-1}
$\quad\quad\quad NO_3^-$-N	0.3–2 mg L^{-1}

variety of chemical forms. Also, the volatility of some N compounds restricts the way in which it can be quantitatively extracted into solution. The determination of inorganic N, mainly ammonium and nitrate, in soils is often useful, because, despite their usually low levels, these inorganic forms are readily available for plant uptake. Generally, N is determined by classical procedures such as various digestion methods (*e.g.* Kjeldahl) followed by distillation. The distillate is analysed by titrimetry or spectrophotometry. Specific ion selective electrodes are also available for ammonium and nitrate ions.

The methods presented here include determination of total organic N (also called Kjeldahl N) and exchangeable NH_4^+, NO_3^- and NO_2^- (also called *extractable* or *available*). The latter are extracted with KCl, which is the accepted method for determining these species. The determination of water-soluble NH_4^+, NO_3^- and NO_2^- in soil is described in Section 5.7.

5.9.2 Organic Nitrogen

Most of the N in the soil, sediments and many sludges is in the organic form. Total organic nitrogen (TON) is one of the most common nutrient element determinations in soil. The method described below is based on the classical Kjeldahl digestion followed by distillation. This method is widely used in soil analysis.

5.9.2.1 Methodology. Organic-N is converted to NH_3-N by digesting with sulfuric acid containing potassium sulfate in order to raise the reaction temperature. This digestion can be carried out in Kjeldahl flasks heated on commercially available Kjeldahl heating units (Figure 4.10). A catalyst such as Cu, Hg or Se is required. Ammonia is then distilled and collected in an acid solution. Analysis is completed by methods used for NH_3 determination in water samples. This method does not strictly give the TON, although it is often assumed that it does. The actual result is the sum of organic-N and ammonium-N in the sample. Most of the nitrate-N and nitrite-N are lost during the digestion. However, since the concentration of inorganic-N forms in soil is considerably less that that of organic-N (*i.e.* organic-N $\gg$ inorganic-N), the result may be assumed to represent both organic-N and total-N. Some prefer to call the N determined in this method *Kjeldahl-N*.

5.9.2.2 Materials
- Materials required for ammonia distillation and analysis (see Section 4.10.2)
- Kjeldahl digestion flasks or similar, 100 mL
- Kjeldahl heating block or similar
- Copper sulfate
- Potassium sulfate
- Sulfuric acid, concentrated

5.9.2.3 Experimental Procedure.
Weigh 5 g of air-dried, sieved (< 2 mm) sample and place in the digestion flask. Add 5 g of K_2SO_4 and 1 g of the copper sulfate. Add 10 mL concentrated H_2SO_4 while gently swirling the flask. Heat gently on the digestion rack in a fume cupboard until frothing subsides. Raise the heat to 380 °C and boil until the digest becomes colourless or pale green. Boil gently for another 3 h. After cooling, add about 40 mL of water and allow to stand for a few minutes until the particles settle to the bottom. Transfer the supernatant into a 100 mL volumetric flask. Repeat this process of washing and standing several times using smaller aliquots of water. Finally make up to the mark with water. Distill this extract by following the procedure outlined for ammonia anlysis in water (see Section 4.10.2) and analyse the distillate receiving solution by one of the methods specified in the same experiment. You may also analyse the distillate receiving solution using the spectrophotometric method described below in Section 5.9.3 for exchangeable ammonium. Calculate the concentration of NH_3-N in the sample extract. Calculate the concentration of organic-N in the soil, sediment or sludge from:

$$\text{Organic-N (mg N kg}^{-1}) = C \times V/M$$

where C is the concentration of NH_3-N in the distillate solution after dilution ($mg\,L^{-1}$), V is the volume of distillate solution after dilution (mL) and M is the weight of sample (g). Repeat the digestion, distillation and analysis using all the same reagents but without the soil sample. Use this blank to correct the results if necessary.

Notes

1. If you follow the procedure given in Section 4.10.2 you will make up the volume of distillate collected in acid to 100 mL, which is the same volume that you made the sample extract to before distillation.
2. You may increase or decrease the amount of sample you digest. In this case, also increase/decrease the volume of acid and mass of the other reagents correspondingly.
3. If bumping is encountered during the Kjeldahl digestion, it can be reduced by adding glass beads or boiling chips.

5.9.3 Exchangeable Ammonium

Exchangeable N species are widely determined in soils since these measure the potential N that the soil can supply to the plant. Exchangeable NH_4^+ is extracted from soil, sediment, dust or sludge samples with 2M KCl. The extract is filtered and the filtrate analysed by spectrophotometry using the indophenol blue method.

5.9.3.1 Materials
- Spectrophotometer
- Solutions A and B used for NH_4^+ determination in Section 2.5.2
- Potassium chloride solution, 2 M, prepared by dissolving 150 g of KCl in 800 mL of water and making up to 1 L
- Ethylenediaminetetraacetic acid (EDTA). Dissolve 6 g of the disodium salt of EDTA in 80 mL water and adjust the pH to 7. Mix well and dilute to 100 mL.
- Stock NH_4^+ solution, $1000\,mg\,L^{-1}$ or $100\,mg\,L^{-1}$ (see Section 2.5.2)
- Mechanical shaker

5.9.3.2 Extraction Procedure. Weight 10 g of soil or other solid sample in a 250 mL bottle or flask and add 100 mL of 2 M KCl solution. Stopper and shake on a mechanical shaker for 1 h. Thereafter, let the solution stand for about 30 min or until the suspension settles and you obtain a clear supernatant. You may filter this through a Whatman no. 42 filter paper if necessary. The same procedure is used for extracting exchangeable NO_3^- and NO_2^-.

5.9.3.3 Analytical Procedure. Take 1 mL of the clear supernatant and transfer to a test tube. Add 1 mL of the EDTA solution and mix. Add 5 mL of solution A followed by 5 mL of solution B, cover with parafilm and shake vigorously. Place test tube in rack and place in a warm water bath at 37 °C for exactly 15 min, or 30 min at room temperature. Measure the absorbance in a 1 cm cell at 625 nm with water in the reference cell. Prepare a series of calibration standards with concentrations in the range of 0.1–10 μg mL^{-1} by dilution of the stock NH$_4^+$ solution. Take 1 mL of each of these and add to a series of test tubes. Add 1 mL of EDTA solution to each tube followed by 5 mL solution A and 5 mL of solution B and proceed as with the sample. Also analyse a blank consisting of 1 mL water + 1 mL EDTA + 5 mL solution A + 5 mL solution B. Subtract the absorbance of this reagent blank from all the other absorbance readings and plot a calibration graph of absorbance against NH$_4^+$ concentration of the calibration standards, and read off directly the concentration in the sample extract. Convert to μg N mL^{-1}. Calculate the concentration of NH$_4^+$ in the sample as:

$$NH_4^+ - N \, (mg \, N \, kg^{-1}) = C \times V/M$$

where C is the concentration of NH$_4^+$ in the extract (μg N mL^{-1}), V is the total volume of extract (mL) and M is the weight of the sample (g). You can convert this to units of mg N $(100\,g)^{-1}$ as recommended by the USDA.

Notes

1. You may vary the mass of sample and total volume of extract, as well as the volume of the extract aliquot taken for analysis (*i.e.* 0.5, 1, 5 mL). If you vary the volume of aliquot, prepare standards over a suitable concentration range so that the aliquot volume of each standard analysed corresponds to that of the sample.
2. If analysis cannot be carried out immediately, you may filter the extract through a Whatman no. 42 filter paper and store in a refrigerator.

5.9.4 Exchangeable Nitrate

5.9.4.1 Introduction. Indirect and direct methods are used to determine nitrate in soil. Two main types of indirect methods include:

- *Reduction of nitrate and nitrite to ammonium.* Devarda's alloy is used to reduce nitrate and nitrite to ammonium, which is then distilled

Table 5.9 *Distillation methods for speciation of inorganic-N. Either NaOH or MgO can be used for the distillation*

Nitrogen species	Method
NH_4^+	Distillation
NO_3^-	Removal of NH_4^+ by distillation → destruction of NO_2^- by sulfamic acid → distillation with Devarda's alloy
$NO_3^- + NO_2^-$	Removal of NH_4^+ by distillation → distillation with Devarda's alloy
$NH_4^+ + NO_3^- + NO_2^-$	Distillation with Devarda's alloy

and analysed in the same way as NH_3. There are several modifications of this method which can be used to determine the different inorganic N components, as shown in Table 5.9. Individual ionic concentrations, total inorganic N or a combination of two different ions can easily be determined. Since the concentration of NO_2^- is normally minimal under aerobic conditions, reaction with sulfamic acid is usually not required. Only NH_4^+ and NO_3^- are present in most samples. However, NO_2^- may be present under anaerobic conditions in sediments and some sludges.

- *Reduction of nitrate to nitrite.* Nitrite is then determined by spectrophotometry.

Direct methods for nitrate determination include spectrophotometry and ion selective electrodes.

5.9.4.2 Materials
- Same as for the distillation and analysis of ammonium in Section 4.10.2
- Devarda's alloy, finely powdered. If it gives high blanks in the nesslerisation procedure, purify by heating at 120–150 °C for 1 h.
- Potassium chloride solution, 2 M, prepared by dissolving 150 g of KCl in 800 mL of water and making up to 1 L

5.9.4.3 Experimental Procedure. Extract the soil, or other solid sample, using KCl in the same way as for NH_4^+ above in Section 5.9.3. Transfer the extract into a 250 mL round-bottom flask and distill according to the method described in Section 4.10.2 for distillation of NH_3 in water after addition of NaOH. This removes NH_3 from the sample extract, which you may determine in the distillate receiving solution. Replace the receiving flask with a new one containing fresh receiving solution. Add 0.5 g Devarda's alloy and 50 mL water to the residue in the distillation flask. Connect the distillation flask to the rest of the apparatus and heat until boiling. Reduce heat and distill slowly. Proceed as described in Section 4.10.2. Determine NH_4^+ in the distillate receiving solution by one of the methods specified in the experiment or by the spectrophotometric method given above for extractable NH_4^+. Convert to $mg\,N\,kg^{-1}$ sample as shown above in Section 5.9.3. Run blanks through the same procedure. You can convert this to units of mg $N\,(100\,g)^{-1}$ as recommended by the USDA.

If nitrite is not present in the sample this will give you NO_3^--N in the sample, otherwise it will give you the sum of $NO_2^- + NO_3^-$. If NO_2^- is present in the sample determine its concentration by following the procedure given below in Section 5.9.5, then calculate NO_3^- by difference. In any case, you may add the concentration of NO_3^- (or $NO_3^- + NO_2^-$ if NO_2^- is present) to that of NH_4^+ as determined during the first distillation stage to get the total inorganic N.

Note

You may determine nitrate directly by analysing the extract, without the need for distillation, by ion chromatography as described in Section 2.4.6 or the spectrophotometric method outlined in Section 2.5.2.

5.9.5 Exchangeable Nitrite

5.9.5.1 Introduction. Levels of nitrite in soils are generally not determined since they tend to be very low and transitory; nitrite is readily converted to nitrate under aerobic conditions. However, nitrite determination may be required for sludges or sediments. The amount of nitrite may be used as an indicator of reducing conditions. Since nitrites are unstable, analysis should be carried out immediately, or as soon as possible after collection. If this is not possible, storing in a fridge at $5\,°C$ will reduce the rate of conversion to nitrate, but even so, samples should be analysed within a couple of days if representative results are to be obtained. Fresh soil samples must be extracted with water but it will be difficult to avoid the effects of aeration.

5.9.5.2 Materials
- Same as for NO_2^- determination in Section 4.10.4. and for the extraction of NH_4^+ above in Section 5.9.3

5.9.5.3 Experimental Procedure. Fresh sample, rather than dried sample, must be used and the analysis should be performed as soon as possible after sample collection. Extract 10 g of soil, sediment or sludge as described for ammonium above in Section 5.9.3. Analyse 40 mL of the sample extract by following precisely the procedure for NO_2^- in water in

Section 4.10.4. Calculate the concentration of NO_2^- as $mg\,N\,kg^{-1}$ sample. You may use less than $40\,mL$ of the sample extract if the absorbance reading is off scale. You can convert this to $mg\,N\,(100\,g)^{-1}$ as recommended by the USDA.

5.9.6 Mineralisable Nitrogen

5.9.6.1 Introduction.   The amount of organic N mineralised and N mineralising capacity (NMC) of soil, as well as sludge and sediment used for polluted soil remediation, is dependent on the activity of microorganisms. The mineralisation of N is a key stage in the N cycle. This in turn is controlled by temperature, moisture and many other factors and shows strong seasonal trends. The extent to which organic N can be mineralised is an important property of any soil but it is difficult to measure reliably. Estimating amounts that could become available over a period of time requires specific treatment involving either incubation (biological methods) or reaction with organic matter (chemical methods). Many procedures have been suggested. The value of NMC is important in order to calculate the optimum fertiliser application rates and avoid both excessive N uptake by plants and pollution of surface and ground waters.

During the analysis of N availability, the soil (sludge, sediment) is incubated under conditions intended to simulate those considered to be optimal for microbial activity although in practice a wide variety of conditions have been specified. Temperatures range from 25 to 40 °C and incubation periods from 6 to 210 days, although two to four weeks is common. Control of the moisture content raises considerable problems that may be solved by using tightly closed polyethylene bags. Secondary variations include soil pre-treatment and the use of additives. In all incubation methods the increase in mineral N during incubation (as recorded by extraction and/or distillation) is a measure of mineralisable N. The methods given below are an aerobic and an anaerobic procedure.

5.9.6.2 Methodology.   In this procedure the soil is prepared by drying (40 °C) and sieving (< 2 mm) as described in Section 5.2 for other soil tests. An additive (sand) is used and the moisture is brought up to an optimum level for aerobic mineralisation. The 2 M KCl extraction and distillation as described above are used to recover NH_4^+-N and $(NO_3^- + NO_2^-)$-N before and after incubation.

5.9.6.3 Materials
  • Same as that required for Sections 5.9.3, 5.9.4 and 5.9.5 (and in the

cross-referenced experiments given in these), depending on which approach is taken to determining total inorganic-N
- Sand, acid washed, 30–60 mesh
- A porous film

5.9.6.4 Experimental Procedure.   Weigh 5 g air-dried, sieved soil into a 250 mL flask (A) and similarly weigh 5 g into a soil extraction bottle (B). Add 15 g of sand to both and swirl to mix. Then add 6 mL water to each and mix. Extract bottle B immediately with 100 mL 2 M KCl by shaking for 1 h on a rotary shaker. Filter through no. 44 filter paper. Incubate flask A for 14 days at 30 °C, keeping the flask sealed with a porous film. After 14 days, extract in the same way as bottle B. Run sand and water blanks with both A and B. Determine total inorganic nitrogen in the extract as follows. Determine the total inorganic N $(NH_4^+ + NO_3^- + NO_2^-)$ as explained in Section 5.9.4. As an alternative, the individual ions can be determined separately using some of the alternative methods mentioned above (see Sections 5.9.4, 5.9.5, and Table 5.9) and combined to obtain total inorganic-N in the extract.

Calculate mineralisable N from:

$$\text{Mineralisable nitrogen } [\text{mg N} (100 \text{ g})^{-1}]$$
$$= A [\text{mg N} (100 \text{ g})^{-1}] - B [\text{mg N} (100 \text{ g})^{-1}]$$

where A is inorganic-N after incubation and B is inorganic-N before incubation. Correct to dry weight if necessary.

5.9.7 Nitrogen-mineralising Capacity

In this procedure, soil or other materials are composted in polyethylene bags with increasing rates of nitrogen fertiliser at optimum temperature and moisture conditions.

5.9.7.1 Materials
- Same as that required for Sections 5.9.3, 5.9.4 and 5.9.5 (and in the cross-referenced experiments given in these), depending on which approach is taken to determine total inorganic-N
- Ammonium nitrate solution (1 mL ≡ 1 mg N)
- Polyethylene bags
- Thermostatted cabinet

5.9.7.2 Experimental Procedure.   Take five 50 g batches from each soil (sludge, sediment) sample and place each in a polyethylene bag. The sample can be air-dried and sieved (< 2 mm), or with natural moisture, in which case the water content should be determined (see Section 5.4.4). Add to these batches dissolved N fertiliser (ammonium nitrate salt or some other commonly used form) in varying amounts, corresponding to increasing fertilisation rates planned for the various crops: for grain crops, 0, 0.5, 1.0, 2.0 and 2.5 mL of solution with a nitrogen concentration of 1 mg mL^{-1}; for potatoes, corn, rice and industrial crops, 0, 0.5, 1.0, 3.0 and 4.0 mL; for intensively grown vegetables, 0, 1.0, 3.0, 4.0 and 5.0 mL. Set the total moisture content of the soil (sludge, sediment) between 60% and 80% of the FFMC (full field moisture content) by adding the required amount of water. Mix thoroughly. Close the bags tightly and compost in a thermostatted cabinet for 4 weeks at 18–28 °C. The exact moisture content and composting temperature are determined on the basis of the predominant hydrothermal conditions of the region. After the composting is over, determine the total inorganic N content in moist samples (with samples selected for their water content), or in air-dried samples that have been ground and screened, by any of the methods described earlier (see Sections 5.9.4, 5.9.5, and Table 5.7). The results of the determination of mineral N ($NO_3^- $-N + NH_4^+-N) in mg $(100\,\text{g})^{-1}$ soil (or kg ha^{-1}) are plotted as y against N fertiliser application rates in mg $(100\,\text{g})^{-1}$ soil (or kg ha^{-1}) as x and fitted to a quadratic equation of the form:

$$y = a + bx + cx^2$$

The values of the coefficients a, b and c are obtained during the fitting procedure using any suitable computer program. The nitrogen-mineralising capacity (NMC) of analysed materials is determined by solving a quadratic equation and using the formula:

$$\text{NMC} = \frac{a}{a - f(0.1x)}$$

where $f(0.1x)$ is the solution of the quadratic equation y for 0.1 of the first rate of fertiliser N applied x [*i.e.* 0.1 mg N $(100\,\text{g})^{-1}$ soil or 3 kg ha^{-1} for all crops and 0.2 mg N $(100\,\text{g})^{-1}$ soil or 6 kg ha^{-1} for vegetables]. All quantities are in mg $(100\,\text{g})^{-1}$ or kg ha^{-1}.

5.9.8 Ammonia-N in Fresh Sludge and Sediment Samples

Ammonia-N (*i.e.* $NH_3 + NH_4^+$) can be determined in wet sediment and sludge samples by direct distillation without any prior treatment. Weigh a quantity of fresh wet sample equivalent to about 1 g of dry weight in a weighing bottle. Transfer into a 250 mL round bottom flask and add 100 mL water. Distill as described for water samples in Section 4.10.2, collecting about 60 mL of distillate. Make the distillate up to 100 mL and analyse by titration or the spectrophotometric method suggested in Section 4.10.2. Using the titration, the concentration of ammonia-N in the sample can be calculated from:

$$\text{mg NH}_3\text{-N kg}^{-1} = 140 \times V/M$$

where V is the volume of titrant (0.01 M HCl or 0.005 M H_2SO_4) used up and M is the weight of sample (g). You may also express this on a dry weight basis by independently drying another aliquot of the same sample and determining the moisture content. The same method can be used to analyse fresh soils from waterlogged sites, marshes, *etc.*

5.9.9 Questions and Problems

1. Discuss the biogeochemical cycle of N on the local, regional and global scale. What alterations in the N cycle have been initiated by anthropogenic activity?
2. Depending on its concentration in various compartments, N can be an essential element or a pollutant. Discuss the reasons for this dual role of N in the biosphere.
3. What is the origin of organic N in the soil? Discuss the role of organic N in the biogeochemical cycle of nitrogen.
4. Outline the major sources of N in the environment and discuss the problems of acidification and euthrophication in terrestrial and aquatic ecosystems.
5. Explain how inorganic N can be speciated (*i.e.* individual forms of NH_4^+, NO_3^- and NO_2^- determined) using distillation.
6. What is the nitrogen-mineralising capacity of soils? Discuss its application and use.

5.9.10 Suggestions for Projects

1. Investigate the seasonal variation in inorganic nitrogen (NH_4^+, NO_3^-) by analysing soil samples at the same sites during spring–summer–autumn (temperate climate) or dry–wet season (monsoon climate).
2. Determine the content of total organic N at various relief positions from the foot of a hill to the top and in conjugated flood-plain soils.
3. Compare levels of N compounds (TON, NH_4^+-N, NO_3^--N) in agricultural and virgin soils.

4. Compare concentrations of NH_4^+-N and NO_3^--N in rice paddy-field soils and sediments from irrigation channels.
5. Carry out a survey of total organic C and total organic N over a larger area in various types of soils with different land uses. Calculate the C:N ratio on the basis of this survey. Discuss the mineralisation or immobilisation trends in different soils.
6. Determine levels of TON, NH_4^+-N and NO_3^--N in different types of sludges and discuss your results on the basis of technological processes in use at the plants.
7. Measure N-mineralising capacity in different soils, sludges and sediments. Discuss the applicability of these sludges and sediments for the remediation of polluted, disturbed or eroded soil sites.
8. Carry out a study of some of the links in the N biogeochemical cycle (*e.g.* soil–plant–water system) on a local (field) or on a regional scale. Illustrate and discuss these links on the basis of your results.

5.9.11 Further Reading

P. Bielek and V. Kudeyarov (eds.), "Nitrogen Cycles in the Present Agriculture", Priroda, Bratislava, 1991.

P. Grennfelt and E. Thornelof (eds.), Critical loads for nitrogen, *Nord*, 1992, **41**.

P. Gundersen and V. Bashkin, Nitrogen cycling, in "Biogeochemistry of Small Catchments. SCOPE 51", ed. B. Moldan and J. Cerny, Wiley, New York, 1994, pp. 255–284.

W. De Vries "Soil Response to Acid Deposition at Different Regional Scales: Field and Laboratory Data, Critical Loads and Model Predictions", DLO Winand Staring Centre, Wageningen, The Netherlands, 1994.

C. A. Black, "Methods of Soil Analysis", American Society of Agronomy, Madison, 1965, pp. 1300–1390.

V. Bashkin and V. Kudeyarov, Determination of the capacity of soils to mineralize nitrogen as an indication of their nitrogen regime. 1. Method of determining the capacity of soil to mineralize nitrogen, *Sov. Soil Sci.*, 1988, 117–125.

V. Bashkin, V. Kudeyarov and T. Kuznetzova, Technique of soil sample pretreatment for analysis of ^{15}N : ^{14}N isotope ratio, *Commun. Soil Sci. Plant Anal.*, 1986, **17**, 115–123.

5.10 PHOSPHORUS

5.10.1 Introduction

Phosphorus is an essential macronutrient for living organisms. It is a constituent of organic compounds with important structural and meta-

Table 5.10 *Concentration ranges of P in some environmental samples (dry samples are reported on a dry weight basis)*

Sample	P content
Mineral soils	0.02–0.15%
Organic soils	0.01–0.2%
Soil extracts	3–80 mg kg^{-1}
Sludges	0.8–11%
Plant materials	0.05–0.3%
Animal tissues	0.3–4% excluding bone
Freshwater	0.005–0.5 mg L^{-1}
Rainwater	0.002–0.05 mg L^{-1}

bolic functions. The reaction of glucose with adenosine triphosphate to produce a phosphate ester is an important reaction in the human body. Phosphate is also a constituent of nucleic acids and certain lipids. Inorganic phosphate ion plays a key role in cell energy metabolism. Also, inorganic phosphate is a major structural component of bone in vertebrates. Typical concentration ranges of P in the environment are given in Table 5.10.

Phosphorus in the environment occurs almost entirely as phosphate and both inorganic and organic forms are of major importance in plant–soil–water interactions, and in the general phosphorus biogeochemical cycling in natural systems. The majority of agricultural soils usually cannot meet crop demands for P, and fertilisation is required. However, the effectiveness of phosphate fertilisers may be limited by adsorption to certain minerals and by transfer to the organic form. Phosphate from fertilisers can affect micronutrient availability to plants and reduce their migration within landscapes.

The main anthropogenic sources of P are the production of white phosphorus, fertilisers and detergents. Agricultural wastes from feedlots and surface runoff from fields contribute significantly to the eutrophication of fresh and marine waters (see Section 4.1.4).

Different forms of P can be analysed: total P, organic P and P bound up in different soil fractions. The latter are extracted into solution, and different extractants can be used. Extractable forms of P in soil are studied more widely than the extractable forms of most other elements. The main reasons for this include the vital role that P plays in metabolism and its major role in soil and surface water eutrophication. In general, the conventional extractants measure the availability index (or P intensity), which is a measure of the extent to which a soil can release phosphate. Total organic and inorganic P can be determined using dry ashing or wet ashing methods.

Phosphate-P in soil extracts is determined almost exclusively by spectrophotometry using the same method as that for phosphate in water (see Section 4.11.3). When carrying out the extraction procedures, determine the moisture content at the time of weighing.

5.10.2 Total Phosphorus, Total Organic Phosphorus and Total Inorganic Phosphorus

5.10.2.1 Methodology. A sample is ignited at 550 °C to oxidise all the organic matter before extracting into H_2SO_4 and analysing. This measurement gives the total phosphorus content. A non-ignited sample is also extracted into H_2SO_4 and analysed. This gives the inorganic phosphorus content. The organic phosphorus content is determined by difference:

Organic P = P in ignited sample − P in non-ignited sample

5.10.2.2 Materials
- Muffle furnace for igniting at 550 °C
- Porcelain crucible
- Sulfuric acid, 0.5 M
- Filtration apparatus (*e.g.* Buchner funnel) and acid-resistant filter paper

5.10.2.3 Experimental Procedure. Weigh two 1.0 g portions of a solid (soil, sediment, *etc.*) sample. Transfer 1.0 g into a porcelain crucible and place in the muffle furnace. Slowly increase the temperature over a 1 h period until it reaches 550 °C. Leave crucible in muffle furnace for 1 further h at 550 °C. Remove the crucible, cool and transfer the ignited soil into a 100 mL polypropylene bottle. In another bottle place the second 1.0 g portion of the soil sample. Add 50 mL 0.5 M H_2SO_4 to each of the bottles, cap and shake on a rotary shaker for 16 h. Filter through the acid-resistant filter paper. Pipette a 10 mL aliquot of the extract into a 50 mL volumetric flask and proceed to Section 5.10.7 below. Calculate the concentrations of phosphorus as indicated in Section 5.10.7 below. The concentration in the ignited sample corresponds to the total phosphorus, the concentration in the non-ignited sample corresponds to the inorganic phosphorus and the difference corresponds to the organic phosphorus.

Note

You will have to neutralise the aliquot of extract that you will analyse according to Section 5.10.7. In this case you will add 5 drops of the specified indicator and add 5 M NaOH dropwise until a yellow colour appears.

5.10.3 Truog's Extraction

This is widely used for all but the most calcareous soils.

5.10.3.1 Extracting solution
- Truog's reagent: 0.001 M H_2SO_4, buffered at pH 3, prepared by diluting 200 mL 0.05 M H_2SO_4 with water to 10 L and adding 30 g $(NH_4)_2SO_4$. Mix well.

5.10.3.2 Procedure.
Weigh 2.5 g air-dried, sieved (<2 mm) soil into a polyethylene bottle. Add 500 mL Truog's reagent. Cap the bottle and shake for 30 min on a rotary shaker. Filter through a Whatman no. 44 paper. Include blank determinations. Proceed to Section 5.10.7 below to analyse the sample extracts and blanks.

Note

Since the amount of extractable phosphorus increases as a result of drying, it may be preferable to extract fresh samples. In this case double the sample weight to be taken. For wet, peat soils an even larger amount (25 g) will be needed.

5.10.4 Olsen's Extraction

This extractant is appropriate for chalk and limestone material owing to the pH (8.5). It also extracts some organic matter, but organic phosphorus will not be estimated by the recommended spectrophotometric procedure. Constant temperature conditions are essential for Olsen's extraction.

5.10.4.1 Extracting Solution
- Olsen's reagent: 0.5 M $NaHCO_3$ buffered at pH 8.5. Dissolve 210 g $NaHCO_3$ in water in a large aspirator and add 100 mL M NaOH. Dilute to 5 L and mix well. Check that the pH is 8.5 ± 0.05.

5.10.4.2 Procedure.
Weigh 5 g air-dried, sieved (<2 mm) soil into a polyethylene bottle and add 100 mL Olsen's reagent. Cap the bottle and shake for 30 min on a rotary shaker. Filter through no. 40 or 44 filter paper. Include blank extractions as before. Proceed to Section 5.10.7 below to analyse the sample extracts and blanks.

5.10.5 Acetic Acid Extraction

This is less widely used for P than some other extractants; however, it can give useful data, especially for sandy soils.

5.10.5.1 Extracting Solution
- Acetic acid, 2.5% v/v, prepared by diluting 250 mL glacial acetic acid to 10 L and mixing

5.10.5.2 Procedure. Weigh 5 g air-dried soil into a bottle and add 200 mL extractant. Shake for 1 h on a rotary shaker. Include blank determinations as before. Filter through Whatman no. 40 paper into polyethylene bottles and reject the first 5–10 mL. Run blanks with extracing solution only. Proceed to Section 5.10.7 below to analyse the sample extracts and blanks.

Notes

1. The note under Truog's extraction applies, with the exception that Olsen's reagent is not recommended for peat type soils.
2. The procedure used for cations with a 25 : 1 ratio is less suitable for phosphorus, except for sandy soils.
3. Acetic acid must not be used as an extractant for calcareous soils.
4. The extract must be neutralised with 10% v/v H_2SO_4 (use 0.1% nitrocresol in alcohol as indicator) before colour development in the molybdenum blue procedure.

5.10.6 Water Extraction

This method determines the phosphate levels in water extracts of the soil that are available to plants. This can be used as an index of availability.

5.10.6.1 Procedure. Weigh 5 g of air-dried soil in a polyethylene bottle and add 50 mL of laboratory grade water. Shake continuously for 5 min. Repeatedly filter through a Whatman no. 42 filter paper until you obtain a clear extract. Analyse as in Section 5.10.7 below but do not add the indicator or neutralise the extract. Also do not add indicator or neutralise the calibration standards used for this determination.

5.10.7 Analysis

5.10.7.1 Materials. As in Section 4.11.3 plus:

- Hydrochloric acid, 5 M
- Sodium hydroxide, 5 M
- Nitrocresol, 0.25% (w/v)

5.10.7.2 Experimental Procedure. Pipette 10 mL of the sample extract into a 50 mL volumetric flask. Add 5 drops of 0.25% nitrocresol and neutralise with either 5 M HCl or 5 M NaOH if necessary. Dilute to just under 40 mL and add 8 mL of the colour developing reagent (see Section 4.11.3). Make up to the mark and determine the concentration by following the procedure for orthophosphate outlined in Section 4.11.3. Follow precisely the procedure given in Section 4.11.3 for analysis of orthophosphates in water with the following modification. Prepare the calibration standards by dilution of the standard phosphate solution but include a volume of the extracting reagent comparable to that in the sample aliquot (*i.e.* 10 mL in this case). Neutralise the standards before adding the colour reagent. Analyse and calculate the concentration of phosphate-P in the sample extract and convert to $mg\,P\,kg^{-1}$ soil according to:

$$PO_4\text{-}P\,(mg\,kg^{-1}) = C \times V/M$$

where C is the concentration of PO_4-P in the extract ($\mu g\,P\,mL^{-1}$), V is the total volume of extract (100 mL) and M is the weight of the sample (5 g).

Notes

1. You may analyse more or less of the sample extract as the need arises, in order to fit the sample absorbance readings within the range of the calibration graph.
2. You may use a cell with a longer path length (*e.g.* 5 cm) to increase sensitivity.

5.10.8 Questions and Problems

1. What are the main environmental problems caused by P?
2. Explain the principle of the dry ashing method for determining the inorganic and organic fractions of P.
3. What is the purpose of ascorbic acid in the molybdenum blue method?
4. What are the main sources of P in municipal sludges and effluents?
5. Why are P tests so widely used?
6. Which of the different forms of P do you think is most available to plants, and why?
7. A soil sample was analysed and the total P content found to be 0.05%. The soil bulk density was $0.9\,g\,cm^{-3}$. How much P is present in 1 ha of soil? Assume a soil depth of 20 cm.

5.10.9 Suggestions for Projects

1. Determine the content of phosphate in various sludges from different plants (sewage works, paper mill, *etc.*).
2. Collect samples from soils at various sites: agricultural fields, animal feedlots, urban areas, *etc.*, and determine the phosphate content. Attempt to identify the sources of P at the different sites.
3. Typically, the bottom sediments of small water reservoirs (ponds, small lakes) are formed by surface runoff from surrounding watershed areas. Determine the content of phosphate in soils and bottom sediments and comment on the sources and inputs of P to the reservoirs.

5.10.10 Further Reading

J. Murphy and J. P. Riley, A modified single solution method for the determination of phosphate in natural waters, *Anal. Chim. Acta*, 1962, **27**, 31–36.

S. R. Olsen and L. E. Sommers, Phosphorus, in "Methods of Soil Analysis, Part 2. Chemical and Microbiological Properties", 2nd edn., ed. A. L. Page, ASA-SSSA, Madison, 1982, pp. 403–430.

S. E. Allen (ed.), "Chemical Analysis of Ecological Materials", 2nd edn., Blackwell, Oxford, 1989, pp. 134–142.

F. S. Watanabe and S. R. Olsen, Test of an ascorbic acid method for determining phosphorus in water and $NaHCO_3$ extracts from soil, *Soil Sci. Soc. Am. Proc.*, 1965, **29**, 677–678.

5.11 SULFUR

5.11.1 Introduction

Sulfur is an essential element, and most of the S in living organisms is found in amino acids and lipids. It is ubiquitous in the environment, where it can be found in four different oxidation states: -2 in H_2S, 0 in elemental S, $+4$ in SO_2 and $+6$ in H_2SO_4. Deposits of elemental sulfur, various sulfide ores and some sulfates are industrially exploited. Sulfides such as H_2S may form under reducing conditions in marsh soils, sediments and various sludges. Sulfate ions are readily available and plant deficiencies are not commonly observed. Wet deposition is a major source of sulfate in soils. Between 10 and $30\,kg\,S\,ha^{-1}$ is supplied by rainfall annually. Another source of sulfate in agricultural soils is superphosphate fertilisers, which are used as a major source of P. These contain about 12% S.

Environmental problems of S pollution are related to acid deposition

Table 5.11 *Concentration ranges of S in some environmental samples (dry samples are reported on a dry weight basis)*

Sample	S concentration
Mineral soils	0.03–0.5%
Organic soils (peat)	0.03–0.4%
Soil extractions	100–500 mg kg^{-1}
Plant material	0.08–0.5%
Animal tissue	0.2–0.8%
Rainwaters (SO_4^{2-}-S)	0.1–30 mg L^{-1}
Freshwater (SO_4^{2-}-S)	0.4–300 mg L^{-1}

and long-range transport of SO_2 originating from fossil fuel combustion. Sulfate is relatively harmless to plants and can be accumulated, but sulfur dioxide gas can be toxic, as shown in experiments. Although some of the research into the effects of SO_2 pollution on vegetation has been inconclusive for both crops and forests, other work indicates that air pollution contributes to forest decline (see also Chapter 2, Sections 3.3 and 6.1.1).

Sulfur is usually present in soils at between 0.03% and 0.5% on a dry basis, depending on the parent rock. Sulfur contents of up to 5% may be found in arid regions with gypsum deposits and in coastal marshes where pyrite (FeS_2) accumulates. Most of the S in soils from humid and semi-humid regions is in organic form ($>95\%$), but the ratio of organic/inorganic S varies considerably with soil type and depth. Under aerobic conditions, most of the inorganic S is present as sulfate, while sulfide and elemental S predominate in anaerobic conditions. Concentrations of S in the environment are given in Table 5.11.

Soil S analysis is generally used in conjunction with plant S tests in order to relate critical concentrations to crop yield. A critical levels of 30 mg S kg^{-1} soil has been reported for rice, corn and cabbage. Different extraction procedures have been used for measuring available sulfate, including water, various electrolyte solutions (NaCl, KCl, $CaCl_2$), sodium bicarbonate, ammonium acetate, HCl, *etc.* Total S is extracted by fusion or by wet ashing. The most effective wet ashing technique is digestion in a mixture of HNO_3 and $HClO_4$. Obviously, sulfuric acid digestion cannot be used.

5.11.2 Methodology

Sulfur is extracted by fusion and determined as sulfate by turbidimetry.

5.11.3 Experimental Procedure

Extract the sample using one of the fusions described in Section 5.3.1 and dilute to volume. Also prepare a blank as specified in the procedure. Determine the sulfate concentration in the sample and blank extracts by following the instructions in the procedure for water analysis as given in Section 4.14. If the sample extract contains sulfate beyond the range of the calibration graph, dilute to fit within the curve. Also analyse the blank and subtract if necessary. Evaluate the concentration of sulfates in the extract and then calculate the concentration in the sample from:

$$\text{mg S kg}^{-1} = C \times V/M$$

where C is the concentration of SO_4-S in the sample extract (mg S L^{-1}), V is the volume of extract (mL) and M is the weight of sample that was extracted (g).

Notes

1. Before fusion, mix soil sample (0.5 g) with 2.5 g Na_2CO_3 and preheat at 450 °C for 30 min.
2. Addition of 0.2 g $NaNO_3$ or 0.1 g Na_2O_2 improves the flux.
3. You may use any wet digestion method that does not involve H_2SO_4.

5.11.4 Further Reading

S. E. Allen (ed.), "Chemical Analysis of Ecological Materials", 2nd edn., Blackwell, Oxford, 1989, pp. 150–154.

5.12 EXCHANGEABLE CATIONS (Ca^{2+}, Mg^{2+}, K^+ AND Na^+) AND CATION EXCHANGE CAPACITY (CEC)

5.12.1 Introduction

Clay and humus in soils and sediments consist of colloidal particles with very large surface areas and these have electrical charges associated with their surfaces. Most of the clay particles are negatively charged, except in tropical acid soils where they may be charged positively. The surface charges are usually neutralised by electrostatic attraction of ions having opposite charge. Ions which are held electrostatically on the surface of the soil colloids can be replaced by other ions from the soil solution. The process of replacement is referred to as *ion exchange*, and it can involve either *anion exchange* or *cation exchange*. The rate of ion exchange is extremely fast; the ions are replaced almost instantaneously. In order to maintain electroneutrality in soils, exchange reactions occur in equivalent amounts:

$$\text{Cation exchange: clay-Ca} + 2H^+ \rightleftharpoons 2H\text{-clay} + Ca^{2+}$$

$$\text{Anion exchange: clay-SO}_4 + 2OH^- \rightleftharpoons 2OH\text{-clay} + SO_4^{2-}$$

The ions which are bound to clay mineral surfaces by weak electrostatic forces and that can be replaced by ion exchange processes are called *exchangeable ions* (anions or cations). The cation (or anion) exchange capacity of a soil is defined as its capacity to adsorb and exchange cations (or anions). Cations which are generally held by electrostatic forces on soil particles are Ca^{2+}, Mg^{2+}, K^+, NH_4^+, with increasing H^+ and Al^{3+} in acid soils, and Na^+ in saline soils. Cation exchange capacity (CEC) is more widely determined in environmental analysis than anion exchange capacity. Anion exchange capacities are small compared to CEC. The cation exchange capacity is a quantitative measure of all the cations adsorbed on the surface of soil or sediment particles:

$$\text{CEC} = \Sigma \text{ exchangeable cations (meq) per 100 g soil}$$

CEC is usually expressed in milliequivalents (meq) per 100 g of soil (or sediment). The USDA, Soil Survey Division, uses meq per 100 g of clay as a unit. From a chemical point of view it is important to distinguish between ions held by means of electrostatic attraction that can participate in ion exchange reactions, and substances adsorbed onto the surfaces of soil particles by chemical bonds (chemisorbed). The number of base (Ca, Mg, K, Na) cations held on the soil in comparison to the total number of sites available is called *base saturation*:

$$\% \text{ Base saturation} = \frac{\Sigma \text{ exchangeable Ca, Mg, K, Na}}{\text{CEC}}$$

Hydrogen saturation is the amount of exchangeable H^+ divided by the CEC. The greater the charge of a cation, the more forcefully it is held by the clay particle. The strength of the attraction between soil particles and different cations decreases on going from Al^{3+} to H^+.

Adsorption and ion exchange processes play an important role in soil physics and chemistry, fertility, nutrient retention in soils, nutrient uptake by plants, fertiliser and lime application, surface and ground water quality, *etc.* Cations held on soil particles are generally available to plants by exchange with H^+ ions produced by the respiration of plant roots. Nutrients added to the soil as fertilisers are retained by the colloidal surfaces and prevented from leaching. Cations which may pollute groundwaters may be filtered by the adsorptive and ion exchange action of soil particles. The adsorption complex (clay and humus) serves the soil as a store of nutrients. It is also a significant contributor to the buffering capacity of soils. Furthermore, the cation exchange capacity can be used to determine the amount of lime that needs to be applied to reduce acidification.

Exchangeable cations may also affect the physical properties of the soil. Calcium and Mg in exchangeable form can promote good soil structure and tilth. Potassium is an essential nutrient. On the other hand, large amounts of Na in exchangeable form may have a harmful effect on soil structure. Excessive Na may cause the soil to disperse, with the consequent destruction of soil structure and loss of pore space. A dispersed soil is plastic and sticky when wet. When dry, a dispersed soil is hard, and hence it is much less permeable to air and water. The concentration and distribution of exchangeable cations in soil can also be used as a measure of the rate and degree of leaching and weathering. These parameters are especially important in determining soil susceptibility to acidification.

There are two general approaches to determining CEC:

- Measure the amount of exchangeable cations (Ca^{2+}, Mg^{2+}, K^+, Na^+, H^+) displaced from the soil by an added cation (*e.g.* NH_4^+ from ammonium acetate).
- Measure the amount of ion (*e.g.* NH_4^+) adsorbed by the soil from an added solution (*e.g.* ammonium acetate) while displacing naturally present exchangeable cations (Ca^{2+}, Mg^{2+}, K^+, Na^+, H^+). The amount of added cation adsorbed by the soil is determined by displacing with another cation (*e.g.* K^+ from KCl) and its concentration taken as the CEC.

The two procedures can produce different results. In the first procedure, Ca^{2+}, Mg^{2+}, K^+ and Na^+ are determined using one analytical method, whereas H^+ is determined with a different method. A potential source of error in the second procedure is the difference in adsorption characteristics of the displacing ion and the naturally present exchangeable ions. For example, the displacing ion may be adsorbed to a greater or lesser extent than the naturally present cation it displaces.

5.12.2 Materials

- 250 mL round-bottom flask with splash head and condenser

- Ammonium acetate, 1 M, pH 7.0. Weigh 77.1 g of ammonium acetate in a 1 L beaker and add 800 mL water. Stir until completely dissolved. Adjust the pH to 7.0 by dropwise addition of glacial acetic acid while stirring. Transfer the solution into a 1 L volumetric flask and make up the mark with water.
- Potassium chloride, pH 2.5. Weigh 100 g of KCl in a 1 L flask and add 800 mL water. Shake until dissolved. Add 3.0 mL 1 M HCl and make up to the mark with water.
- Sodium hydroxide, 40%
- Boric acid (H_3BO_3). Dissolve 20 g in water and dilute to 1 L.
- Ethanol, 95%
- Hydrochloric acid, 0.01 M
- Mixed indicator. Dissolve 0.1 g methyl red and 0.1 g bromocresol green in 250 mL ethanol. Store in an amber-coloured bottle.
- Borate buffer solution. Dissolve 9.5 g $Na_2B_4O_7 \cdot 10H_2O$ in water and dilute to 1 L. Place 500 mL of this solution in a 1 L volumetric flask. Add 88 mL NaOH and make up to the mark with water
- Atomic absorption spectrometer or flame photometer
- Mechanical shaker
- Filter paper (Whatman no. 44)
- Filter funnels

5.12.3 Exchangeable Cations (Ca^{2+}, Mg^{2+}, K^+, Na^+)

This involves replacing the exchangeable cations in a soil and measuring their concentration.

5.12.3.1 Experimental procedure. Weigh 5.0 g of soil and place in a 100 mL polyethylene bottle. Add 25 mL ammonium acetate solution and shake the mixture for 1 h. Filter the supernatant directly into a 100 mL volumetric flask through a filter paper held in a funnel inserted in the neck of the flask, but leave the soil in the bottle (*i.e.* do not pour any soil into the filter funnel). Use a pipette to withdraw the supernatant if you need to. Add 20 mL 95% ethanol to the bottle and shake. Allow to settle and filter the supernatant into the same 100 mL flask as before. Repeat this washing, shaking and filtering procedure twice more, each time letting the soil remain in the bottle. Make up the extract to 100 mL with laboratory water and determine the concentrations of exchangeable cations (Ca^{2+}, Mg^{2+}, Na^+ and K^+) by AAS as described in Section 2.5.1 (you may use a flame photometer instead).

Convert the concentrations to meq per 100 g of soil as follows:

$$\text{mg Ca g}^{-1} \text{ soil} = 0.1 \times C_{Ca}/5$$

where C_{Ca} is the concentration of Ca in the soil extract in $mg\,L^{-1}$.

$$mg\,Ca\,(100\,g)^{-1}\,soil = 100 \times 0.1 \times C_{Ca}/5 = 20 \times 0.1 \times C_{Ca}$$

and

$$meq\,Ca\,(100\,g)^{-1}\,soil = 2 \times C_{Ca}/20.0$$

Similarly:

$$meq\,Mg\,(100\,g)^{-1}\,soil = 2 \times C_{Mg}/12.2$$
$$meq\,Na\,(100\,g)^{-1}\,soil = 2 \times C_{Na}/23.0$$
$$meq\,K\,(100\,g)^{-1}\,soil = 2 \times C_{K}/39.1$$

5.12.4 Cation Exchange Capacity (CEC)

The CEC is determined by measuring the amount of NH_4^+ retained by the soil while replacing the exchangeable cations in the previous procedure.

5.12.4.1 Experimental Procedure. Mix the extracted soil from the previous procedure, Section 5.12.3, with 25 mL KCl (pH 2.5) solution. Shake for 30 min and filter into a 100 mL volumetric flask. Leach the soil on the filter with successive small portions of KCl solution into the flask and make up to the mark. Transfer 50 mL of the filtrate into a 250 mL round-bottom flask. Dilute with 50 mL of water. Add borate buffer and proceed to distill according to the procedure for ammonia in water (Section 4.10.2). Analyse the distillate using either titrimetry or the spectrophotometric method recommended in Section 4.10.2. If using titrimetry, calculate the NH_4^+ in the soil from:

$$mg\,NH_4^+\text{-}N\,kg^{-1} = 140 \times F \times V/M$$

where V is the volume of titrant (0.01 M HCl or 0.005 M H_2SO_4) used up (mL), M is the weight of the sample (g) and F is the inverse of the fraction of the extract taken for distillation (*i.e.* in above case, $F = 1/\text{fraction}$ distilled $= 1/0.5 = 2$). This can be expressed as $meq\,kg^{-1}$ by dividing by the atomic mass of N (14). Convert this to $meq\,(100\,g)^{-1}$ to give you the CEC:

$$CEC\,[meq\,(100\,g)^{-1}] = NH_4^+\text{-}N\,[meq\,(100\,g)^{-1}] = F \times V/M$$

Since:

$$CEC = \Sigma \text{ exchangeable cations } [\text{in meq } (100 \text{ g})^{-1}]$$

you may calculate the exchangeable H^+ as:

$$\text{Exchangeable } H^+ = CEC - \Sigma \text{ exchangeable Ca, Mg, Na and K}$$

using the concentration of exchangeable bases determined in the first procedure. You may also calculate the % base saturation as:

$$\% \text{ Base saturation} = \frac{\Sigma \text{ exchangeable Ca, Mg, K, Na} \times 100\%}{CEC}$$

Note

You may vary the volume of the KCl extract used for distillation, depending on how much NH_4^+ is adsorbed in the sample. In that case, adjust the calculation by changing factor F in the above equation.

5.12.5 Questions and Problems

1. What is the cation exchange capacity and why is it an important determination in soils and sediments?
2. What is the difference between cation and anion exchange?
3. What is the difference between adsorption and ion exchange?
4. Describe the role of exchangeable cations in soil physical properties?
5. Describe the two main approaches to CEC determination. What are their advantages and disadvantages?
6. What is the meaning of the term *base saturation*? How can you determine this value?

5.12.6 Suggestions for Projects

1. Carry out a survey of CEC in soils and sediments from sites with different uses (surface water reservoirs, agricultural land, forests, *etc.*). Calculate the values of base saturation and explain the actual and potential influence of acid deposition on soil and water acidification.
2. Investigate the relationships between CEC values and soil purification from pollutants in a model experiment. For example, shake soils with different CEC values (sandy and clayish, low and high in organic matter, cultivated and virgin, *etc.*) with a solution containing a polluting substance (heavy metal salts, NH_4 compounds, *etc.*). After

shaking, filter the mixture and determine the amount of pollutant in the filtrate. Explain the results.
3. Measure the pH, CEC and base saturation values in soils from intensively cultivated crop fields. Calculate the amount of alkaline materials (limestone, dolomite) required to increase the pH values by 1 unit.

5.12.7 Further Reading

K. H. Tan, "Soil Sampling, Preparation and Analysis", Dekker, New York, 1983, pp. 114–127.
J. D. Rhoades, Cation exchange capacity, in "Methods of Soil Analysis, Part 2, Chemical and Microbiological Properties", ed. A. L. Page *et al.*, ASA-SSSA, Madison, 1982, pp. 149–157.
W. H. Hendershot, H. Lalande and M. Duquette, Ion exchange and exchangeable cations, in "Soil Sampling and Methods of Analysis", ed. M. R. Carter, Lewis, Boca Raton, 1993, pp. 167–176.

5.13 HEAVY METALS

5.13.1 Introduction

Some heavy metals are pollutants with harmful influences on natural ecosystems and human health (*e.g.* Hg, Pb), while others are essential nutrients (*e.g.* Zn, Fe, Cu, Co, Mo). However, even these micronutrients can become harmful if present in excessive amounts. Heavy metal pollution of terrestrial ecosystems is of concern for a number of reasons. Pollutants in the soil may be absorbed through the roots together with soil water in which they are dissolved, and they may either cause injury to the plants or pass through the food chain when these plants are eaten. Also, metals present in atmospheric aerosols, rainwater or fogwater may be deposited onto plant surfaces. Unlike organic pollutants, which are broken down to a greater or lesser extent depending on their reactivity, metals cannot be degraded and will remain in the soil permanently unless they are leached out. However, when they are leached out, heavy metals move into surface and ground waters and may eventually end up in drinking water. The pH of the soil is a controlling factor in the leaching of many metals from soils: the greater the acidity, the more of the metal will leach out. Of particular concern is the leaching of toxic aluminium from soil in acidic conditions. Aluminium can have damaging effects both on terrestrial and aquatic ecosystems, and the major concern over acid rain is related to the leaching of aluminium.

Anthropogenic sources of heavy metals include mines, smelters and numerous other industrial and agricultural activities. Heavy metals have been used in many pesticide preparations precisely because of their toxic properties. Emissions of air pollutants from metal-ore smelting, including heavy metals such as Ni, Cu, Co and Ti, together with SO_2, have been implicated as the cause of serious damage to terrestrial and aquatic ecosystems in the Sudbury area in Ontario (Canada) and in the Kola peninsula (Russia). Urban dusts contain many heavy elements emitted by industrial and other urban activities. Urban aerosols or resuspended street dusts may be deposited on vegetable gardens in urban areas. Urban street dust may also be picked up by children while playing outdoors and ingested. Of major concern are Pb and Cd. Sewage sludges and sludges from various industrial processes can have extremely high levels of heavy metals (Table 5.3) and when these sludges are disposed of in the environment the toxic metals are released. A common practice is to spread sewage sludge on agricultural land owing to its high nutrient content; however, the harmful effects of heavy metals in sludge may cancel its beneficial effects. For this reason, the heavy metal content of sludges used for agricultural purposes is carefully controlled. Landfill sites are used to dump solid wastes from various sources: domestic refuse, sewage sludge, furnace ash and fly ash from power stations and industrial boilers, incinerator residue, sludge from paper mills, foundry waste, and so on. These wastes generally have a high concentration of heavy metals, the proportion of the different metals varying from source to source. Metal pollutants can easily be mobilised from poorly maintained landfills. Concentrations of some heavy metals in landfill soils are compared to those of typical soils in Table 5.12.

Table 5.12 *Concentrations of heavy metal pollutants in soils*

Metal (mg kg^{-1})	Typical soils	Landfill soils
Cd	<0.01–8	16–50
Co	0.3–200	–
Cr	0.9–1500	28
Cu	<1–390	440–11 000
Hg	<0.01–5	1.8–39
Mn	<1–18 300	–
Mo	0.1–28	–
Ni	0.1–1520	–
Pb	<1–890	640–34 000
V	0.8–1000	–
Zn	1.5–2000	4500–9500

Metalloids such as B, Si and As are usually included in discussions of heavy metal pollution. Although these are not strictly metals, they are analysed by the same methods (*e.g.* AAS) and are therefore grouped together with the transition metals. Metals that are of major interest in soils are discussed in more detail below.

5.13.2 Aluminium

Aluminium is the third most abundant element in the earth's crust and it is a major component of many common minerals. It is also one of the most frequently analysed metals in soils, sediments, sludges, rocks, minerals and plant and animal tissues. *Aluminosilicates* (kaolin, feldspar, mica, *etc.*) contain significant quantities of Al, but Al is also found in nonsilicate minerals (*e.g.* gibbsite). Mineral soils contain between 1–12% Al, while organic soils (peat) contain between 0.05–0.5% Al. Despite its abundance in soil, only a small fraction is mobile. However, significant quantities of Al may be released in acid soils. Current concerns about acid deposition are focused on long-term soil and water acidification, mainly because of the leaching of soluble Al ions. Mobile Al has an important influence on soil pH. Aluminium is present in different forms depending on the solution pH (Figure 5.6). At pH <5, Al^{3+} is the dominant species and it hydrolyses to release hydrogen ions:

$$Al^{3+} + H_2O \rightleftharpoons AlOH^{2+} + H^+$$

$$AlOH^{2+} + H_2O \rightleftharpoons Al(OH)_2^+ + H^+$$

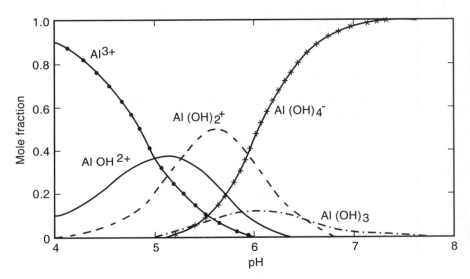

Figure 5.6 *Mole fraction of Al(III) in equilibrium as a function of pH*

The concentration of Al^{3+} decreases with increasing pH as hydroxy-Al complex ions are formed. At pH > 5, some Al precipitates as the neutral $Al(OH)_3$ complex and as the pH increases further the aluminate ion $Al(OH)_4^-$ becomes the predominant species. Aluminium can be held by electrostatic attraction to negatively charged clay particles and it can contribute to the cation exchange capacity (CEC) of the soil. Some mobile Al is organically bound and may be less toxic than the more soluble inorganic forms.

Aluminium is not essential for plant growth, although a beneficial role has been claimed for some plant species. Instead, increasing attention has been given over the years to the potential toxicity of mobile Al to plants growing on acid soils, and tolerance mechanisms have been studied. Aluminium levels in acid soils may be high enough to be phytotoxic, and this is usually remedied by liming the soil. Aluminium also affects nitrogen uptake, and conversely, inorganic nitrogen in the soil may influence the uptake of Al by a particular plant species. The major cause of soil acidification in cultivated soils is the acidifying effect of nitrogen fertilisers. The leaching of Al from soils into groundwaters could be a potential concern, as an association between Al and Alzheimer's disease has been demonstrated. However, the issue of whether Al is the causative agent is hotly debated, and is far from resolved.

5.13.3 Arsenic

Arsenic is found at levels between 1 and $20\,mg\,kg^{-1}$ in rocks and from 0.2 to $40\,mg\,kg^{-1}$ in soils. Soils to which As has been applied in the past can have concentrations $> 500\,mg\,kg^{-1}$, while soils overlying sulfide deposits can have levels as high as $10\,000\,mg\,kg^{-1}$. Although As compounds can be extremely toxic to plants, animals and humans, natural levels of As in soils are generally not considered as being hazardous. The toxicity of As depends on the chemical form. Prior to the 1970s, inorganic As compounds were used extensively in agriculture as pesticides, defoliants and even as animal food additives.

5.13.4 Cadmium

Cadmium is widely used in the manufacture of paints, plastics, batteries and in metal plating. Cadmium is a major concern in sewage treatment works as sewage sludge containing high levels of Cd cannot be used for fertilising soils because of the possible accumulation by plants. Resulting levels of Cd in plant tissue could be hazardous to animals and humans consuming the plants as food. High Cd levels are found in shale-derived soils, and around phosphate and Zn mines. Concentrations of Cd as high as $500\,mg\,kg^{-1}$ have been measured in soils near to mines and smelters. Cadmium can cause high blood pressure, kidney damage and sterility among males. Long-term exposure can cause bone to become brittle. In Japan, in the 1950s, about 100 people died from Cd poisoning caused by the consumption of rice grown on Cd contaminated soils. Since then, the condition has become known as the "itai itai" disease ("it hurts it hurts" in Japanese) because of the pain experienced by sufferers.

5.13.5 Chromium

Major industrial sources of chromium in the environment are nonferrous metal plants, steel works, organic chemicals and petrochemicals, paper and pulp production (paper mills, pulp, board mills, paperboard, building paper), petroleum refining, power plants, textile mills, leather tanning, electroplating, motor vehicles, cement, fertilisers, asbestos products, paints, dyes, fungicides, corrosion resistant or hardened steels, catalysts and more. It is released into the environment either directly during various manufacturing and treatment processes, or when products containing Cr are disposed.

Naturally occurring Cr is ubiquitous in soils and vegetation, although the concentrations are generally very low, except over serpentine deposits. Levels of Cr in some soils originating from parent materials rich in chromite can be relatively high. Average Cr concentrations in sludges tend to be between 100 and $1000\,mg\,kg^{-1}$ but in extreme cases levels of almost $100\,000\,mg\,kg^{-1}$ have been measured. Concentrations of Cr in soils are generally in the range from < 1 to $1000\,mg\,kg^{-1}$, although at some sites levels as high as $10\,000\,mg\,kg^{-1}$ have been reported. Chromium is toxic to animals and humans, but less so to plants. Hexavalent Cr(VI) is a suspected human carcinogen.

Inhalation and ingestion are the main routes of human exposure to Cr. The toxicological effects of Cr poisoning include lung and kidney damage, inactivation of various enzymes and skin disorders. Inhalation of chromium oxide fumes and dust as well as hexavalent Cr salts can lead to lung disease (*e.g.* bronchitis, pulmonary oedema).

5.13.6 Lead

Average concentrations of Pb in soil are between 15 and $25\,mg\,kg^{-1}$. Lead is naturally present in galena (PbS), and soils derived from the weathering of this mineral may have high Pb contents. Major anthropogenic sources of Pb include the use of Pb as a petrol additive, Pb mining and smelting, printing, Pb paint flakes, sewage sludge and the use of pesticides containing Pb compounds (*e.g.* lead arsenate). Lead in suspended atmospheric aerosol is deposited onto the earth's surface and retained in the upper 2–5 cm of the soil. Soils in the vicinity of mines and smelters may have concentrations $> 10\,000$ mg kg^{-1}. Numerous studies have shown a clear impact of road traffic on levels of Pb in the environment. Dust particles can be swept onto roadside soils and vegetation. Concentrations of Pb in soil and grass are found to decrease dramatically with distance from the roadside, as shown in Figure 5.7, and high correlations with traffic density have been found in several urban areas [see Figure 1.8(c)]. Lead can be mobilised from soils in surface run-off following heavy rains. Smaller particles may be blown

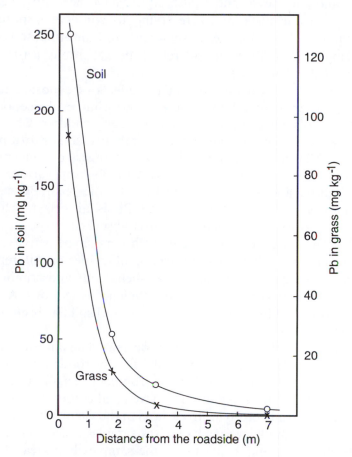

Figure 5.7 *Lead (Pb) in soil and grass as a function of distance from the roadside*

back into the atmosphere by wind. Generally, the downward movement of Pb through the soil as a result of leaching is very slow. Leaching can be accelerated when the solution pH decreases due to acidic rainfall, when substances capable of forming soluble complexes with Pb are present in the soil, or where the Pb concentration in the soil is extremely high. Lead compounds are very persistent; even if all emissions were suddenly to cease, Pb pollution would remain around us for centuries. Very high Pb levels of several thousand $mg\,kg^{-1}$ can be found in soils treated with Pb-containing pesticides. Recently, the recycling of Pb–acid batteries has created problems in the local environment around recycling plants. Most of these plants are located in developing countries of Asia and Latin America and they process batteries imported from industrialised nations. Levels of Pb as high as $60\,000$–$70\,000\,mg\,kg^{-1}$ have been measured in soils in the vicinity of Pb-battery recycling plants in the Philippines, Thailand and Indonesia, and health effects have been observed. This appears to be one instance where trying to conserve resources and minimise pollution has gone seriously wrong. In California, soil containing $1000\,mg\,kg^{-1}$ of Pb is considered to be hazardous waste and its disposal is strictly regulated.

Lead is a well-known poison, but the effects of exposure to lower levels have been contentious. There is growing evidence of "sub-clinical" Pb poisoning, especially among young children who play in polluted parks, gardens and streets. Contaminated soil or dust particles may be transferred to children's hands and ingested accidentally or intentionally (pica). Humans are exposed to Pb from various sources and road dusts and soils can contribute to the total Pb exposure. Approximately one half of the inhaled Pb is absorbed, whereas a smaller fraction of Pb ingested in food is absorbed. Exposure to low Pb levels can cause nervous system disorders, hyperactivity, hypertension, behavioural changes and learning difficulties in children. Some have gone as far as to blame anti-social behaviour and criminality on sub-clinical Pb poisoning, although the evidence is tenuous. A correlation between Pb in blood and Pb in air, dust and soil has been observed in many studies.

Although atmospheric Pb levels have decreased significantly in many countries over the last 10 years as a result of the introduction of unleaded petrol, Pb can remain in the urban environment, usually in dusts and soils. The threat from Pb pollution in developed countries has far from receded. A 1992 study of Pb in street dusts in the vicinity of two London schools found mean Pb levels in half of the samples collected to be higher than $1000\,mg\,kg^{-1}$, and a quarter of the samples had mean Pb levels higher than $2000\,mg\,kg^{-1}$ (Evans *et al.*, 1992). In the US, concern continues to be expressed about Pb pollution. Some countries and local authorities have adopted standards for Pb in soils intended for residential use, and these are as follows (in $mg\,kg^{-1}$): US guideline 500, Minnesota proposed emergency level 500, Canada 375–500 depending on soil type, UK 500 for re-developed industrial sites, Netherlands clean-up value 600. Unlike for other environmental media, standards for pollutants in dust particles have not been widely introduced. In 1980 the Greater London Council suggested an action level of $5000\,mg\,kg^{-1}$ for Pb in dust, and recommended that a guideline value of $500\,mg\,kg^{-1}$ should be adopted. Maximum permissible levels of Pb in dust and soil of between < 100 and $1000\,mg\,kg^{-1}$ have been recommended by experts. Standards $< 100\,mg\,kg^{-1}$ have been suggested in order to protect children from pica Pb.

5.13.7 Manganese

Total Mn concentrations in soils are highly variable (< 20 to $> 3000\,mg\,kg^{-1}$), but only some of this is available to plants. Soluble Mn^{2+} ions are the most readily available of all the different forms of Mn in soil. Deficiency and toxicity are both possible, depending on the Mn content of soil and the prevailing conditions. The availability of Mn is strongly influence by the pH and Eh. Reducing and acidic conditions could lead to excessive Mn^{2+} which could have toxic effects. On the other hand, Mn deficiency may occur at high pH in aerobic conditions.

5.13.8 Mercury

The abundance of Hg in the earth's crust is relatively low; however, higher levels are encountered where cinnabar (HgS) and other Hg-containing minerals are found. There are several sources of Hg in the environment, both natural and anthropogenic. Natural sources include volatilisation of Hg owing to its high vapour pressure, followed by its deposition on the earth's surface. Industrial and agricultural uses of Hg and its compounds, especially organomercury compounds, have caused a serious pollution problem.

A number of industrial processes utilise Hg and its compounds, and mercurial waste is considered as the main source of this pollutant. Anthropogenic sources of mercury include chlor-alkali plants, mines, smelters, pulp mills (although less than formerly), coal combustion, plastic and drug manufacture, timber preservation, pesticides, gold and silver extraction, batteries, instrumentation, paint, amalgams, catalysts and sewage sludge. The use of Hg as a seed dressing to inhibit the growth of fungi and its use in pesticides are being restricted. There have been many recorded cases of serious Hg pollution, with lethal consequences. Severe cases of Hg poisoning were reported in Minamata, and other places, in the 1950s (see Section 4.16.5).

The transformation of inorganic Hg to highly toxic organic forms can occur in sediments and is of particular concern. Methylation of Hg has also been demonstrated in soils.

5.13.9 Nickel

Nickel emissions to the atmosphere are mostly anthropogenic; industrial sources account for more than 80% of the total emission. Nickel levels in soils are generally between < 50 and $100 \, \text{mg} \, \text{kg}^{-1}$, but very high levels may be found in some areas, particularly in soils over serpentine deposits. Concentrations of Ni as high as $5000 \, \text{mg} \, \text{kg}^{-1}$ have been reported in some soils. Plants seem to be more sensitive to nickel toxicity than animals, although both can be affected by Ni pollution. Nickel carbonyl, $Ni(CO)_4$, is extremely toxic and any emission is especially hazardous to mammals and humans. Nickel pollution from metal smelting has been reported in Canada, Russia, Australia, Cuba and other countries.

Nickel is introduced into the terrestrial environment as solid waste from metallurgical industries or as deposition of atmospheric emissions. Another major source of Ni pollution is sewage sludge when applied to land. In spite of Ni accumulation in soils, uptake by plants is not sufficient to be of concern in the food chain. Application of phosphate fertilisers to cultivated land is also a source of nickel and this could lead to elevated concentrations since nickel, together with other heavy metals, is found in phosphate minerals in variable amounts. Coal contains varying amounts of Ni (< 2–$150 \, \text{mg} \, \text{kg}^{-1}$ in European and Canadian coals; < 1–$90 \, \text{mg} \, \text{kg}^{-1}$ in Australian bituminous coals; < 2 to $> 300 \, \text{mg} \, \text{kg}^{-1}$ in bituminous coals from the Appalachian region of the USA) and coal burning electric power stations are a source of nickel emissions. Nickel is especially concentrated in the fly ash upon combustion.

5.13.10 Silicon

After oxygen, Si is the most abundant element in the earth's crust. Silicon is widely distributed in nature in various forms. In rocks, Si can be present as quartz and numerous silicates (*e.g.* feldspar), and it is a major component of clay minerals that result from the weathering of rocks.

Silicon is often found together with Al, Fe, Mg and other metals present in the silicate structure. Since clay minerals possess negative surface charges, they are largely responsible for the cation exchange capacity (CEC) of soils. The Si content in soil is greatly influenced by the nature of the parent material. Concentrations of soluble Si in soil water vary between 1 and $40 \, \text{mg} \, \text{L}^{-1}$.

5.13.11 Zinc

Zinc is an essential trace element for plants, animals and humans as it is associated with many enzymes and with certain other proteins. Zinc is relatively more abundant in the earth's crust than some other metals (*e.g.* Cu); however, there are not many minerals that contain Zn. There is only one common sulfide (ZnS), but it forms minerals which are worked in many parts of the world. Clay minerals in soils can also adsorb some Zn. Anthropogenic sources of Zn in the environment include printing processes, construction materials, metals (iron, steel and brass coated with zinc), fertilisers, batteries, sewage sludge, animal wastes in the form of manure (dairy, feedlot, swine or chicken), Zn-containing pesticides (*e.g.* Zineb, Mancozeb and Ziram), atmospheric deposition and coal combustion. The latter can contribute to Zn input through deposition of atmospheric emission and when the residue of coal combustion (furnace ash and fly ash) are disposed of in landfills. Zinc concentrations in soils typically range from 1 to $2000 \, \text{mg} \, \text{kg}^{-1}$, but at some sites levels as high as $10\,000 \, \text{mg} \, \text{kg}^{-1}$ have been reported.

Although levels of Zn in soils are higher than those of copper, molybdenum and other micronutrients, plant requirements are also greater. Zinc deficiencies are known in crop plants, especially in peaty, sandy and chalky soils. High levels of Zn are normally toxic to plants. Concentrations of Zn in natural waters are generally low; however, industrial effluents and drainage from mining areas can act as sources of pollution.

The major health concern in the general population is marginal or deficient Zn intake rather than its toxicity. Zn is generally considered as being of low toxicity due to the wide margin between usual environmental concentrations and toxic levels. However, high levels of Zn are undesirable as it may lead to Cu deficiency by inhibiting Cu absorption. The daily dietary allowances for Zn as recommended in different countries and by the International Commission on Radiological Protection (ICRP) are as follows (in $\text{mg} \, \text{d}^{-1}$): USA (adults and growing children over one year old), 10; UK, 14.3; Japan, 14.4; India, 16.1; Italy, 4.7–11.3; ICRP, 13.0.

5.13.12 Other Metals

Iron, cobalt, copper, molybdenum and selenium are considered to be essential elements, and deficiencies of these metals can be of concern. Iron is the fourth most abundant element in the earth's crust and it is a major constituent of many soils. If it is present in insoluble form, Fe deficiency in plants may occur. On the other hand, if high levels of soluble Fe are present it may act as a phytotoxin. Bacteria in the roots of legumes require Co for the symbiotic fixation of nitrogen. Cobalt deficiency in ruminant animals can result if the levels of Co in plants are too low to meet their dietary needs. Cobalt levels in soils range from 0.2 to 30 mg kg^{-1}. Both Mo deficiencies in crops and toxicities in foraging animals have been reported. Deficiencies of Mo are possible at levels ≤ 0.1 mg kg^{-1}, while toxic effects are observed in cattle feeding on plants with Mo levels > 10 mg kg^{-1}. Molybdenum levels in soils range from 0.1 to 40 mg kg^{-1}. At low levels, Se is an essential element and deficiencies in animals occur if they feed on plants with Se concentrations ≤ 0.05 mg kg^{-1}. However, at high concentrations (> 4 mg kg^{-1}) Se can be toxic as it prevents proper bone formation in animals. Selenium levels in soils range from < 0.1 to 100 mg kg^{-1}. Although Se may be present at trace amounts in soil, it may be accumulated by plants and crops in quantities that could pose a hazard to animals and humans. Selenium pollution may result from its industrial uses: rubber production, glass, pigments, metal alloys and electronics. Copper is one of the most essential elements for living organisms. Copper concentrations in soils are generally < 100 mg kg^{-1}, while in sludges, levels as high as 17 000 mg kg^{-1} have been reported.

Other metals which may be toxic if present at high enough levels include antimony, barium, beryllium and thallium. These have various industrial uses and may be emitted into the environment with wastes.

5.13.13 Analysis

Heavy metals in soils, sediments, sludges and dusts are usually determined by atomic absorption spectrometry (AAS) (both flame and flameless) and atomic emission–inductively coupled plasma (ICP) spectrometry, after acid digestion of the samples. Anodic stripping voltammetry (ASV) and other electrochemical techniques have been used for some of the metals. Spectrophotometric methods based on colour development after addition of specific reagents have also been used. Conventional atomic absorption spectrometry is suitable for most metals; however, for some metals (Hg, As, Se) the method has to be modified. Cold vapour flameless AAS is generally used for Hg (see Section 5.13.19), while hydride generation has been used for As and Se. As the toxity of the metal varies with the form, sometimes it is of interest to know the concentrations of the individual metal-containing compounds rather than the total metal. Speciation of individual metal compounds requires more sophisticated methods, such as X-ray methods for inorganic compounds, and gas chromatography–mass spectrometry (GC-MS) or GC-AAS for organometallic compounds. Digestion of solid samples prior to AAS analysis generally involves HF and $HClO_4$, since these are capable of breaking down silicates and releasing metals into solution. However, other acids have also been widely used for digestion of soils: HCl, HNO_3, aqua regia and various acid mixtures. These other acids may give the total metal concentration, depending on the metal and soil type. Any of the digestions given below is suitable for extracting most heavy metals.

5.13.14 Methodology

Heavy metals are extracted into acid solution and analysed by AAS.

5.13.15 Hydrofluoric Acid–Perchloric Acid Digestion

For determining total metals, the most effective extraction procedure is that involving hydrofluoric acid (HF) and perchloric acid ($HClO_4$). However, these acids are extremely dangerous and facilities for handling them may not be available in all laboratories. If you have access to such facilities, and you require great accuracy in your work, follow the HF–$HClO_4$ extraction procedure given in Section 5.3.2. Then proceed to analyse the metals by AAS as described in Section 4.16.

5.13.16 Nitric Acid Digestion

If you are not able, or willing, to use the HF–$HClO_4$ extraction, you may extract the solid samples into nitric acid (HNO_3) according to the procedure outlined here. This procedure, although not as effective at extracting all the metals, has been widely used for heavy metals and reasonably reliable results can be obtained. This method will not measure the metals that are associated with silicates. However, it will give reliable measurements of the heavy metals in the soil originating from industrial sources as non-silicates, and is therefore suitable for environmental pollution analysis. There are many variations of the HNO_3 method, with some using concentrated HNO_3, $HNO_3 + H_2O_2$ or

$HNO_3 + H_2SO_4$. Different soil/acid ratios have been used, different extraction times, *etc.* Some extract simply by shaking the soil/acid mixture without heating; however, most heat without stirring. The official US EPA procedure for metals in soils uses $HNO_3 + HCl + H_2O_2$ as in Section 5.13.18 below. The aqua regia method given below in Section 5.13.17 is also effective for heavy metal determination, and it is gradually replacing the $HClO_4$ method in soil science because of safety concerns. Bear in mind that nitric acid is also very hazardous and you must carry out digestions in a fume cupboard. Wear safety glasses and protective gloves at all times when handling concentrated acids.

5.13.16.1 Materials
- Atomic absorption spectrometer
- Hot plate with magnetic stirrer
- Nitric acid, concentrated, high purity
- Nitric acid, 0.25 M
- Filter paper, Whatman no. 541, acid washed
- Stock metal solutions, $1000 \, mg \, L^{-1}$ (see Table 4.19)
- Working metal solutions prepared by dilution of the stock solution

5.13.16.2 Experimental Procedure.
Grind some air-dried and sieved ($< 2 \, mm$) soil, sediment, sludge or dust sample. Weigh 1.0 g and place inside a 100 mL tall-form beaker. Add 30 mL of 1:1 HNO_3 (10 mL water + 10 mL concentrated HNO_3) and boil gently on a hotplate until the volume is reduced to approximately 5 mL, while stirring on a magnetic stirrer. Add a further 10 mL of 1:1 HNO_3 and repeat. Cool and filter the extract through a Whatman no. 541 filter paper and wash the beaker and filter paper with successive small portions of 0.25 M HNO_3. Transfer the filtrate and the washings to a 50 mL volumetric flask and dilute to the mark with water. Prepare a series of calibration standards over the required range (see Table 4.20) by diluting aliquots of the working metal solution with 0.25 M HNO_3. Analyse by AAS using the flame and wavelength specified in Section 4.16 for the metal you wish to determine and use background correction. Construct a calibration graph of absorbance (or peak height if a chart recorder is used) against metal concentration and read off the concentration in the sample extract. Calculate the concentration in the solid sample as:

$$\text{Concentration in soil (mg kg}^{-1}) = C \times V/M$$

where C is the concentration of metal in the extract ($\mu g \, mL^{-1}$), V is the volume of extract (mL) and M is the weight of sample (g). Analyse also a dilute nitric acid blank to confirm that there is no contamination in the reagent.

Note

You may use a simpler version of this procedure that doesn't involve heating. Weigh 2 g of air-dried and sieved soil in a polyethylene bottle and add 30 mL 1 M HNO_3. Shake for 2 h and filter. This method is quite effective for many heavy metals, including Pb, Ni and Cd.

5.13.17 Aqua Regia Digestion

This method is replacing the $HClO_4$–HNO_3 digestion in many soil analysis laboratories as it is safer to use.

5.13.17.1 Materials. Same as above in Section 5.13.16.1 plus:

- Aqua regia. Add 130 mL concentrated HCl to 120 mL water and mix. Add 150 mL of this solution to 50 mL concentrated HNO_3 and mix. Prepare and handle this solution with great care.
- Kjeldahl flask or similar digestion flask

5.13.17.2 Experimental Procedure. Grind some air-dried and sieved (<2 mm) soil, sediment, sludge or dust sample. Weigh 1.0 g and place inside a Kjeldahl flask. Add 15 mL of aqua regia and swirl to wet the sample. Stand overnight. The next day, place the flask in the heating block and heat at 50 °C for 30 min. Raise the temperature to 120 °C and continue heating for 2 h. Cool and add 10 mL 0.25 M HNO_3. Filter through a Whatman no. 541 filter paper. Wash the flask and filter paper with small aliquots of 0.25 M HNO_3. Transfer the filtrate and washings to a 50 mL flask and make up to the mark with 0.25 M HNO_3. Make up all standards with 0.25 M HNO_3. Analyse and calculate the results as indicated above in Sections 5.13.16.2 and 4.16.

5.13.18 Nitric Acid–Hydrogen Peroxide Digestion

This method was originally developed by the US EPA for the determination of heavy metals in soils, sediments, and sludges. This method does not measure the true total metal concentration since metals bound up in the crystal lattice of silicate minerals are not released. However, it gives a reasonable measurement of those metals that are available in the soil.

5.13.18.1 Materials. Same as in Section 5.13.16 plus:

- Hydrochloric acid, concentrated

- Hydrogen peroxide, 30%
- Nitric acid, concentrated

5.13.18.2 Experimental Procedure. Weigh 2 g of air-dried and sieved (< 2 mm) soil and place in a 150 mL beaker. Add 10 mL 1 : 1 HNO_3 (*i.e.* 5 mL water + 5 mL concentrated HNO_3), cover with a watch glass and heat on a hot plate for 15 min. Cool and add 5 mL concentrated HNO_3. Heat for another 30 min. Repeat with another aliquot of 5 mL concentrated HNO_3 but do not cover the beaker completely. Heat until the volume is reduced to 5 mL. Cool, and add 2 mL water and 3 mL 30% H_2O_2. Cover the beaker and heat gently. If effervescence is vigorous, remove from the hot plate. Repeatedly add 1 mL aliquots of 30% H_2O_2 and heat until effervescence subsides. Add 10 mL water and 5 mL concentrated HCl, cover with watch glass and heat for 15 minutes without boiling. Cool and filter through a Whatman no. 541 filter paper into a 50 mL volumetric flask. Wash the watch glass and beaker with water and filter into flask. Make up to the mark with water. Analyse and calculate the results as indicated above in Sections 5.13.16.2 and 4.16.

Notes for 5.13.16.2, 5.13.17.2 and 5.13.18.2

1. If the sample is too concentrated, you may dilute it to fit it within the range of the calibration curve.
2. If the sample is dilute, you may make up the extract to a lower final volume (*e.g.* 10 or 25 mL) rather than 50 mL. You may also increase the quantity of sample that you extract (*e.g.* 5 g), but you may also have to increase the volume of acid. Anyway, you can always bring the volume down by heating.
3. Analyse blanks of acid in order to confirm that they are free of metals. You may prepare a blank by following the same procedure but without the soil sample. Use this to correct the results for any background contamination in the acids.

Notes for specific metals

(a) Lead. When analysing Pb, use the line at 217.0 nm rather than the resonance line at 288.3 nm for samples with low Pb concentrations.

(b) Aluminium
1. Use a high-temperature nitrous oxide–acetylene flame. Particular care is necessary when using nitrous oxide–fuel mixtures and the instructions of the instrument manufacturer concerning gas pressures, lighting and extinguishing the flame should be followed carefully.

2. Include an ionisation buffer (2% potassium chloride) to eliminate interference by P and Cl.
3. Greater sensitivity can be achieved by including an organic solvent such as isopropanol in the sample and standard solutions.

5.13.19 Cold Vapour Mercury Analysis

Direct AAS analysis of Hg is generally not employed owing to low sensitivity and signal noise in most conventional flames at the most intense line of 185 nm. Mercury in environmental samples is best determined using the flameless cold vapour technique.

5.13.19.1 Methodology.   Mercury compounds are reduced by Sn^{2+} to metallic Hg, which is then vapourised in a stream of air and swept through an absorption cell placed in the path of the hollow cathode lamp beam (Figure 5.8). The 253.7 nm line is always used rather than the most intense line. The cell has quartz windows which are transparent to radiation of 253.7 nm. Most manufacturers of AAS instruments offer accessories for cold vapour determination.

5.13.9.2 Materials
- Atomic absorption spectrometer
- Absorption cell, 11–15 cm long, 2.5 cm diameter

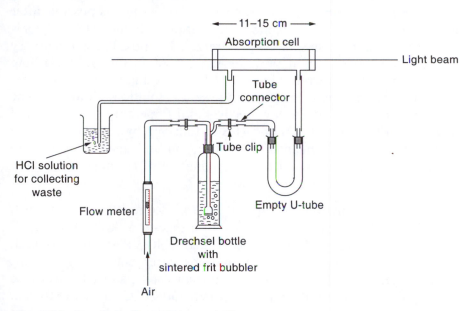

Figure 5.8 *Accessories for cold vapour AAS*

- Air pump or air cylinder with regulator
- Flowmeter
- U-tube
- Clear vinyl plastic tubing (Tygon tubing or similar)
- Drechsel bottle with sintered frit, 250 mL
- Nitric acid, concentrated
- Potassium permanganate solution. Dissolve 60 g $KMnO_4$ in water and dilute to 1 L.
- Potassium persulphate solution. Dissolve 50 g $K_2S_2O_8$ in water and dilute to 1 L.
- Sodium chloride–hydroxylamine solution. Dissolve 50 g NaCl and 50 g $(NH_2OH)_2 \cdot H_2SO_4$ in water and dilute to 1 L.
- Tin(II) chloride solution. To 500 mL of water add 12.5 mL concentrated HCl. Dissolve 50 g $SnCl_2$ in this mixture and dilute to 1 L.
- Sulfuric acid, concentrated
- Water bath and ice bath
- Plastic tubing clips
- Tubing connectors
- Stock Hg solution, 1000 mg L^{-1}. Dissolve 135.4 mg $HgCl_2$ in 70 mL water, add 0.15 mL concentrated HNO_3 and dilute to 100 mL with water.

5.13.19.3 Experimental Procedure. *(a) Sampling and storage.* Samples for mercury determination should be analysed as soon as possible after collection to minimise loss due to volatilisation. If immediate analysis is not possible, store fresh samples in a freezer. The possibility of mercury loss through volatilisation prevents the use of fusion or dry-ashing techniques to bring mercury into solution. Acid digestion methods are therefore normally used for preparing the test solution, although some acid treatments result in low recoveries of mercury. The method described below uses HNO_3 and H_2SO_4 followed by further oxidation with potassium permanganate. The digestion and analysis procedure described here is suitable for the determination of Hg in soils, sludges, sediments and plant and animal tissue.

(b) Mercury extraction. Weigh 1.0 g of soil, sediment, sludge or dust, or 2–5 g of plant material, and place in a digestion flask. Add 10 mL concentrated H_2SO_4 and 5 mL concentrated HNO_3. Place in an ice bath to prevent loss of Hg by volatilisation. After cooling, add 2 mL concentrated HCl. When analysing calcareous soils, add HCl dropwise while shaking. Place the flask in a water bath (50–60 °C) and leave for 2 h. If the suspension is not clear, leave for another hour. Remove flask and

place in an ice bath. Slowly add 5 mL KMnO$_4$ solution while stirring. Continue adding 5 mL of KMnO$_4$ solution at a time until a persistent purple colour remains. Add 5 mL 5% K$_2$S$_2$O$_8$ solution and leave for 5 h in the ice bath to oxidise any organomercury compounds. Transfer to a 100 mL volumetric flask, make up to the mark and stopper. This can be stored for up to 4 days in a refrigerator at 4 °C.

(c) Analysis. Install absorption cell by supporting it on the burner head and align in the path of the light beam to give maximum transmittance. Connect all the pieces of equipment using vinyl plastic tubing, as shown in Figure 5.8. The U-tube serves to prevent any solution from entering the cell. Insert the exhaust tube into dilute HCl to collect the Hg vapour. At the end of the experiment this should be transferred to the Hg waste collection bottle for treatment and disposal according to the regulations in your laboratory. Mercury should not be emitted into the atmosphere through the exhaust hood. Connect a length (*ca.* 5 cm) of flexible vinyl plastic tubing to the glass tube inlet and outlet of the Drechsel bottle and connect these to the rest of the assembly by means of tubing connectors. Use tubing clips attached to these short plastic tube extensions at the inlet and outlet of the bottle to open or close the flow of air through the bottle. Disconnect the Drechsel bottle and head together with the short lengths of plastic tubing from the assembly. Apply tubing clips to each length of extension tubing so as to be able to seal completely the Drechsel bottle when necessary. Remove the head of the Drechsel bottle and transfer the 100 mL extract to it. Add enough NaCl–hydro-xylamine sulfate solution to reduce excess KMnO$_4$, and then immediately add 5 mL tin(II) chloride solution. Insert the bottle head with the absorption tube into the Drechsel bottle with the tubing clips on the extensions still in the closed position, thus sealing the contents. Shake for 2 min and connect the Drechsel bottle to the assembly by connecting the plastic tubing extensions to the tubing connectors. Stir the flask and release the tubing clips, allowing air to bubble through the solution in the bottle at a rate of 2 L min^{-1}. Volatilised mercury is swept into the absorption cell and the absorbance will increase to a maximum. When the recorder returns to the baseline, replace the Drechsel bottle containing the sample with one containing water and continue blowing air through in order to clean the system. Repeat with the next sample or standard.

Prepare a series of 100 mL standards with Hg concentrations in the range of 1–10 μg L^{-1} by diluting the stock solution. Analyse the 100 mL standards in the same way as the samples. Also analyse a 100 mL water blank. Construct a calibration graph of peak height *versus* concentration of Hg in the standards. Read off the concentration in the sample and calculate the concentration in the solid sample from:

$$\text{Concentration in soil (mg kg}^{-1}) = C \times V/M$$

where C is the concentration in the sample extract (μg L^{-1}), V is the volume of the sample extract (L) and M is the weight of the sample (g). Note that V is expressed in L; if you express it in mL, your result will come out in ng g^{-1} or μg kg^{-1}.

Notes

1. If the concentration in the extract is too high, dilute an aliquot of the extract to 100 mL and then analyse.
2. To increase the sensitivity you may increase the amount of sample that you extract.
3. For samples with a high organic content you may have to double the amount of H_2SO_4 and HNO_3 used in the digestion.
4. If the sample contains high levels of chloride, add an extra 25 mL of the sodium chloride–hydroxylamine sulfate reagent and purge briefly with air before adding the tin(II) chloride reagent.

5.13.20 Questions and Problems

1. Explain the difference between "Zn as an essential nutrient" and "Zn as a pollutant".
2. Which other heavy metals play this dual role, acting both as essential nutrients and as toxic pollutants?
3. What are the principal sources of Zn in the environment?
4. Explain the similarity between Zn and Cd from a biogeochemical point of view.
5. What are the usual concentrations of Cr and Ni in various environmental media (soil, sludge, sediments, water, *etc.*)?
6. What are the main anthropogenic sources of Cr and Ni?
7. Describe the toxicological properties of Cr. What is the most toxic form of this element?
8. The total content of Al in many soils is very high but it only becomes toxic in acid soils. Explain this in terms of the speciation of Al with pH.
9. It is well known that the content of free Al^{3+} is highest in old tropical and subtropical soils as well as in northern podzols. What are the reasons for this accumulation?
10. If Fe, Al, Cd and Pb in soil were determined by using both the HF–$HClO_4$ and the HNO_3 digestions, for which metals would you expect

to see the greatest difference in results between the two methods, and for which the least? Explain.

11. What are the main natural and anthropogenic sources of Hg in the environment?

12. Explain the origin and environmental fate of organomercury compounds. Why are they considered as the most poisonous pollutants?

13. What is the function of the absorption cell in cold vapour Hg analysis?

5.13.21 Suggestions for Projects

1. Carry out surveys of various heavy metals in soil over a large area and prepare a map of heavy metals distribution. Calculate correlation coefficients between pairs of metals and identify those metals that are closely associated with each other. What could be the reason for these relationships?

2. Determine the Zn content in various horizons of mineral and organic soils. Illustrate the relationship between humus content, mechanical soil composition and Zn distribution.

3. Determine the concentration of various heavy metals in sludges from different sources, and calculate inputs into an agricultural field on the basis of your measurements and typical sludge application rates.

4. Determine the Cr content of sludges sampled from different industrial plants. Investigate the relationships between the different technological processes and Cr concentrations.

5. Compare the levels of Cr in various layers of clay and sand soils sampled in urban and rural areas.

6. Sample the soil at varying distances from a coal-burning electric power plant (0.1, 0.5, 1, 3, 5, 15, 25 and 50 km). Determine the content of Al^{3+} and soil pH. Explain your results.

7. Sample acid soil (pH <4.5) and measure the content of free Al^{3+}. Add increasing amounts of $CaCO_3$ (10, 20, 40, 60, 100 mg per 100 g of soil) to fresh soil or soil rewetted to its natural moisture content, mix well and put in polyethylene bags. Keep the bags closed for two to three weeks. Measure the resulting content of free Al^{3+}. Explain the experimental results.

8. Measure the content of free Al^{3+} in sediment and water of a water body. Compare the results with pH values and the content of exchangeable base cations. Explain the results from the point of view of Al speciation.

9. Carry out a survey of Pb in dust over an area (*e.g.* school, campus or

town), or you may survey Pb in dust inside car parks, petrol station forecourts, outside schools, *etc.* Prepare a map of lead distribution in the area and identify any sites which may have levels above those recommended by experts as being safe.

10. Measure the concentration of Pb and other heavy metals (*e.g.* Cd) in house dust and in street dust. Comment on your results.
11. The variation of Pb in soil and plants with distance from the roadside could be investigated at various sites.
12. Measure the concentrations of Pb and other heavy metals (Cd, Ni, Cr, *etc.*) in road dust, soils and plants in urban and rural areas and compare.
13. Investigate the relationship between traffic density (as given by vehicle counts) and Pb in dust, soil and vegetation by sampling at various sites with differing traffic densities (quiet country lanes, busy city centre streets, highways, *etc.*).
14. Study the variability of heavy metals in dust or soil at a particular site by taking many samples over a particular area and calculating the mean, standard deviation and coefficient of variation.
15. Analyse heavy metals in street dust and soil soon after a rain event and after a long dry period at several sites. What do the results tell you?
16. Determine levels of Hg in soils and sediments of urban and rural areas and compare.

5.13.22 Reference

E. Evans, M. Ma, L. Kingston and S. Leharne, The speciation pattern of lead in the vicinity of two London schools, *Environ. Int.*, 1992, **18**, 153–162.

5.13.23 Further Reading

D. C. Adriano, "Trace Elements in the Terrestrial Environment", Springer, New York, 1986, pp. 156–180, 298–328, 421–469.

S. E. Allen (ed.), "Chemical Analysis of Ecological Materials", 2nd edn., Blackwell, Oxford, 1989, pp. 81–84, 201–215.

N. J. Bunce, "Environmental Chemistry", 2nd edn., Wuerz, Winnipeg, 1994, chap. 5.

B. E. Davies (ed.), "Applied Soil Trace Elements", Wiley, New York, 1980, pp. 1–154, 287–352.

R. G. Burau, Lead, in "Methods of Soil Analysis, Part 2, Chemical and Microbiological Properties", 2nd edn., ed. A. L. Page *et al.*, ASA-SSSA, Madison, 1990, pp. 347–365.

T. J. Ganje and D. W. Rains, Arsenic, in "Methods of Soil Analysis, Part 2,

Chemical and Microbiological Properties", 2nd edn., ed. A. L. Page *et al.*, ASA-SSSA, Madison, 1990, pp. 385–402.

J. Kubota and E. E. Cary, Cobalt, molybdenum, and selenium, in "Methods of Soil Analysis, Part 2, Chemical and Microbiological Properties", 2nd edn., ed. A. L. Page *et al.*, ASA-SSSA, Madison, 1990, pp. 485–500.

H. M. Reisenaur, Chromium, in "Methods of Soil Analysis, Part 2, Chemical and Microbiological Properties", 2nd edn., ed. A. L. Page *et al.*, ASA-SSSA, Madison, 1990, pp. 337–346.

D. E. Baker and M. C. Amacher, Nickel, copper, zinc, and cadmium, in "Methods of Soil Analysis, Part 2, Chemical and Microbiological Properties", 2nd edn., ed. A. L. Page *et al.*, ASA-SSSA, Madison, 1990, pp. 323–336.

R. P. Gambrell and W. H. Patrick, Jr., Manganese, in "Methods of Soil Analysis, Part 2, Chemical and Microbiological Properties", 2nd edn., ed. A. L. Page *et al.*, ASA-SSSA, Madison, 1990, pp. 313–322.

R. V. Olson and R. Ellis, Jr., Iron, in "Methods of Soil Analysis, Part 2, Chemical and Microbiological Properties", 2nd edn., ed. A. L. Page *et al.*, ASA-SSSA, Madison, 1990, pp. 301–312.

R. Barnhisel and P. M. Bertsch, Aluminium, in "Methods of Soil Analysis, Part 2, Chemical and Microbiological Properties", 2nd edn., ed. A. L. Page *et al.*, ASA-SSSA, Madison, 1990, pp. 275–300.

C. T. Hallmark, L. P. Wilding and N. E. Smeck, Silicon, in "Methods of Soil Analysis, Part 2, Chemical and Microbiological Properties", 2nd edn., ed. A. L. Page *et al.*, ASA-SSSA, Madison, 1990, pp. 263–273.

J. A. Risser and D. E. Baker, Testing soils for toxic metals, in "Soil Testing and Plant Analysis", 3rd edn., ed. R. L. Westerman, SSSA, Madison, 1990, pp. 275–298.

D. C. Martens and W. L. Lindsay, Testing soils for copper, iron, manganese, and zinc, in "Soil Testing and Plant Analysis", 3rd edn., ed. R. L. Westerman, SSSA, Madison, 1990, pp. 229–264.

J. W. B. Stewart and J. R. Bettany, Mercury, in "Methods of Soil Analysis, Part 2, Chemical and Microbiological Properties", 2nd edn., ed. A. L. Page *et al.*, ASA-SSSA, Madison, 1990, pp. 367–384.

CHAPTER 6

Plant Analysis

6.1 INTRODUCTION

Not only do plants serve as a source of nutrition for animals and humans, they play a myriad of other roles fundamental to the environment and society. Plants provide an essential link between the atmosphere and terrestrial ecosystems with both beneficial and potentially harmful consequences. Through *photosynthesis*, plants fix inorganic carbon dioxide from the atmosphere and convert it into organic matter which can be used as a source of energy by plants, animals and humans, while at the same time replenishing the air with that most vital substance of all, oxygen:

$$6CO_2 + 6H_2O + h\nu \rightarrow C_6H_{12}O_6 + 6O_2$$

This *primary production* is the source of food on which all animals, including humans, are dependent. Organic matter can be used to produce energy required for various essential activities (reproduction, growth, *etc.*) by living organisms through respiration:

$$C_6H_{12}O_6 + 6O_2 \rightarrow 6CO_2 + 6H_2O + energy$$

Respiration releases carbon dioxide and water vapour back to the atmosphere, thus balancing the chemical action of photosynthesis. This equilibrium forms the basis of the main nutrient cycle in the biosphere, the carbon cycle. Terrestrial plants, together with their aquatic counterparts, namely phytoplankton, play a vital role in maintaining conditions on the earth that are conducive to life. Phytoplankon supply approximately 75% of the oxygen in the biosphere, while land-based plants provide the remainder.

Civilisation, as we know it, arose when humans first learned to utilise plant growth for their needs by mass cultivation of crops, and to this day agriculture remains the most indispensable of all activities, ensuring the continued survival of our race. Furthermore, we have the plants to thank not only for meeting our biological energy requirements, but also those of our modern society. The massive amounts of energy used up daily throughout the world are almost exclusively provided by what were once living plants. Fossil fuels, originating from long-dead plants and other organisms, supply approximately 90% of the world's energy.

On the other hand, many of today's environmental problems are closely linked with plants. The very combustion of fossil fuels, over the past 150 years, has released into the atmosphere huge quantities of carbon dioxide, which had taken nature millions of years to convert, through the photosynthetic action of plants and other biogeochemical processes, from atmospheric CO_2 into coal, oil and gas. By simply comparing the timescales of anthropogenic and natural processes, one cannot escape the conclusion that it is only a matter of time before nature's delicate balance is broken. Agricultural practices themselves are responsible for numerous environmental problems:

- Greenhouse gas emissions of methane (CH_4) from rice fields and cattle farming.
- Deforestation, whether by means of fire, or otherwise, which contributes to the increase in the atmospheric content of CO_2 and other greenhouse gases.
- Eutrophication caused by N and P fertilisers.
- Acidification by N fertilisers and SO_2 and NO_x from biomass burning.
- Desertification by poor farming practices.
- Accumulation of toxic substances.

Desertification is currently threatening some marginal areas of the world, but it is by no means a new phenomenon. In fact, desertification was the consequence of early human agriculture, and also the cause of the decline of many early civilisations. Saharan Africa and the Middle East were all once productive crop-growing regions that were reduced to deserts as a result of poor agricultural practices and deforestation.

Many of the polluting wastes of our industrial society are introduced into food chains *via* plants. Once in the food chain, these pollutants, many of them highly toxic, can pass on from one animal species to another, eventually to be returned, often in a highly concentrated form, to the source of the pollution, man himself. Animals and humans can obtain this pollution either directly from plants, or from other animals. Important soil/air/water interactions are illustrated in Figure 6.1.

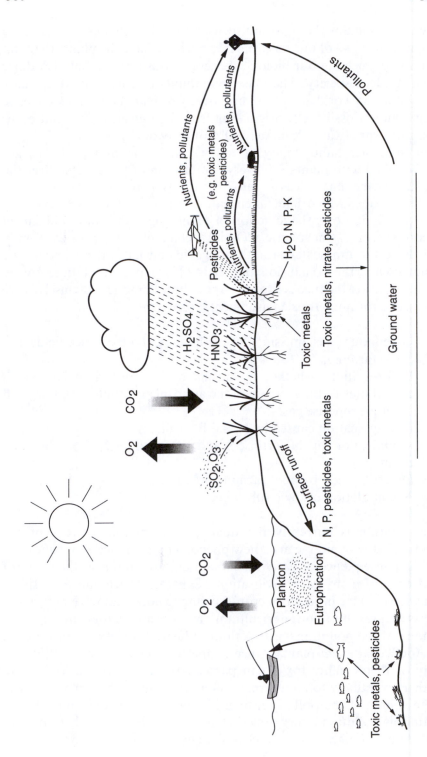

Figure 6.1 *Plant/soil/air/water/food chain interactions*

6.1.1 Plants and Pollution

Pollutants can impact on plants from the atmosphere, as in the case of SO_2 injury to leaves, or through the roots of plants, as in the case of soluble Al. Plant pollution is of concern for two reasons:

- Pollutants may have direct or indirect phytotoxic impacts on the plants themselves, leading to a decline in crop yields and threatening our food supplies. For example, SO_2 pollution may cause chronic and acute injury to plants, alone, or in synergism with NO_2 or ozone, while acidification of soils by rainfall originating from the very same SO_2 may mobilise toxic Al from the soil and cause injury to plants.
- Plants may act as a vehicle for transferring pollutants into the food chain. For example, Cd is readily accumulated by plants, and levels which may not be harmful to the plants themselves could pose a significant threat to animals and humans that consume plants.

The latter is of great concern to public health authorities since pollutants which enter the food chain can eventually affect human health. Heavy metals and pesticides are major pollutants in this respect. There has been major concern about various pesticides (insecticides, herbicides, fungicides, rodenticides) that are applied in agriculture ever since Rachel Carson brought the issue to public attention in 1962 in her book "Silent Spring". Although persistent chlorinated hydrocarbons (*e.g.* DDT, aldrin, chlordane, dieldrin), which are stable to chemical breakdown, have been banned in most developed countries, many developing nations still continue to use them. More degradable pesticides, such as organophosphates and carbamates, have been developed but these tend to be more toxic to invertebrates. Pesticide residues in plants may be transported through food chains or they may be washed off from the soil and contaminate surface and ground waters. Heavy metals can end up in soils from various sources, of greatest concern being the application of soil sludge to agricultural soils. Under appropriate conditions these could be absorbed by the plant roots, either causing direct injury to plants or being accumulated and passing into the food chain. Both organochlorine compounds and toxic metals may be accumulated as they pass through the food chain, the concentration increasing at each successive trophic level.

Direct effects of air pollutants on plants can vary from subtle to severe depending on:

- Nature of pollutant
- Concentration of pollutant
- Plant species
- Exposure time

There is also variation in sensitivity within plant species, depending on the environmental conditions (temperature, nutrient availability, soil moisture, *etc.*). Phytotoxic air pollutants include SO_2, NO_2, O_3, peroxyacetyl nitrate (PAN) and formaldehyde. Both acute and chronic effects have been observed. Furthermore, two or more pollutants may act in combination to produce a greater effect than the sum of individual effects in what is known as *synergism*. Plants exposed to air pollution exhibit both physical and biochemical responses, including changes to photosynthesis and metabolism. Effects of major air pollutants on plant leaves are summarised in Table 6.1.

Growth retardation and other physiological responses have been observed at SO_2 levels as low as 10 ppbv; however, ambient levels of SO_2 are unlikely to cause losses in crop yields of more than 10%. On the other hand, O_3 pollution may cause significant reduction of crop yields at ambient concentrations.

A major concern with regard to vegetation is the effects of acid rain, acid fog and acid mists. These can have more significant effects than direct pollutant impacts. Major acidifying substances in rainwater are SO_4^{2-} and NO_3^- (see Section 2.1). Soil acidification releases potentially toxic levels of soluble Al^{3+} (see Section 5.13.2) which can cause root injury or inhibit the uptake of certain nutrients (Ca, Mg). Forest decline has been an increasing problem in Europe and North America over the last 20 years and acid rain is generally considered to be the main

Table 6.1 *Effects of air pollutants on plant leaves*

Pollutant	Symptom	Injury threshold (ppmv)	Exposure time
SO_2	Bleach spots, bleached areas between veins, chlorosis	0.3	8 hours
NO_2	Irregular collapsed lesions, white or brown, on intercostal tissue	2.5	4 hours
O_3	Flecking, stippling, bleached spotting, pigmentation, conifer needle tips become brown and necrotic	30	4 hours
PAN	Glazing, silvering or bronzing on lower surface of the leaf	0.01	5 weeks

Adapted from Stern *et al.* (1994)

causative agent, although other factors may also be involved. Soil acidification may also be caused by ammonia fertilisers.

Apart from pollution, another major concern, especially from the standpoint of agronomy, is the availability of essential nutrients to plants. It is this availability that guarantees high crop yields needed to feed the world's growing population. Apart from C, O and H, which are supplied by air and soil water, the following elements are essential for the growth of most plants: N, P, K, Ca, Mg, S, B, Cl, Fe, Zn, Mn and Mo. Sodium and Si are essential for some plants while Ni and Co are essential for legumes that symbiotically fix atmospheric nitrogen. Three major nutrients, N, P and K, are commonly deficient in soils. Soils can also be deficient of some micronutrients, notably S, B, Zn and Fe. Nutrient deficiency is usually remedied by applying fertilisers. Fertilisers are generally water-soluble salts of nutrients. However, all fertilisers can be toxic if applied in excessive amounts, and especially so to plant roots and germinating seeds. The application of fertilisers has to be carefully controlled, and decisions regarding which fertilisers to apply, where and when to apply them, and in what quantities, are made on the basis of soil tests and plant analysis.

6.1.2 Plant Analysis

Plant analysis is the determination of chemical substances in a specific plant part. Substances determined in plant analysis include:

- Macronutrients (*e.g.* N, P, K)
- Micronutrients (*e.g.* Zn, Mo)
- Biologically important organic compounds (*e.g.* amino acids, hormones)
- Pesticides (*e.g.* Dieldrin, Aldrin, Malathion)
- Heavy metal pollutants (*e.g.* Pb, Cd)

Plant analysis is widely employed in agronomy, where it is used mainly, but not exclusively, to measure the essential nutrient content of plant tissue and evaluate the fertility of crops. Typical nutrient contents in plants are shown in Table 6.2. Plant analysis is generally, but not necessarily, used in conjunction with soil testing as a diagnostic tool to identify nutrient deficiencies and recommend suitable treatment. Normally, the results of plant analysis are compared against certain *critical* values of nutrients. These critical values are usually concentrations that correspond to 10% reduction in growth. The relationship between nutrient concentration and plant growth is illustrated in Figure 6.2. The

Table 6.2 *Typical concentrations of nutrients in crops, expressed on a dry weight basis*

	Concentration (g kg^{-1})					
Crop	*N*	*P*	*K*	*Ca*	*Mg*	*S*
Oilseed rape	36	7	10	4	2.5	10
Potato, tuber	14	1.8	22	0.9	0.9	1.4
Ryegrass	25	3	18	4	1.2	1.2
Cereals, grain	20	4	6	0.6	1.5	1.5
Cereals, straw	7	0.8	8	3.5	0.9	1.1

Adapted from Archer (1988).

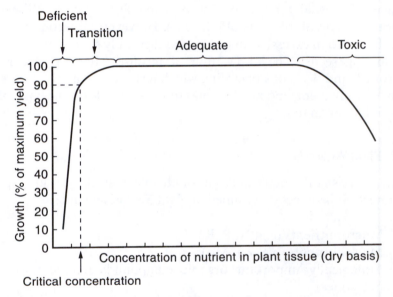

Figure 6.2 *Schematic graph of the relationship between percentage of maximum yield and nutrient concentration in plant tissue*

critical value is assigned within the transition zone, so that symptoms of nutrient deficiency appear below the critical value but not above it. There is a range of nutrient concentrations which is considered adequate. Very high concentrations indicate that the nutrient is acting as a phytotoxin. Depending on the concentration, nutrient levels are generally classified as deficient, adequate or toxic. Plant analysis is therefore an invaluable tool in the management of crop production, the objective of which is to increase crop yields. Plant analysis usually requires expert interpretation before the most appropriate cost-effective measures can be recommended.

Plant analysis consists of the following stages:

Sampling → Washing → Drying → Grinding → Digestion/extraction → Analysis

These will be discussed in turn. Analysis of extracts is carried out using the same techniques as for other environmental samples discussed in previous chapters (AAS, titrimetry, *etc.*).

6.1.3 Questions and Problems

1. Why is absorption of pollutants by plants a major concern?
2. Which air pollutants are most damaging to plants?
3. Describe the different routes by which pollutants could affect plants.
4. Describe the process of bio-accumulation.

6.1.4 References

J. R. Archer, "Crop Nutrition and Fertilizer Use", Farming Press, Ipswich, 1988.
A. C. Stern, R. W. Boubel, D. B. Turner and D. L. Fox, "Fundamentals of Air Pollution", 2nd. edn., Academic Press, Orlando, FL, 1984.

6.1.5 Further Reading

W. N. Richardson and T. Stubbs, "Plants, Agriculture, and Human Society", Benjamin/Cummins, Reading, MA, 1978.

6.2 SAMPLING AND SAMPLE PREPARATION

Obtaining a plant tissue sample which is representative of the general population is both important and difficult. General field variation of biological materials greatly exceeds any introduced during analysis in the laboratory. Therefore, it is essential to evaluate the field variation if the results are to be meaningful and serve the objectives of the analysis. The elemental content of a plant may vary in different parts of a plant, it may vary from plant to plant (even in plants belonging to the same species), it may vary with the season, and even with the time of day. Major considerations when planning a sampling strategy are:

- Which plant part to sample?
- When to sample?
- Which chemical substance to determine?

The latter has an influence on the former two. Obviously, the number of samples required to obtain representative results also has to be considered. If plant-to-plant variations are large, intensive sampling is required. Numbers of plants sampled in field surveys vary between 10 and 100 per hectare.

6.2.1 Plant Part

The selection of the plant part for analysis is largely determined by the aims of the study and the type of plant. For small herbs and grasses it is often sufficient to sample the entire aerial growth, but larger species may require sampling of leaves or other tissues. For monocotyledons it is recommended to sample the leaf blade above the sheath junction. The current year's growth of leaves is often an adequate indicator of nutrient or pollution status for woody species. Plant tissues that are either young or past maturity are not sampled.

6.2.2 Time of Sampling

Time of sampling is crucial because nutrient levels in all active plant tissues fluctuate with time. Different sampling times for different elements are rarely practical; however, plants suspected of suffering from nutrient deficiency should be sampled immediately. Nutrient variations and sampling recommendations for tree leaves are given below:

6.2.2.1 Diurnal Variation. The best time to sample trees is around midday.

6.2.2.2 Seasonal Variation. Seasonal changes in concentration are generally due to the movement of nutrients into a component during growth, and movement in the opposite direction when senescence approaches, although individual nutrients differ in their mobilities. Although some general trends in nutrient and other element levels may be derived from reported studies, these are far from consistent. In some studies, N, P, K and S in the leaves of many deciduous and conifer trees in temperate climates were observed to peak in early spring, level out in the summer and decline in autumn. In other studies these were found to exhibit a maximum in summer. Nevertheless, most studies show a sharp decline in September to October. However, in some studies, no decrease in N, P and K was observed in the autumn, and in others still, an increase from October to November was observed. For Ca, Mg and Si the trend is generally reversed: these tend to increase steadily during the growing season. In some studies, the Ca increase towards autumn was more pronounced, and in others the Mg remained relatively stable. Iron, Al, Zn and sometimes Mn show little initial change and then usually increase to a peak in early autumn with a sharp decline thereafter. Other elements have less well-defined trends. Nutrient levels in green stems follow the seasonal trend of the leaves. The total dry biomass of deciduous leaves rises to a steady maximum from July to September and then drops in the autumn. Therefore, a sampling period including the end of July and early August is recommended. Similar trends were also reported for conifer trees; however, the age of foliage is an important consideration with these species. Generally, the concentrations of many elements in conifer needles tend to increase through the year and sampling in autumn or early winter is recommended. Since needles of differing ages show different seasonal trends, the sample should be representative of a full year's growth. There are no clear seasonal trends in plant composition in tropical climates; however, in monsoon climates, increasing element contents are generally observed at the end of the wet season.

6.2.2.3 Year-to-year Variation. Much less is known about year-to-year variations than about seasonal variations. Significant variation in N, P, K, Ca and Mg was observed in a six-year study of conifer trees. However, some other studies showed relatively minor differences in the nutrient levels. Fluctuations are mainly influenced by the prevailing climate and the onset and length of the growing season.

6.2.3 Methods of Sampling

Plants can be sampled using equipment employed in agriculture and forestry (*e.g.* high-level pruners for sampling tree leaves and small branches, saws for thicker stems, incremental corers for stem samples, shears and secateurs for grassland and herbage). For small-scale work, such as student practicals and projects, grass and leaves can simply be plucked by hand. Polyethylene gloves should be worn when handing samples, especially in warm weather since perspiration could be a source of interference. The sampling of roots can be tedious because they need to be separated from soil particles.

Recommendations regarding time of sampling and which plant parts to sample are available for individual crops. Some general plant sampling rules are summarised below:

- Sample randomly at least 20 plants throughout the study area.
- Each sample should consist of at least 100 g of fresh plant tissue.

- Sample areas exhibiting specific problems separately from the main area.
- Sample mature leaves exposed to sunlight just below the growing tip on stems and main branches.
- Sample at the beginning of the plant's reproductive stage, or just before.
- Do not sample plants that are past full maturity.
- Do not sample tissue that is covered with dust or soil.
- Do not sample tissue from plants with visible insect damage.
- Do not sample damaged or diseased tissue.
- Do not sample dead plants.

Relevant site data should be recorded at the time of sampling, *e.g.* sources of pollution, type of land use, vegetation density, height and form of growth, associated species, nature of soil, drainage, fertilizer application and local topographical features. The national grid reference should be noted so that information regarding geological features can be obtained if required, or a hand-held GPS (global positioning system) could be used.

6.2.4 Sample Preparation

6.2.4.1 Transport and Storage. Fresh sample should be taken to the laboratory as soon as possible. Fresh plant material should be packed loosely in polyethylene, paper or cloth bags, and kept for several hours. Prolonged storage should be avoided, particularly at room temperature. If labile constituents are to be analysed, the samples should be maintained in a condition as close as possible to that in the field. In the case of some specialised analyses, it may be necessary to freeze the sample in the field using solid carbon dioxide (in acetone) or liquid nitrogen and store in vacuum containers.

If it is not possible to take the sample directly to the laboratory, samples should be dried in an oven at $\leqslant 60\,^{\circ}\text{C}$ to prevent spoilage. They can also be air-dried by spreading out over an area for 1 or 2 days, but contamination by soil and dust has to be prevented. Dried samples should be packed loosely in paper or cloth bags and taken to the laboratory or stored in a refrigerator. It is preferable to grind the sample at this stage if possible. Air-dried and ground plant samples can be stored for long periods at room temperature in well ventilated conditions. Samples of organic soils and surface litterfall can be handled in the same way as plant samples, particularly for total nutrient analysis.

6.2.4.2 Washing.   It is necessary to wash the samples in order to remove the surface dust or soil particles which may interfere with some analyses. Also, leaves may be contaminated with spray residues (*e.g.* pesticides). Plant materials should be washed quickly in order to avoid leaching some of the chemical components, and water or a weak detergent solution can be used. Wiping with a damp cloth may be satisfactory for many samples. The recommended procedure is to sponge each plant tissue sample with cotton wool wetted with a 0.1% detergent solution (phosphate-free), and then rinse with water. Other methods of cleaning the tissue include brushing or wiping. You may have to consider which procedure to adopt depending on the objective of your analysis. For most determinations you should follow the recommended decontamination procedure; however, if you are, for example, studying the deposition of pollutants onto plant surfaces, then the sample should not be washed.

6.2.4.3 Drying.   Samples should be dried as soon as possible after washing in order to minimise biochemical changes. Also, dry samples are easier to homogenise, and results of analysis are reported on a dry weight basis. The drying temperature should be sufficiently high to destroy enzymes responsible for decomposition yet low enough to prevent thermal decomposition. Drying temperatures as high as $105\,^{\circ}C$ have been used. Thermal decomposition has been observed at temperatures $> 60\,^{\circ}C$. Oven drying at temperatures between 60 and $80\,^{\circ}C$ for 24 hours is recommended.

6.2.4.4 Comminution.   It is necessary to reduce the particle size of the dried material in order to obtain a representative sample suitable for further treatment and analysis. Various mechanical devices are used for cutting, grinding, macerating, homogenising or emulsifying fresh plant tissue. Samples may be macerated in a blender in the presence of water or an organic solvent which are afterwards removed or recovered for analysis. Emulsifiers break down the plant tissue into much finer particles. Direct analysis of the emulsion is possible if it is sufficiently homogeneous. Devices commonly used to grind dried material include various types of ball mills, hammer mills and cutting mills. Contamination by heavy metals (*e.g.* Cu, Zn, Al, Fe, Na) from various parts of these grinders is possible. Ground samples are usually passed through a 0.5 or 1 mm sieve. If grinding devices are not available, you may finely chop the dried plant material with scissors or a food chopper. The technique of coning and quartering described in Section 5.2.3 may be used to obtain subsamples of fresh or dried plant materials for analysis.

The ground samples should be placed in clean glass bottles and dried again for 24 h at $65\,^{\circ}C$. After drying, the bottles should be sealed and

stored in a refrigerator for long-term storage. Plastic and metal containers may also be used for soil samples depending on the analyte that is to
be determined (*i.e.* metal sample containers may not be suitable if some
heavy metals are to be determined). Containers may be stored in a dry
and cool place for short periods.

6.2.5 Questions and Problems

1. Why is sampling of plants more problematic than that of other
   environmental samples?
2. What are the main objectives of plant analysis?
3. What are the reasons for drying plant samples?

6.2.6 Suggestions for Project

Design a standardised data form for plant sampling and analysis. This
could consist of columns for entering the analytical results for the
various components, as well as other relevant information such as time
and place of sampling, sampling method, site description, visual observations, *etc.* These forms could be photocopied to be used during
monitoring programmes.

6.2.7 Further Reading

R. D. Munson and W. L. Nelson, Principles and practices in plant analysis, in
    "Soil Testing and Plant Analysis", ed. R. L. Westerman, SSSA, Madison,
    1990, pp. 359–387.
J. B. Jones, Jr. and V. W. Case, Sampling, handling, and analysing plant tissue
    samples, in "Soil Testing and Plant Analysis", ed. R. L. Westerman, SSSA,
    Madison, 1990, pp. 389–427.

6.3 DIGESTION AND EXTRACTION

6.3.1 Introduction

Organic substances in plants are extracted into a solvent (hexane, ether,
etc.). This can be done by shaking the mixture, or a Soxhlet apparatus
may be used instead. In the Soxhlet extraction, fresh solvent is continuously refluxed through the sample held in a porous thimble.

Ashing procedures are used to prepare plant samples for elemental
analysis. Both dry ashing, involving combustion of sample, and wet
ashing, involving digestion with strong acids, have been used to destroy

the organic matter and dissolve the analytes. Various acids and acid mixtures have been recommended for wet ashing, including HCl, HNO_3, $HClO_4$, H_2SO_4/HNO_3, $HNO_3/HClO_4$, $H_2SO_4/HClO_4$ and $H_2SO_4/HNO_3/HClO_4$. In some procedures, H_2O_2 is added to promote oxidation. Extraction into water and salt solutions is also employed in some tests. Volatile elements such as Hg and As may be lost during dry ashing and some wet ashing procedures. Acid-washed filter papers should be used for filtering extracts if trace levels of heavy metals are to be determined. The dry ashing and wet ashing procedures outlined below are the most effective methods for elemental analysis. Since the wet ashing method utilises $HClO_4$, the dry ashing method should be used in preference. The dangers of using $HClO_4$ were pointed out in the section on soil extraction (see Section 5.3) and the precautions outlined there should be taken if this acid is to be used. In other experiments we give details of other, more benign, wet ashing procedures.

The final residual acid after dilution is generally 1% HCl or H_2SO_4 and calibration standards used for subsequent analysis (*e.g.* in AAS) should be made up in 1% acid to match the sample extracts. Some trace metals can be analysed directly in the digested sample extract (5% acid) without further dilution. Blanks should always be prepared, especially for trace metal analysis.

6.3.2 Dry Ashing

This involves complete combustion of all the organic matter in a muffle furnace followed by dissolution of the chemical components in HCl. According to most recommended procedures, samples should be ashed at 500 °C for between 4 and 10 h. Nitric acid should be included at the dissolution stage to ensure complete oxidation, as in the procedure outlined below. This method can be used to prepare samples for the analysis of K, Na, Ca, Mg, P, S, Al, Fe, Zn and other heavy metals. According to some authors, addition of HCl and dehydration of silica by further heating are not desirable for Ca, Cu, Fe, Mn, Mg, K and Zn. Their recommended procedure is to simply heat the sample in the furnace for at least 4 h, cool and dissolve the ash in 2.5 mL 6 M HNO_3. Some authors recommend using a temperature of 550 °C rather than 500 °C used in the procedure below; however, according to others, losses by volatilization could occur at temperatures > 500 °C. Like with all the other extraction methods, there are obviously many variations, and each analyst seems to have his/her own preferences. Minor modifications of the procedure are unlikely to have a major impact on the results. What is most important is that the temperature be increased slowly. Ashing times of between 4 and 8 h are usually recommended if uncovered crucibles are used. An additional 2 h is needed if crucibles are covered.

6.3.2.1 Materials
- Crucible, 15 mL, porcelain, tall form
- Muffle furnace
- Water bath
- Filter paper, Whatman no. 541
- Hydrochloric acid; approximately 6 M, prepared by mixing equal volumes of concentrated HCl and water
- Nitric acid, concentrated.
- Nitric acid, 6 M, for alternative procedure

6.3.2.2 Experimental Procedure.
Weigh 0.5 g of oven-dried or air-dried ground and sieved (<1 mm) plant material into an acid-washed porcelain crucible. Place in a muffle furnace and raise the temperature slowly over 2 h to reach 500 °C. Leave inside the furnace for at least 4 h. Remove crucible from furnace and cool. Add 10 mL of 6 M HCl and cover. Heat on a steam bath for 15 min. Add 1 mL HNO_3 and evaporate to dryness. Continue heating for 1 h to dehydrate silica. Add 1 mL of 6 M HCl, swirl, then add 10 mL of water. Heat again on the steam bath to complete dissolution. Cool and filter through a Whatman no. 541 filter paper into a 50 mL volumetric flask and make up to the mark with water. Prepare a blank by repeating the same procedure but omitting the plant sample.

Notes
1. This extract is suitable for the determination of Na, K, Ca and Mg, and often for Fe, Mn, Al, Zn and Cu. For P analysis, first add 5 mL 20% (v/w) magnesium acetate to the sample and evaporate to dryness. Mixed acid digestion may also be used for P.
2. You may leave the crucible in the furnace at 500 °C overnight if this is practical.
3. If greater sensitivity is required, as in the analysis of trace metals, make up to 25 mL rather than to 50 mL or process a larger sample (*e.g.* 2 g). You may have to ash for a longer period.
4. For Co, use 2.5 g of sample and make up to 25 mL.
5. For Cu and Zn, use 0.4 g and make up to 50 mL.

6.3.2.3 Alternative Procedure.
Weigh 1.0 g of oven-dried or air dried ground and sieved (<1 mm) plant material into an acid-washed porcelain crucible. Place in a muffle furnace and raise the temperature slowly over 2 h to reach 500 °C. Leave inside the furnace for at least 4 h.

Remove crucible from furnace and cool. Add 2.5 mL 6 M HNO_3 and ensure that the ash dissolves (stir with a plastic policeman if necessary). Transfer to a 20 mL volumetric flask. Wash the crucible (and plastic policeman if used) with water, transferring washings into the volumetric flask. Make up to the mark with water. Allow any suspended matter to settle to the bottom of the flask. Prepare a blank by repeating the same procedure but omitting the plant sample.

Notes

1. This method is suitable for heavy metals present at trace levels.
2. You may use a greater amount (*e.g.* 2 g) of plant tissue and make up the extract to 10 mL to increase the sensitivity if necessary.

6.3.3 Wet Ashing

Various acid digestion procedures can be employed. The most effective method involves digesting the sample with a mixture of HNO_3, H_2SO_4 and $HClO_4$. This procedure is suitable for preparing samples for the determination of K, Na, Ca, Mg, Zn, Al, Cu, P and also for Fe and Mn if the additional boiling stage outlined below is included. Other digestions, using less hazardous chemicals, are outlined in the text where appropriate.

6.3.3.1 Materials
- Kjeldahl digestion flasks or similar, 50 mL
- Kjeldahl heating unit
- Filter paper, Whatman no. 541
- Perchloric acid, 60%
- Nitric acid, concentrated
- Sulfuric acid, concentrated

6.3.3.2 Experimental Procedure. Weigh 0.5 g of oven-dried or air-dried ground and sieved sample (<1 mm) into a 50 mL Kjeldahl flask or similar digestion flask. Add 1 mL $HClO_4$, 5 mL HNO_3 and 0.5 mL H_2SO_4. Swirl gently and digest at moderate heat first, increasing the heat slowly. Digest for about 15 min after the appearance of white fumes. Let the flask cool (the cold digest is usually colourless or occasionally pink) and add 10 mL of water. If Fe and Mn are to be determined, boil for a few minutes before filtering. Otherwise filter into a 50 mL volumetric flask and make up to the mark with water. Prepare blanks by repeating the same procedure but omitting the plant sample.

Notes

1. Prolonged heating at the white fume stage will lead to drying out and low recoveries.
2. The residue left on the filter paper may be recovered and used for an approximate silica analysis. It is not easy to recover silica adhering to the inside of the digestion flask.
3. For samples high in fat, protein or resinous substances HNO_3 should be added first and the mixture should either be digested or allowed to stand for a while. Perchloric acid should only be added afterwards.
4. Do not exceed 0.5 g of sample when using this procedure. You may, however, use smaller amounts of sample (*e.g.* 0.1 g).
5. You may place a funnel in the mouth of the digestion tube.

6.3.4 Questions and Problems

1. Outline the main principles of wet ashing and dry ashing.
2. Which method do you expect to be more efficient for most of the elements and why?
3. Why are these methods not very effective for metals such as Hg and As? What method would you suggest for such elements?
4. Why is $HClO_4$ hazardous and what precautions should be taken when using this acid?

6.3.5 Suggestions for Projects

1. Prepare extracts of some plant samples using both dry ashing and wet ashing methods. Determine the concentrations of various analytes in the extracts and compare the results by plotting them against each other. Calculate the correlation coefficient and the regression line. Comment on your results.
2. Dry ash some plant samples at different temperatures (*e.g.* 400, 500, 600, 700 °C) and determine the concentrations of several elements in the extracts. Are any of the elements affected by the temperature? Which temperature would you recommend for which element?

6.3.6 Further Reading

J. B. Jones, Jr. and V. W. Case, Sampling, handling, and analysing plant tissue samples, in "Soil Testing and Plant Analysis", ed. R. L. Westerman, SSSA, Madison, 1990, pp. 389–427.

J. E. Richards, Chemical characterization of plant tissue, in "Soil Sampling and Methods of Analysis", ed. M. R. Carter, Lewis, Boca Raton, 1993, pp. 115–139.

6.4 WATER CONTENT AND ASH CONTENT

6.4.1 Water Content

Water, or moisture content, of freshly harvested crops varies between 8 and 80% depending on the plant species. If fresh or air-dried samples are used for chemical analysis, the water content should be determined and a correction applied so that the results can be expressed on the basis of dry weight. Alternatively, air-dried and ground samples can be dried at 105 °C for 3 h before weighing to eliminate the need for a correction.

The same basic procedure is used for both the fresh and air dried samples. The method used for determining water content in plant materials involves measurement of the loss in weight after oven drying to a constant weight at 105 °C.

Materials and experimental procedure are the same as those used for determining water content in soil, sediment, dust and sludge samples (see Section 5.4.4). Thinly spread 1 g air-dried sample or 5–10 g fresh material in a container and determine the moisture content by following the procedure given in Section 5.4.4. Air-dried samples may take about 3 h of drying to reach constant weight, but fresh material may take longer. Express the water content as a percentage. Also calculate the percentage of dry matter.

6.4.2 Ash Content

The term *ash* refers to the residue left after combustion of the oven-dried sample. This is a measure of the total mineral content and it is often determined. Errors in ashing arise through volatilisation losses when too high a temperature is used and from incomplete combustion when the temperature or time are insufficient. Determine the ash content of 1 g of oven-dried plant material by following the procedure given for the determination of loss-on-ignition in soil, sediment and sludge samples in Section 5.4.5. Calculate the ash content from:

$$\text{Ash } (\%) = 100 \times M_2/M_1$$

where M_1 is the weight of the oven dried plant material and M_2 is the weight of the ash residue after combustion.

Notes

1. The presence of blackened particles may indicate incomplete combustion. In this case, moisten with water, dry at 105 °C and heat again
   in the muffle furnace at 500 °C. Some care is needed in interpretation,
   as samples with high contents of Mn may give a dark residue on
   ashing.
2. The temperature should be raised slowly to prevent losses if the
   sample suddenly catches fire.

6.4.3 Silica-free Ash

The ash content is sometimes required on a silica-free basis. This is
obtained by extracting the ash with hydrochloric acid to remove all the
minerals except silica. This method is described below.

6.4.3.1 Materials
- Those required for determining loss-on-ignition in soils (see
  Section 5.4.5)
- Hydrochloric acid, 10% v/v
- Hydrochloric acid, 25% v/v
- Water bath
- Watch glass
- Filter paper, Whatman no. 44

6.4.3.2 Experimental Procedure. Follow the ashing procedure given
above in Sections 6.4.2 and 5.4.5 and determine the ash content. Add
5 mL of 10% HCl to the crucible containing the ash residue and
evaporate to dryness on a water bath. Add a further 0.5 mL of 10%
HCl and evaporate again. Add 5 mL of 25% HCl. Cover with a watch
glass and boil for 30 min on a water bath. Filter through a Whatman no.
44 filter paper, transferring all the residue to the filter. Transfer the filter
to the original crucible and ash for 2 h at 500 °C. Cool in a desiccator and
weigh. Calculate the silica-free ash content from:

$$\text{Ash } (\%) = 100 \times \Delta M / M_1$$

where ΔM is the loss in weight after acidification and M_1 is the weight of
oven-dried sample. ΔM = total ash determined as M_2 in Section
6.4.2 − ash content after extraction with HCl.

6.4.4 Questions and Problems

1. What are the criteria for selecting the appropriate temperature for igniting the sample?

6.4.5 Suggestions for Project

Investigate the effect of temperature and duration of combustion on the ash content of plant samples. Determine the ash content at several temperatures (450, 500, 550, 600 °C, *etc.*) and also at a specific temperature over different periods.

6.5 NITROGEN, PHOSPHORUS AND SULFUR

6.5.1 Introduction

These are three important nutrients required by plants. Nitrogen and phosphorus are required in appreciable quantities, while the needs for S are much less. Deficiencies of S are much less pronounced than those of macronurients such as N and P, and S fertilisation needs are also much lower. Sulfur and P are taken up by plant roots from soil solution as SO_4^{2-} and $H_2PO_4^-$, respectively. Sulfate deposition from the atmosphere and mineralisation of organic sulfur usually provide adequate supplies of S to meet requirements. Therefore, S deficiencies are found only in some soils. Sulfur and P analysis in plant tissue is usually combined with soil S and P tests in studies relating critical soil concentrations to plant yields. Total S and P are usually determined following dry ashing or wet extraction into a mixture of perchloric and nitric acid, while N is usually determined using the Kjeldahl digestion method. In the procedures given below, both P and S are extracted by dry ashing but the S procedure is modified slightly to ensure complete oxidation of S. Phosphorus, as orthophosphate, is determined in the extract by spectrophotometry, while S, as sulfate, is determined by turbidimetry.

6.5.2 Nitrogen

Determine total organic nitrogen (TON) using the Kjeldahl digestion procedure followed by steam distillation as outlined in Section 5.9.2 for soil samples. Analyse NH_4^+ in the distillate using one of the suggested methods. Express your result in g N kg^{-1} as shown for sulfur in Section 6.5.4.

6.5.3 Phosphorus

Dry ash 0.5 g of dry plant material according to the procedure given in Section 6.3.2. Analyse the extract by spectrophotometry using the procedure described in Section 4.11.3 for phosphate determination in water. Calculate the concentration in the plant material according to the equation given for S below in Section 6.5.4. Express the result in g P kg^{-1}.

6.5.4 Sulfur

This is essentially the same as the regular dry ashing procedure except that magnesium nitrate is used to prevent loss of S.

6.5.4.1 Materials
- Muffle furnace
- Hot plate
- Evaporating basin, 20 mL
- Hydrochloric acid, concentrated
- Magnesium nitrate solution, prepared by dissolving 71.3 g $Mg(NO_3)_2 \cdot 6H_2O$ in water and diluting to 100 mL
- Filter paper, Whatman no. 541

6.5.4.2 Experimental Procedure. Weigh 0.5 g dry plant material and place in an evaporating basin. Add 5 mL magnesium nitrate solution, making sure that the entire surface of the sample is covered. Heat on a hot plate to 180 °C, and when dry raise to 280 °C. When the colour changes from brown to yellow, transfer to a muffle furnace at 500 °C and heat for at least 4 h. Remove crucible from furnace and cool. Add 10 mL concentrated HCl and cover with a watch glass. Boil gently for 3 min. Cool and add 10 mL water. Rinse the watch glass into the basin. Filter through a Whatman no. 541 filter paper into a 50 mL volumetric flask and make up to the mark with water. Analyse the extract either by turbidimetry as in Section 4.14. Calculate the concentration of sulfate in the extract and convert to units of mg S L^{-1}. Calculate the concentration in the sample from:

$$\text{Sulfur (g S kg}^{-1}) = 10^{-3} \times C \times V/M$$

where C is the concentration of sulfate in the extract (mg S L^{-1}), V is the volume of the extract (mL) and M is the weight of the sample (g).

6.5.5 Questions and Problems

1. Explain why only a small fraction of S in soil is available to plants.
2. Plants are most likely to experience deficiencies with respect to which nutrient? Discuss the reasons for this.

6.5.6 Suggestions for Projects

1. Determine the levels of N, P and S in plants and the soil on which they grow. Collect samples from various fields fertilised by the application of different fertilisers (mineral phosphorus fertilisers, manure, sludge, *etc.*). Comment on your results.
2. Determine the concentrations of these elements in various parts of the plant (root, stem, leaves).
3. Investigate the seasonal variation in the levels of these elements in different plants.

6.6 POTASSIUM, SODIUM, CALCIUM AND MAGNESIUM

6.6.1 Introduction

Potassium is one of the most important nutrients required for plant growth, and in high yielding crops the content of K may exceed that of N. During rapid growth, crops accumulate K at high rates, and at times the soil's capacity to supply K may be exceeded. The K content of aerial portions of crops varies from about 40 kg K ha^{-1} in flax to 1400 kg ha^{-1} in bananas. Calcium levels in plants are generally less than one half of K levels, and availability of Ca is usually not a problem. Neutral and alkaline soils contain adequate natural supplies of Ca while acid soils receive Ca from the lime applied to raise the soil pH. Magnesium content is half that of Ca, and Mg deficiency has been noted in livestock feeding on forage with inadequate Mg content. Concentrations of K, Ca and Mg in plant tissue are frequently determined in field studies, often together with soil analysis, in order to establish the nutrient status of crops, critical nutrient concentrations, study soil/plant interactions, *etc.*

6.6.2 Experimental Procedure

Dry or wet ashing are used to extract K, Na, Ca and Mg from plant materials and extracts are analysed using flame AAS, flame photometry or inductively coupled plasma (ICP). The most efficient method to use is the dry ashing procedure described in Section 6.3.2. Wet ashing methods involving H_2SO_4, either alone or as a component of a mixture, are unsuitable for Ca owing to the possible precipitation of relatively insoluble $CaSO_4$. The mixed acid wet digestion outlined in Section 6.3.3 is therefore not recommended for Ca, but it may be used for K, Na and Mg. You may use the wet ashing procedure involving HNO_3 as described in Section 5.13.16 for the extraction of heavy metals in soils. This method can be used for Ca as well as for the other alkali and alkaline earth metals. Use dry and ground samples and follow the instructions for the dry or wet ashing procedures. Analyse the elements by AAS as described in Section 2.5.1.

6.6.3 Questions and Problems

1. Why is H_2SO_4 not suitable for the extraction of Ca?
2. Which analytical techniques other than AAS could be used to determine alkali and alkaline earth metals in plant and soil samples?

6.6.4 Suggestions for Projects

1. Determine the level of these elements in plants and soils on which they grow. Collect samples from various fields fertilised by the application of different fertilisers (mineral phosphorus fertilisers, manure, sludge, *etc.*). Comment on your results.
2. Determine the concentrations of these elements in various parts of the plant (root, stem, leaves).
3. Investigate the seasonal variation in the levels of these elements in different plants.

6.7 NITRATE AND NITRITE

6.7.1 Introduction

The nitrate (NO_3^-) content in crops is one of the most important indicators of farm production quality. Nitrate content in food is strictly regulated because of its toxicity, especially to young children. The actual toxin is not the nitrate ion itself but rather the nitrite ion (NO_2^-), which is formed when nitrate is reduced by intestinal bacteria. The health impacts of nitrate and nitrite are mentioned in Section 4.10.3 in the context of drinking water quality. The average human daily intake of nitrate/nitrite is 95 mg d^{-1} in adults. Estimates of the relative contributions of nitrate from drinking water and food to the daily intake vary considerably, depending on how they are calculated. Nevertheless, they show that between 50% and 90% of nitrates in human intake may originate from vegetables, conserved meat (sausages, canned meat, smoked meat, *etc.*) and even milk products.

Vegetables tend to concentrate nitrate ion, especially if they are grown using high application rates of N fertilizers. The concentration of nitrate in vegetables can vary considerably. Lettuce, spinach, cabbage, celery, radish and beetroot can contain as much as 3000–4000 mg kg^{-1}, and these levels could have potential health impacts. The problem of nitrate accumulation seems to be especially severe in leafy vegetables grown in greenhouses under winter conditions, owing to intensive application of N fertilisers and low light levels which retard nitrate utilisation by crops. Another source of nitrate and nitrite in food is their use as food additives. Nitrate and nitrite salts ($NaNO_2$, $NaNO_3$, KNO_2, KNO_3) are added to meats and other food products as a curing salt, colour fixative (preventing the meat turning brown) and as a food preservative to prevent the growth of the dangerous bacterium *Clostridium botulinum*, which produces the highly poisonous botulism toxin. Cured meats, bacon, ham, smoked sausages, beef, canned meat, pork pies, smoked fish, frozen pizza and some cheeses contain nitrate and nitrite additives, typically at levels of 120 mg kg^{-1}. Although without them there would certainly be many deaths due to the growth of toxic microorganisms in meat, excessive intakes of these salts may cause gastroenteritis, vomiting, abdominal pain, vertigo, muscular weakness and an irregular pulse. Long-term exposure to small amounts of nitrates and nitrites may cause anaemia and kidney disorders. The level of these additives is strictly controlled (for example, 500 mg kg^{-1} as $NaNO_3$ in the UK), and the addition of nitrates and nitrites to baby foods is now banned in many countries. The *acceptable daily intake* (ADI) for $NaNO_3$ is 0–5 mg kg^{-1} body weight. For KNO_2 and $NaNO_2$ the ADI is 0–0.2 (temporary), while for KNO_3 the ADI is not specified. The value of ADI $= 0$ refers to baby food. The WHO sets the ADI at 220 mg for an adult.

Since vegetables are a major source of ingested nitrates, the most rational way of reducing the problem is to grow crops with safe levels of nitrates. Most countries do not have actual standards but some kind of guideline or criteria value based on the ADI. Criteria values of nitrate content in the same kind of vegetable may vary broadly from country to country owing to differences in vegetable consumption and in vegetable production practices. For example, the maximum permissible nitrate level (in mg kg^{-1}) in spinach in different countries is: USA, 3600; the Netherlands, 4000; Switzerland, 3000; the Czech Republic, 730; Russia, 2100. For leafy vegetables, the Netherlands and Austria employ a maximum allowance of 4500 mg kg^{-1} while Germany uses a guideline value of 3000 mg kg^{-1}. Standards vary with the type of vegetable, as shown by the maximum permissible nitrate content in different vegetables given in Appendix III.

Nitrate content is also one of the indicators of fodder quality. Numerous cases of cattle poisoning by nitrates present in fodder have been reported in various countries, and many head of cattle were lost from affected herds. Feed beetroot, cabbage, mustard, sunflower, oat, as well as various types of ensilage used as green fodder, may contain potentially toxic levels of nitrate. The toxic level of nitrate for animals is 0.7 mg kg^{-1} of body weight. Among farm animals, cattle and young pigs are the most sensitive to nitrates, while sheep are more resistant. Also, consumption of nitrates at sub-toxic levels by cattle has been reported as a cause of reduced milk production and weight gain, vitamin A deficiency, abortions, stillbirths, cystic ovaries, *etc*. Nitrate-N concentrations of 0.21% in the feed are considered to be toxic to farm animals. At pH values occurring in animal and human stomachs, nitrite ion is converted to nitrosoamines, which may be carcinogenic. For example, dimethylnitrosoamine is carcinogenic to many animal species, although it has not yet been confirmed as a human carcinogen.

Crops high in nitrate not only pose a direct danger to human and animal health, but also cause financial losses to agriculture and the food-processing industry. High nitrate content leads to a low shelf life of vegetables, thereby increasing losses during storage. If the recommended nitrate limits are exceeded, the produce either has to be destroyed or used as animal fodder. Therefore, the analysis of nitrates and nitrites in crops and food is of great importance. Nitrates and nitrites are usually extracted into water or a 0.025 M $Al_2(SO_4)_3$ solution.

6.7.2 Methodology

Nitrates and nitrites are extracted into water from fresh or dry plant materials or food products. Nitrate and nitrite ions are determined either by spectrophotometry or indirectly after reduction to ammonium.

6.7.3 Experimental Procedure

Chop up fresh plant or food material. Weigh 1.0 g and add 50 mL of water in a 100 mL polyethylene or glass bottle. Alternatively, add 40 mL of water to 200 mg of dry ground material. Cap and shake for 30 min. Filter into a flask. Analyse the water extract as soon as possible in order to avoid errors which could arise as a result of nitrite oxidation if samples are stored for long. Determine nitrate and nitrite ions either by ion chromatography as described in Section 2.4.6 or using the methods given in Sections 4.10.3 and 4.10.4. Also see Section 5.9.4 for nitrate and Section 5.9.5 for nitrite ions.

6.7.4 Questions and Problems

1. Discuss the effects of nitrate and nitrite in crops and foods on animal and human health.
2. Which vegetable species are the most intensive accumulators of nitrate?
3. What are the economical implications of excessive nitrate accumulation in crops?
4. Compare daily intakes of nitrates from food (vegetables, meat, *etc.*) and from drinking water. Which is the predominant source and why?

6.7.5 Suggestions for Projects

1. Measure the content of nitrate and nitrite in different agricultural crops that are commonly used in your area and compare with existing guidelines.
2. Measure the content of nitrates in the same vegetables (cabbage, onion, carrot, lettuce, spinach, *etc.*) or fruits (apples, melons, bananas, watermelons, *etc.*) from different agricultural fields or plantations where different fertilisers are applied. Compare the fertiliser application rates with nitrate content.
3. Analyse the nitrate content in different parts of the same crop or fruit species (leaves, stems, roots, tubers, fruits).
4. On the basis of nitrate and nitrite measurements in crops, foods and drinking water, calculate your daily intake of nitrate and nitrite and compare these with recommended values.
5. Discuss the mechanism of nitrate toxicity in humans and animals.
6. Describe the different techniques that could be used to determine nitrate in plants and soils.

6.7.6 Further Reading

P. Bielek and V. Kudeyarov (eds.), "Nitrogen Cycles in the Present Agriculture", Priroda, Bratislava, 1991, pp. 127–168.

6.8 HEAVY METALS

6.8.1 Introduction

By accumulating metals in above-ground tissue, plants can transfer heavy metal pollutants from soils into the food chain, and this accumulation is one of the most serious environmental concerns of the present day, not only because of the phytotoxicity of many of these metals to the crops themselves, but also because of the potentially harmful effects toxic metals could have on animal and human health. The monitoring of heavy metals in crops and other foodstuffs is therefore of great importance in protecting the public from the hazards of toxic pollution. As mentioned in the chapter on soil analysis (see Section 5.13), some heavy metals are also nutrients and only become toxic at high concentrations, while others have no beneficial properties and are exclusively toxic.

Crops differ widely in their sensitivity to heavy metal pollution (Table 6.3) and also in their relative sensitivity to the individual metals or compounds. The soil pH and redox potential also have an important effect on metal toxicity since they determine in what form the metal will be present. Different forms of the same metal may have different toxicities. For example, in slightly acidic soils (pH 5.5–6), Cu could be twice as toxic as Zn, while Ni could be four times as toxic as Zn. Of the heavy metals, Cr, Ni, Cd, Hg and Pb are considered to be the most phytotoxic.

Plants absorb heavy metals from soil solution through the roots. The uptake of metals by plants is influenced by various factors, including type of plant, nature of soil, climate and agricultural practices. There are significant differences in uptake of heavy metals among cultivars; for example, translocation and differential uptake of Zn and Cu among corn

Table 6.3 *Relative phytotoxicity of heavy metals to some crops*

Low	*Moderate*	*High*	*Very high*
Corn	Cucumber	Mustard	Lettuce
Sudangrass	Flatpea	Spinach	Carrot
Smooth bromegrass	Oat	Broccoli	Turnip
	Orchard grass	Radish	Peanut
	Japanese bromegrass	Tomato	
		Alfalfa	
		Korean lespedeza	
		Soybean	
		Colonial bentgrass	
		Perennial ryegrass	
		Timophy	

and lettuce cultivars have been observed. Furthermore, the concentrations of many heavy metals are not uniformly distributed throughout the plant, with different levels found in different tissues of the same plant. In general, the roots contain the highest levels of heavy metals, followed by vegetative tissue, which in turn has higher concentrations than seeds or grain. Management practices aimed at maximising agricultural production may also affect plant uptake of heavy metals (the method and frequency of fertiliser application, source of fertiliser). Disposal of solid wastes (*e.g.* sewage sludge, animal manure, compost) and liquid manure on agricultural fields results in increased accumulation of heavy metals in crops. Irrigation and climate (*e.g.* rainfall and drought) also affect uptake of metals. An additional source of heavy metals is through foliar absorption. Heavy metals present in atmospheric dust particles settle on the leaves and other plant surfaces. Although some metals (*e.g.* Pb) do not penetrate the cuticle of higher plants, others (*e.g.* Cd, Cu, Mn) can be absorbed.

The normal range of heavy metal concentrations in plants is given in Table 6.4, together with critical concentrations. These are concentrations

Table 6.4 *Concentrations of heavy metals in plants and soils and critical concentrations in plants and soils*[a]

Metal	Normal range in plants (mg kg^{-1})	Critical plant concentration A (mg kg^{-1})	Critical plant concentration B (mg kg^{-1})	Normal range in soils (mg kg^{-1})	Critical soil concentration (mg kg^{-1})
As	0.02–7	5–20	1–20	0.1–40	20–50
Cd	0.1–2.4	5–30	4–200	0.01–2.0	3–8
Co	0.02–1	15–50	4–40	0.5–65	25–50
Cr	0.03–14	5–30	2–18	5–1500	75–100
Cu	5–20	20–100	5–64	2–250	60–125
Hg	0.005–0.17	1–3	1–8	0.01–0.5	0.3–5
Mn	20–1000	300–500	100–7000	20–10 000	1500–3000
Mo	0.03–5	10–50	–	0.1–40	2–10
Ni	0.02–5	10–100	8–220	2–750	100
Pb	0.2–20	30–300	–	2–300	100–400
Sb	0.00001–0.2	–	1–2	0.2–10	5–10
Se	0.001–2	5–30	3–40	0.1–5	5–10
Sn	0.2–6.8	60	63	1–200	50
V	0.001–1.5	5–10	5–13	3–500	50–100
Zn	1–400	100–400	100–900	1–900	70–400

[a] Adapted from B. J. Alloway (1995).
A: concentrations above which toxicity is likely, B: concentrations likely to cause a 10% reduction in yield.

that are likely to be either toxic or cause a reduction in yield. Normal soil and critical soil concentrations are also given.

Human intake of heavy metals is the sum total of the contribution of metals from three sources: food, drinking water and inhaled air. Ingestion of foods of plant and animal origin is the primary route by which heavy metals impact on human health, while drinking water and inhalation of airborne particles generally make a smaller contribution to the total intake of toxic metals. Most serious incidents of heavy metal toxicity to humans have involved ingestion of contaminated foods. Health effects of different heavy metals are described in Sections 4.16 and 5.13. The greatest health risk comes from those heavy metals that are less toxic to plants than to humans. Some plants may accumulate certain toxic metals at levels which may be harmless to the plant but could be harmful to animals and humans if ingested. Cadmium is the metal of major concern in this respect, but Hg is also more toxic to humans and animals than to plants. Leafy vegetables are greater accumulators of soil Cd than the edible portions of tomato or radish. Dietary tolerance for Cd is 65 μg d^{-1}. Some countries have adopted maximum admissible levels for a number of heavy metals in specific food products. Guideline values adopted for Cd, Hg and Pb by Germany are given in Appendix III.

The metal content of foods may be altered by processing. Washing, in general, reduces the levels of heavy metals, although the extent of reduction varies with the metal and plant. For example, washing can reduce Pb by as much as 80% in some plants, but Cd by only 20%. On the other hand, packaging may introduce heavy metals into foods. In the past, the use of uncoated soldered tin cans greatly increased the content of Pb, Sn, Zn and Fe in canned food; however, such practices are unlikely nowadays.

6.8.2 Methodology

Fresh or dried plant samples are extracted into an acid solution using either dry or wet ashing procedures. Analysis is by AAS, although ICP-AES may be used instead. Dry ashing is preferable as organic components are effectively oxidised and highly dangerous HClO$_4$ is not required.

6.8.3 Experimental Procedure

Extract the sample by either the dry ashing or the wet ashing procedures described in Section 6.3. The wet ashing procedure employs perchloric acid and extreme caution should be taken when using this acid. Extraction using HNO_3 (see Section 5.13.16), aqua regia (see Section 5.13.17) or HNO_3–H_2O_2 (see Section 5.13.18) are quite satisfactory and can be used instead of the methods based on $HClO_4$. Determine the metal concentration in the extract using flame AAS as described in Section 4.16. As an alternative, you may use graphite furnace AAS or ICP-AES if these are available in your laboratory. Make up standards in acid of strength equivalent to that in the sample extracts. Use background correction in AAS determination. Calculate the concentration of metal in the original sample from:

$$\text{Metal concentration (mg kg}^{-1}\text{ dry matter)} = C \times V/M$$

where C is the concentration of the metal in the sample extract as determined by AAS ($\mu g \text{ mL}^{-1}$), V is the volume of extract (mL) and M is the weight of dry sample (g). Mercury may be determined using the cold vapour procedure (see Section 5.13.19).

6.8.4 Questions and Problems

1. Outline the mechanism of heavy metal accumulation in crops and discuss the various factors which may influence this process.
2. Discuss phytotoxity and bioavailability in various plants with respect to heavy metals.
3. Why are phosphorus fertilisers considered to be the main sources of Cd and Zn enrichment in agricultural soils? Which other heavy metals are of environmental concern in this respect?
4. On the basis of your understanding of soil composition, structure and chemistry, rank the various dry and wet ashing procedures that can be used for heavy metals (see Sections 6.3, 5.13 and 5.3) in order of decreasing recovery of heavy metals (*i.e.* from most efficient to least efficient) and justify your answer.

6.8.5 Suggestions for Projects

1. Carry out a survey of heavy metals in various crops and compare the measured concentrations with critical levels. Report the differences between plant species and cultivars in your locality.
2. Investigate the distribution of heavy metals in food chains, for example in the soil–plant–food system. Calculate the daily intake of different heavy metals and identify sources of the metals on the basis of your results.
3. Measure heavy metal levels in plant samples collected from various

fields fertilised by the application of different fertilisers (mineral phosphorus fertilisers, manure, sludge, *etc.*). Assess any possible relationship between the heavy metal content of plants and the nature of the fertiliser.

4. Determine heavy metal concentrations in various fresh and processed foods of plant and animal origin. Calculate daily intakes and discuss any interesting results you may obtain.

5. Determine the concentrations of some heavy metals in both soils and plants at the same sites. For each site and plant type, calculate the *soil–plant transfer coefficient*. The soil–plant transfer coefficient is defined as the metal concentration in plant tissue above ground divided by the total metal concentration in the soil.

6. Compare heavy metal concentrations in plants growing in urban and industrial areas to those growing in rural areas.

7. Determine the concentration of Pb, and other heavy metals, in grass and other plants with increasing distance from the roadside. Sample near to roads with different traffic densities (busy, moderate, quiet, highway). Compare these results with those of soil analysis at the same sites. Attempt to relate the concentration of Pb, and other heavy metals, to road traffic density.

8. Determine Hg in seaweed, fish and other seafood by means of cold vapour AAS as described in Section 5.13.19.

6.8.6 Reference

B. J. Alloway (ed.), "Heavy Metals in Soils", 2nd edn., Blackie, London, 1995.

6.8.7 Further Reading

S. E. Allen (ed.), "Chemical Analysis of Ecological Materials", 2nd edn., Blackwell, Oxford, 1989, pp. 201–215.

D. C. Adriano, "Trace Elements in the Terrestrial Environment", Springer, New York, 1986, pp. 1–45.

J. A. Risser and D. E. Baker, Testing soils for toxic metals , in "Soil Testing and Plant Analysis", ed. R. L. Westerman, SSSA, Madison, 1990, pp. 275–298.

Safety

Safety must be the primary concern of anyone involved in practical work of any kind. Hazards could arise: (a) in the field (during sampling) and (b) in the laboratory (during analysis). While it is impossible to give advice regarding every possible hazard, common sense should be the guiding principle.

Both physical and chemical hazards have to be considered. Environmental work can vary considerably from site to site; you may have to sample from boats, river banks, platforms, *etc.*, in different weather conditions. While collecting samples you should take all the necessary precautions to protect yourself and others from possible accidents. Other than the obvious physical hazards of outdoor work (falling objects, slipping, drowning, *etc.*), chemical hazards may also be encountered during sampling. Many environmental samples have to be treated as toxic wastes due to the potentially high content of hazardous pollutants. When working at industrial sites, landfill sites, sewage treatment plants, and so on, you should use the appropriate safety equipment (safety helmet, boots, gloves, ear mufflers, *etc.*) and follow procedures recommended by staff working at the site. In any case, use your common sense and do not take any risks. When in doubt; don't! The following rules relate to field work:

- Never work alone in a dangerous location.
- Assess the level of hazard before going into the field (*e.g.* obtain weather forecasts, time of tide).
- Leave a record of the planned field trip with someone in your laboratory, giving date and time of departure, location of field work, any potential hazards that may be anticipated and expected time of return.
- Wear clothing appropriate to the field work (safety helmet, safety boots, *etc.*).
- Take the necessary equipment (ear mufflers, first-aid kit, life-jacket, flash-light, *etc.*).

Be aware of the special hazards and appropriate procedures specific to the location. Intertidal flats, marshes, rocky shores and cliffs are particularly dangerous. Working on boats and diving can also present special hazards.

With regard to safety in the laboratory, the following rules should be followed:

- Treat all chemicals with respect. Most chemicals are hazardous. They can be flammable, corrosive or toxic.
- Read the safety manual available in your laboratory.
- Acquaint yourself with the safety features of your laboratory: location of fire extinguishers, fire exits, safety showers, first aid kits, telephones, emergency phone numbers, *etc*.
- Read the labels on all containers. Acquaint yourself with various hazard labels. Obtain the toxicity information on all the reagents that you will use and acquaint yourself with specific emergency procedures for each reagent in case of an accident. You can obtain this information from safety charts, books (see "Further reading" at the end of this Appendix) and the Material Safety Data Sheets (MSDS) on the Internet. MSDS can be obtained from various university and chemical manufacturer's sites on the Internet. A particularly useful site is that of the International Programme on Chemical Safety (IPCS) run by NIOSH/WHO at:

http://www.cdc.gov/niosh/ipcs/icstart.html

This site lists *International Chemical Safety Cards* (ICSC) in English, Finnish, French, German, Malay, Japanese, Spanish and Swahili. A safety card is available for each hazardous chemical, giving details of the type of hazard, exposure, symptoms, advice on hazard prevention, storage, spillage, disposal, occupational safety limits, physical properties, *etc*.

- Do not eat, drink or smoke in the laboratory.
- Always wear safety glasses and a lab coat.
- Never pipette by mouth.
- Never work alone or without supervision.
- Keep your work space clean and tidy.
- Label all containers.
- Tie long hair in a pony tail.
- Use a fume cupboard when handling chemicals that produce hazardous vapours (*e.g.* concentrated acids, ammonia).
- When diluting acids, always add acid to water and never water to

acid. Add the acid gradually, little by little, and allow to cool before adding subsequent portions of the acid.

- Use rubber gloves when handling concentrated acids.
- In the event of an acid spill, rinse immediately with copious amounts of water.
- Report all accidents, however minor, to the laboratory instructor.
- Do not deviate from the set experimental procedure without first consulting the laboratory instructor.
- Be aware of electrical hazards when operating electrically powered equipment.
- Dispose of chemical waste according to procedures established in your laboratory. Some of the waste may be poured down the drain as long as it is flushed with adequate tap water. Organic solvents and heavy metals should not be flushed down the sink and must be stored in specially labelled bottles to be professionally disposed of later on.
- Plan your work, think about what you are doing, be watchful, aware and considerate. Always be attentive and careful. Do not daydream or fool around in the laboratory. Remember: practical work may be fun, but it can also be dangerous!

The above list is by no means exhaustive and you should add to it rules appropriate to the specific work that you are doing. Many institutions offer regular in-house courses in first aid, fire prevention and fire fighting. You should consider attending some of these courses.

Further Reading

S. G. Luxon, "Hazards in the Chemical Laboratory", 5th edn., Royal Society of Chemistry, Cambridge, 1992.

D. A. Pipitone, "Safe Storage of Laboratory Chemicals", Wiley, New York, 1984.

IUPAC/IPCS/WHO/UNEP/ILO, "Chemical Safety Matters", Cambridge University Press, Cambridge, 1992.

Laboratory Practice

CLEAN ROOM

It is not uncommon to observe interference by gases (*e.g.* CO_2, NH_3) and dust in ordinary laboratories when analysing environmental samples. For very accurate environmental analysis, clean-room facilities should be used. Clean rooms are special laboratories with the following features:

- Laboratory air is filtered.
- Special fume-cupboards without metallic or wooden fittings are used.
- Metallic installations are removed from the laboratory.
- Pipes and other metallic parts are painted with non-abrasive plastic paint.
- Teflon (PTFE) labware is widely used.
- Laboratory staff are required to wear disposable gloves.
- Laboratory procedures are clearly outlined.

Converting a laboratory into a clean room can be quite expensive and beyond the budget of many teaching laboratories. Designating a part of the laboratory for environmental work and maintaining a higher level of cleanliness than in the rest of the laboratory is generally sufficient for student practicals and projects. Some other precautions that can be taken in any laboratory to improve the quality of analytical work include:

- Use a clean bench with all unnecessary items removed.
- If possible, use a separate refrigerator for storing environmental samples. Do not store environmental samples in refrigerators used for food, drinks or standards.
- If possible, handle samples and standards on different benches. Never use the same glassware for standards and samples. Preferably

use one set of glassware (pipettes, flasks, *etc.*) exclusively for samples and another for standards in order to avoid cross-contamination.

- Always place stoppers upside down on the bench to avoid contamination.
- Always keep your workspace clean and tidy. This is supposed to be standard laboratory practice, but a visit to many laboratories will reveal that this is far from the truth.
- Ideally, disposable gloves should be used.
- Use your knowledge of general chemistry and common sense to avoid contamination.

REAGENTS

Reagents of varying grades of purity are available. Typical grades of reagents are listed in Table A1. Obviously, the price of the reagent is related to its purity; the higher the purity, the greater the cost.

Many manufacturers use their own classifications of reagent purity which differ from those in Table A1. Classifications adopted by some manufacturers are given in Tables A2, A3 and A4.

Table A1 *Typical grades of reagents, their purity and application*

Reagent grade	Purity	Application
Technical grade	Lowest purity (70–90%)	Widely used in industry. Can be used where purity is not critical (*e.g.* cleaning solutions, desiccating reagent). Not suitable for environmental analysis.
USP (United States Pharmacopoeia)	Low purity	Used in medical, pharmaceutical and food analysis. Not suitable for environmental analysis.
Reagent grade	Moderate purity	Widely used in chemical analysis. Can be used in environmental analysis if higher purity reagents cannot be obtained, and depending on the sample being analysed and the analyte in question.
Primary standard grade	High purity (>99%)	Recommended for environmental analysis.
Special purpose reagent (*e.g.* HPLC grade)	Variable, but generally high purity	Many of these reagents may be used for environmental analysis.

Table A2 *Fluka reagent classification*

Purity grade	Assay
techn.	Fluctuating
pract.	>90%, mostly >95%
purum	>97%
puriss.	>99%; impurities present at sub-ppm levels

Table A3 *Merck reagent classification*

Reagent	Comments
ACS	Reagents comply with the standards of the American Chemical Society
ISO	Reagents tested according to the specification of the International Organisation for Standardisation
LAB	Purity properties required for applications in the laboratory
Extra pure	>98% purity
Suprapur	>99.5% purity; impurities present at sub-ppm levels

Table A4 *BDH reagent classification*

Reagent grade	Comments
General Purpose Reagent (GPR)	Purity greater than technical or commercial grade. Suitable for analytical work.
AnalaR	Pure reagents widely used in analytical work.
ARISTAR	Exceptionally pure reagents produced to meet the requirements of highly sensitive analytical work. Tested by the most up-to-date methods of trace analysis, in most cases down to the ppb range. Impurities present at sub-ppm levels.
Hi Per Solv	High purity solvents and reagents for high performance liquid chromatography (HPLC).
ConvoL	Concentrated volumetric solutions.
AVS	Analar volumetric solutions.
SpectrosoL	Solvents, reagents and standards of high and consistent purity for spectroscopic applications. Impurities present at sub-ppm levels.

While many reagent grades can be used for general chemical analysis, environmental analysis requires reagents of the highest purity. This is especially true when analysing substances that may be present at trace levels in environmental samples. It is not uncommon that the amount of impurity in the reagent may exceed the concentration of the analyte in the sample. In order to minimise contamination, highest purity reagents (primary standard grade, puriss., Suprapur or ARISTAR) should be used where high accuracy is required such as in research projects. Special purpose, reagent grade, purum or AnalaR chemicals may be suitable for general laboratory work. Most reagent grade and primary standard grade chemicals have the percentage purity and maximum limits of impurities listed on the labels. This can help you decide if the reagent is suitable for a particular analysis. Reagent blanks should be frequently analysed to determine the background level of impurities which may interfere with a specific analysis.

Concentrated volumetric solutions (*e.g.* ConvoL) are suitable where high accuracy is required (*e.g.* titrations). These are made from AnalaR grade reagents and are supplied in sealed ampoules. When diluted according to the enclosed instructions these give accurately standardised solutions of specific strength. They are convenient to use as they require no weighing, pipetting or standardisation. These solutions are quick to prepare, and highly accurate ($\pm 0.2\%$ error).

LABORATORY WATER

Owing to the low concentrations of trace constituents in some environmental samples, the water used for preparing reagent solutions, dissolving or diluting samples, *etc.*, can act as a source of contamination. The term *laboratory water* is used to designate water used in chemical analysis. Laboratory waters can be distilled, doubly distilled, deionised or ultrapure waters produced by special water purification systems involving ion exchange, adsorption on charcoal and reverse osmosis (*e.g.* NANOpure, Milli-Q). These systems are usually equipped with a meter that monitors the purity of the water. Ultrapure water produced by such systems, and having a specific resistance of 18 MΩ cm (equivalent to a conductivity of 0.06 μS cm^{-1}), is recommended for environmental analysis. If this is unavailable, deionised or doubly distilled water should be used. Distilled water can be used if none of the above are available. Laboratory water blanks should be frequently analysed, together with samples and standards.

DISPOSAL OF WASTE

In many countries it is prohibited to dispose of laboratory waste by pouring it into the sewage system *via* the drain. Waste is usually collected, stored and passed on to a professional waste disposal firm. Many chemicals cannot be readily degraded by wastewater treatment plants and they enter the environment in an unaltered state, contributing to the pollutant load. The disposal of laboratory waste *via* the drain should therefore be discouraged.

Empty Winchester bottles can be used for collecting waste liquids. These should be properly labelled so that the type of waste can be identified. It is common practice to make one bottle available for *halogenated organic solvents* and another for *non-halogenated organic solvents*. Bottles for inorganic wastes should be labelled according to the nature of the waste (*e.g.* Hg). Acidic and alkaline wastes should be neutralised before disposal. Aqueous solutions may be poured down the drain after dilution, unless regulations in your country prohibit this practice, but, as mentioned above, this practice should be actively discouraged.

CLEANING OF GLASSWARE

Various cleaning agents are available for cleaning plastic bottles and glassware. One of the best is Decon 90, a biodegradable, non-phosphate detergent. Otherwise, soaking glassware in a sulfuric acid–dichromate mixture is adequate. To prepare this mixture, moisten 20 g of powdered sodium or potassium dichromate with water inside a glass container to form a thick paste. Add about 500 mL of concentrated sulfuric acid and stir. Store in glass-stoppered glass bottles. If some of the salt remains undissolved, leave it at the bottom of the bottle and use the supernatant liquid. Use the solution repeatedly as long as it remains red in colour. When the solution turns green in colour, discard, as this indicates that it is spent. Exert extreme caution when using this solution. Avoid spilling at all costs, but if any is spilled, wash off the spillage immediately with plenty of water. Do not allow this solution to come into contact with metallic or enamelled containers. Soaking glassware overnight in this solution should suffice. After soaking, rinse thoroughly with tapwater. Then rinse the glassware with laboratory water to ensure that any traces of cleaning agent are removed. Finally, dry in a drying cabinet and store in a clean place. Other cleaning reagents (*e.g.* nitric acid, hydrochloric acid) may be required for specific analyses. If specific cleaning procedures are recommended, follow them closely.

USE OF PIPETTES, BURETTES AND GRADUATED GLASSWARE

Two types of volumetric glassware are available: those calibrated "*to deliver*" and those calibrated "*to contain*". Volumetric flasks calibrated "to contain" should be used for preparing volumetric solutions. These are calibrated to contain the specified volume of solution when filled to the mark etched on the neck. Pipettes and burettes calibrated "to deliver" should be used; these are calibrated to deliver the specified volume when filled to the etched graduation mark. A small amount of liquid remains in the tip after the pipette is emptied. This residual solution should NOT be blown out when using a pipette calibrated "to deliver". Various automatic pipettes, capable of delivering between 1 μL and 10 mL, are commercially available. When used properly, these pipettes can give excellent precision. Two types of automatic pipettes are available: fixed-volume and variable-volume. Fixed-volume pipettes deliver the specified volume of solution. Variable-volume pipettes can be set to deliver the desired volume over a specified range (*e.g.* 10 to 100 μL). The solution is contained in a disposable plastic tip and this should be replaced between use. Always follow the manufacturer's instructions when using automatic pipettes. Some burettes have a glass stopcock that requires lubrication. These should be avoided, and burettes with a Teflon (PTFE) valve should be used.

Figure A1 shows how to read the meniscus when using graduated glassware, including pipettes, burettes and volumetric flasks. The eye should be level with the meniscus and the minumum in the curvature is taken as the reading. A white card can be placed behind the vessel while taking the reading.

Volumetric flasks ranging in volume from 5 mL to 5 L are available. You will generally be using 25, 50 or 100 mL flasks for preparing calibration standards, and 500 mL or 1 L flasks for preparing stock standard solutions and other reagents. Volumetric flasks should be filled to just under the graduation mark and made up to the mark by adding the solution dropwise from a dropping pipette or a wash bottle. If a dry

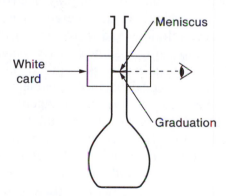

Figure A1 *How to read the meniscus of volumetric glassware*

reagent, or if several different solutions, are to be introduced into the flask, then only fill the flask about half-full with solvent and swirl the contents to dissolve. Add more solvent and swirl again and proceed as above. After filling to the mark the flask should be stoppered and shaken by inverting, while holding the stopper in with the thumb, to mix the contents thoroughly.

Pipettes capable of delivering a single volume have only one graduation mark. Pipettes which can be used to deliver variable volumes have a graduated scale printed on them. Pipettes should be used as follows. Use a rubber bulb to draw several portions of the solution into the pipette in order to wet the walls, but never draw into the bulb itself. Then draw a volume of solution to a level above the etched mark, making sure that there are no bubbles of air in the solution. Slowly allow the solution to drain out of the pipette until the meniscus reaches the etched mark. Dispense the solution into the receiving vessel and rest the pipette tip against the inner wall of the vessel for a few seconds. Do not blow out the small volume of solution remaining in the tip into the receiving vessel.

The most common burettes have a 50 mL capacity and are graduated to 0.1 mL. Microburettes with a capacity of 5 mL are also available. Burettes should be rinsed with the titrant several times before use. Pour about 5 mL of the titrant into the burette with the stopcock closed and rotate to wet the walls. Drain the titrant and repeat several times. Then fill the burette with titrant through a small funnel. Fill to above the zero mark and open the stopcock to remove the air in the tip. Then adjust the level to just below the zero mark and record the initial volume.

WEIGHING

Weighing can be done on a *laboratory balance* or on a more precise *analytical balance*. Modern laboratories employ electronic balances with a digital display of the weight. High accuracy is not required when preparing many reagent solutions and in this case the reagent may be roughly weighed on a laboratory balance, which weighs to the nearest 0.1 g. However, when high accuracy is required, as when preparing standard solutions, an analytical balance capable of weighing to the nearest 0.00001 g, or better, should be used. Reagent salts should first be dried in an oven at 105–110 °C for 1–2 h and then cooled to room temperature in a desiccator before weighing. Weigh the required reagent "by difference" as follows. Place a glass weighing bottle or plastic weighing boat on the balance, and press the "tare" control on the balance (this zeros the balance). Using a spatula, transfer the required amount of dried reagent to the weighing vessel and record the weight. If you are not using the "tare" control, then record the weight of the empty weighing vessel before adding the reagent and calculate the weight of reagent by difference from the final reading (bottle + reagent).

PREPARATION OF STANDARD SOLUTIONS

After weighing the required amount of dried reagent in a weighing boat or bottle, tip it into a beaker for dissolution. Rinse the weighing vessel with small aliquots of the solvent and transfer the washings to the beaker. Add some more solvent (less than the total volume required) to the beaker and stir or heat, as required, to dissolve. Once the reagent has completely dissolved, pour into the volumetric flask through a small funnel placed in the neck of the flask. Do this by pouring down a glass rod as shown in Figure A2. Rinse the beaker several times with aliquots of the solvent and pour washings into the volumetric flask. Also, rinse the funnel. Stopper the flask and mix the solution by shaking and inverting several times. Finally, make up to the mark as indicated above. Stopper and shake again.

FILTRATION

Filtration is the separation of particles from a solutions by means of a filter paper. Filtration can be achieved by gravity or by means of a vacuum. It is commonly used in water analysis to separate dissolved and

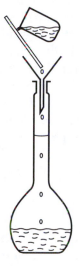

Figure A2 *How to pour a solution into a volumetric flask*

insoluble substances, and in air analysis to collect suspended particulate matter.

Gravity filtration is achieved by means of a filter paper held in a funnel. A circular filter paper is folded in half and then in half again. A cone is formed by pulling three sides of the filter paper away from the fourth side. The filter cone is inserted in a funnel and wetted with water so that it adheres to the sides of the funnel. The solution is then poured into the funnel and the filtrate collected in a vessel.

Vacuum filtration is accomplished by applying a vacuum to the filtering flask. A Buchner funnel or specialised filtering apparatus can used. The vacuum can be obtained by means of an air pump or by a water aspirator connected to a water tap. Details of vacuum filtration are given in Section 4.1.13.

DIGESTION

Some analyses require very little sample preparation, and the sample may be analysed directly (*e.g.* pH in water). However, many analyses require sample digestion in order to convert the analyte to a form suitable for analysis. For example, soil and dust samples need to be dissolved before they can be analysed by methods suitable for analysis of solutions. While water may be used for dissolving water-soluble ionic compounds, strong acids are often required for dissolving less-soluble species such as heavy metals. Laboratory reagents used for sample dissolution are listed in Table A5 together with their properties and typical uses.

Digestion procedures are described in the text where required. These should be followed closely and extreme caution should be exercised when performing any of the acid digestions. Acid digestion and evaporation of liquids is carried out on a heating plate inside a fume cupboard. Safety glasses and rubber gloves should be worn. Evaporation should be performed in a beaker covered with a watch glass with a stirring rod inserted between the watch glass and the top of the beaker wall. This should permit the vapours to escape while protecting the solution from contamination. Boiling chips or glass beads can be added to the solution to minimise bumping.

TITRIMETRIC ANALYSIS

Basic Principles

Titrimetric analysis, also called *volumetric* analysis, is based on the measurement of the volume of a reagent required to react with an

Table A5 *Laboratory reagents used for sample dissolution*

Reagent	Properties	Uses
Laboratory water (H_2O)	Highly polar, clear, colourless liquid. Very safe to use.	Dissolves polar and ionic compounds.
Hydrochloric acid (HCl)	Concentrated HCl is 38% solution. Evolves irritating fumes. Must be handled in a fume cupboard. Dangerous to use.	Dissolves metals, metal oxides and carbonates (*e.g.* Fe, Zn, Fe_2O_3, $FeCO_3$).
Nitric acid (HNO_3)	Concentrated acid is 70% solution. Evolves thick white and brown fumes. Reacts on contact with skin and clothing. Must be handled in a fume cupboard. Dangerous to use.	Dissolves metals and some organic samples (*e.g.* wastewater).
Sulfuric acid (H_2SO_4)	Concentrated acid is 96% solution. Evolves heat when mixed with water. Reacts on contact with skin and clothing. Must be handled in a fume cupboard. Dangerous to use.	Dissolves some organic samples (*e.g.* Kjeldahl method) and oxides of Al and Ti. Not very useful for metals as most metals form insoluble sulfates.
Perchloric acid ($HClO_4$)	Commercially available as 72% solution. Hot acid gives explosive reactions with organic materials and easily oxidised inorganic compounds. Must be handled in a specialised fume cupboard. Very dangerous to use.	Dissolves difficult organic samples and metals.
Hydrofluoric acid (HF)	Concentrated acid is 50% solution. Must be stored in Teflon or plastic containers as it dissolves glass. Causes serious and painful burns of the skin. Very dangerous to use.	Dissolves silica-based materials (*e.g.* soils, rocks, sand).
Aqua regia	1:3 by volume mixture of concentrated HNO_3 and HCl. Dangerous to use.	Dissolves highly unreactive metals (*e.g.* Au).

analyte in solution. It is a simple and inexpensive method requiring basic laboratory equipment available in all laboratories: burettes, pipettes, flasks and balances. Many analyses that can be performed by titrimetry can also be carried out by more sensitive instrumental techniques (*e.g.*

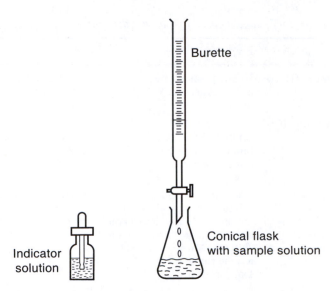

Figure A3 *Equipment required for titration*

spectrophotometry). However, because of its simplicity, titrimetry is still widely used in many environmental laboratories. The experimental setup for carrying out a titration is shown in Figure A3. The reagent solution, which is added from a burette, is called a *titrant*. The elements of a titration analysis are as follows:

$$\text{Analyte} + \text{Reagent} \rightarrow \text{Product}$$

The amount of added reagent needed to consume all of the analyte is measured and converted to the mass or concentration of the analyte based on the stoichiometry of the reaction. The amount of reagent is determined by measuring the volume of the reagent solution added. The total consumption of the analyte (*i.e.* completion of reaction) is signalled by a visual sign, such as a colour change in the solution. Usually this colour change is caused by the addition of a substance called an *indicator*. The point where the titration reaction is complete is called the *equivalence point*. The *end point* is the point at which the indicator signals that the reaction is complete, by changing colour. Often, the indicator is not able to signal the precise moment at which the analyte has completely reacted with the titrant. Therefore there is usually some discrepancy between the end point and the equivalence point. In general, the accuracy of the titration does not suffer significantly from this difference. In cases where the difference is appreciable, some other method of determining the equivalence point has to be used (*e.g.* ion selective electrode).

On the basis of the types of reactions involved in a titration we can subdivide titrations into four classes:

- Neutralisation, or acid–base, titrations.
- Complexation titrations.
- Precipitation titrations.
- Oxidation–reduction, or redox, titrations.

A graphic illustration of the variation in pH of solution during a typical acid–base titration is shown in Figure A4. This is called a *titration curve*. Initially, the solution is acidic and the pH increases slowly as base is added from a burette. In the vicinity of the equivalence point the rate of pH change is rapid. Thereafter, the pH increases slowly. The pH of the equivalence point for a strong acid–strong base titration (*e.g.* HCl–NaOH) is 7. Similar titration curves can be drawn for other titration reactions but in this case it is $-\log_{10}[\text{ion}]$ that is plotted instead of the pH as the *y*-axis. When a pH, or other ion selective electrode, is used instead of an indicator, the entire titration curve is plotted and the end point is determined as the point of inflection on the titration curve.

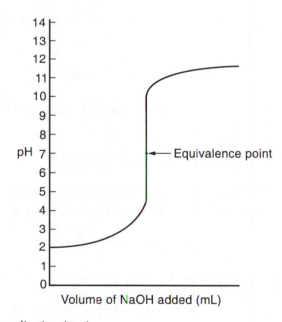

Figure A4 *Neutralisation titration curve*

According to the way the titration procedure is carried out we can subdivide titrations into three classes:

(a) Direct titrations.

$$\text{Analyte} + \text{Titrant} \rightarrow \text{Product}$$

The concentration of the analyte is determined directly from the volume of titrant used up.

(b) Indirect titrations. Excess reagent is added to convert all of the analyte to another product:

$$\text{Analyte} + \text{Reagent} \rightarrow \text{Product}$$

The product of the reaction is then titrated and the concentration of the original analyte determined from the volume of titrant used to react with the product:

$$\text{Product} + \text{Titrant} \rightarrow \text{Product}^*$$

(c) Back titrations. Excess reagent is added to covert the analyte to another product:

$$\text{Analyte} + \text{Reagent} \rightarrow \text{Product}$$

The amount of reagent remaining in solution after reaction with the analyte is then titrated:

$$\text{Excess reagent} + \text{Titrant} \rightarrow \text{Product}^*$$

The concentration of the original analyte is determined from the volume of the original reagent added and the volume of titrant used up:

$$\text{Analyte} = \text{Reagent added} - \text{Titrant}$$

In this titration, the amount of excess reagent added to the analyte in the initial step must be accurately measured.

The first step in a titration involves preparation of the titrant. The titrant is prepared by dissolving an accurately measured weight of a pure solid reagent in water. The reagent should be weighed on a balance to four significant figures in order to avoid diminishing the accuracy during the course of the titration. Often, titrants are prepared by dilution of a more concentrated solution.

Standardisation

In many cases, it is not possible to obtain the necessary accuracy in titrant concentration simply by preparing the titrant because the solid reagent may not be pure or because it may be hygroscopic (*e.g.* sodium hydroxide). In this case, it is necessary to determine the concentration of the titrant accurately by performing a separate determination, called a *standardisation* procedure. In a standardisation procedure the titrant is usually titrated with a second, so-called *primary standard* (*i.e.* the "titrant" becomes the "analyte" and the "primary standard" becomes the "titrant" in this procedure). To *standardise* a solution means to determine its concentration accurately by reaction with the primary standard. The primary standard must meet the following criteria:

- It must be pure (*i.e.* its purity must be 100%). Reagents which are not 100% pure may be acceptable so long as the purity is accurately known.
- It should be stable at temperatures in a drying oven.
- It should not be hygroscopic (*i.e.* it should not absorb water from humid air).
- The reaction with the reagent that is being standardised should be quantitative and fast.
- It should have a high molecular weight.

Performing a Titration

When titrating, place the tip of the burette into the neck of the conical flask and introduce the titrant in increments of about 1 mL. Stir or swirl the flask constantly while adding titrant to ensure mixing. Decrease the volume of increments as the titration progresses. In the vicinity of the end point, add titrant dropwise and observe the indicated colour change. Record the volume of titrant used to the nearest 0.01 mL. First carry out a rough titration to determine the end point approximately, then carry out several titrations carefully following the above procedure and use the mean value when calculating the concentration of analyte.

A volume of solution in a 50 mL burette can be read to four significant figures. The graduations on the burette allow for reading three figures with certainty and the last figure has to be estimated. For example, the reading from a burette reported as 21.57 mL implies that 21.5 mL can be read with certainty from the graduations, while the fourth digit (7) is estimated as the reading lies somewhere between graduations for 21.5 mL and 21.6 mL.

SPECTROSCOPY

Spectroscopic methods of analysis are based on the interaction of electromagnetic radiation with atoms and molecules making up matter. The various branches of spectroscopy can be classified according to the portion of the electromagnetic spectrum inducing energy changes in matter. This classification is shown in Table A6. Although many spectroscopic techniques are used in environmental analysis (e.g. γ-ray, X-ray, infrared, Raman), we shall only consider UV/visible spectroscopy as this is more widely used than the other techniques.

Electromagnetic radiation is a form of energy propagated in waves which can be characterised in terms of wavelength or frequency:

$$v = c/\lambda$$

where c is the speed of light in vacuo (2.997924590×10^8 m s^{-1}), λ is the

Table A6 *Various branches of spectroscopy*

Type of spectroscopy	λ	v (Hz)	E (J mol^{-1})	Type of change
γ-ray	<100 pm	$>3 \times 10^{18}$	$>10^9$	Change in nuclear configuration
X-ray	100 pm–10 nm	3×10^{16}–3×10^{18}	10^7–10^9	Change in distribution of inner electrons
UV/Vis.	10 nm–1 μm	3×10^{14}–3×10^{16}	10^5–10^7	Change in distribution of outer electrons
IR	1 μm–100 μm	3×10^{12}–3×10^{14}	10^3–10^5	Change in configuration due to vibration
MW	100 μm–1 cm	3×10^{10}–3×10^{12}	10–10^3	Change in orientation due to rotation
ESR	1 cm–100 cm	3×10^8–3×10^{10}	10^{-1}–10	Change of electron spin in magnetic field
NMR	100 cm–10 m	3×10^6–3×10^8	10^{-3}–10^{-1}	Change of nuclear spin in magnetic field

wavelength and v is the frequency. The energy of each wave is a function of the frequency and it is defined as:

$$E = hv$$

where h is Planck's constant $(6.62620 \times 10^{-34}$ J s). The energy of the electromagnetic wave must correspond to the difference in energy between two energy levels in an atom or molecule for interaction to take place:

$$E_0 + hv \rightarrow E^*$$

where E_0 is the lowest energy level, or so-called *ground state*, and E^* is the higher energy level, or so-called *excited state*. Atoms and molecules can exist only in distinct energy levels which are said to be *quantised*. Absorption of radiation of the appropriate frequency can cause the atom or molecule to jump to a higher energy level, as shown by the above equation. Conversely, when atoms or molecules return from a high energy level to a lower energy level, radiation of appropriate frequency is emitted (Figure A5):

$$E^* \rightarrow E_0 + hv$$

In quantum mechanics, hv is called a *quantum*, and a quantum of light energy is called a *photon*. According to quantum theory, changes in energy states take place by means of loss or gain of quanta and not continuously.

Normally, atoms and molecules prefer to occupy the lowest energy level possible, although there will always be some atoms and molecules in higher energy levels. In fact, an equilibrium exists between the higher and lower energy level and this equilibrium is influenced by the temperature.

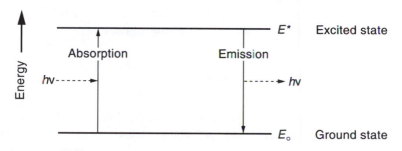

Figure A5 *Simple energy level diagram showing absorption and emission of radiation*

The ratio of atoms or molecules in a higher energy level to those in a lower energy level is given by:

$$N_{upper}/N_{lower} = \exp(-\Delta E/kT)$$

where ΔE is the energy separation between the two energy levels in J, T is the temperature in K and k is the Boltzmann constant (1.38062×10^{-23} J K^{-1}).

In *absorption spectroscopy*, electromagnetic radiation is directed at the sample (gaseous, liquid or solid) and the radiation passing through the sample is measured with a detector. The incident radiation is attenuated (*i.e.* intensity is reduced) on passage through the sample. In *emission spectroscopy* the radiation emitted by a sample is detected. The output of the detector is recorded and referred to as a *spectrum*. Spectra can be of three types: line, band or continuous. Atomic spectra are quite simple and consist of lines. When UV or visible light strikes atoms it induces electrons to rise to higher electronic energy levels and each line in the spectrum corresponds to a particular electronic transition. On the other hand, band or continuous spectra are observed with molecules. These reflect not only the transitions of electrons between different molecular electronic energy levels but also transitions between different molecular vibrational energy levels. Molecular bonds vibrate, much as a spring vibrates, and this gives rise to vibrational energy levels which complicate the appearance of molecular spectra. Although both atomic and molecular spectroscopy involve interaction with UV/visible radiation, the term "UV/visible spectrophotometry" generally refers to molecular spectroscopy. Atomic UV/visible spectrophotometry is generally called *atomic absorption spectroscopy* (AAS) or *atomic emission spectroscopy* (AES).

The intensity of monochromatic radiation (*i.e.* light of a single wavelength) decreases exponentially as the concentration of the absorbing substance increases. The relationship between absorbance and concentration is defined by the Beer–Lambert law:

$$A = \log_{10}(I_0/I) = \varepsilon cl$$

where I_0 is the intensity of the incident beam, I is the intensity of the transmitted beam (Figure A6), A is the absorbance, ε is the molar absorptivity in units of L mol^{-1} cm^{-1}, c is the concentration in mol L^{-1} and l is the path length in cm. The absorptivity of a species is

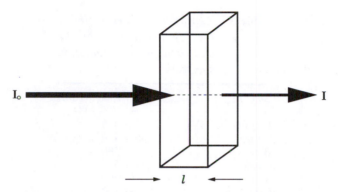

Figure A6 *Decrease in the intensity of incident light I_0 upon passing through a sample cell of length l containing an absorbing solute at concentration c*

constant at a given wavelength. The transmittance, T, is defined as the fraction of incident radiation transmitted by the sample:

$$T = I/I_0$$

and the absorbance can be related to the transmittance by:

$$A = -\log_{10}T$$

A series of calibration standards are analysed and a calibration graph of absorbance against concentration prepared. The concentration of the unknown is determined from the calibration graph.

UV/visible Spectrophotometry

Parts of the electromagnetic radiation spectrum relevant to UV/visible spectroscopy are summarised in Table A7. Those techniques that involve the use of a photomultiplier tube, a special type of light detector, are commonly termed *spectrophotometric* and the corresponding instrument is called a *spectrophotometer*. Techniques involving only visible light are sometimes called *colorimetric* and the corresponding instrument is called a *colorimeter*. Some colorimeters involve matching visually the colour of the developed sample with a set of standards.

UV/visible spectrophotometry is the most widely used technique in environmental analysis because it is:

- *Widely applicable.* Many inorganic and organic chemicals absorb in

Table A7 *Approximate wavelengths of the UV/visible part of the electromagnetic radiation spectrum*

Colour	Wavelength (nm)
Ultraviolet (UV)	200–400
Violet	400–450
Blue	450–500
Green	500–570
Yellow	570–590
Orange	590–620
Red	620–760
Infrared	>760

the UV or visible region and many non-absorbing species can be converted to absorbing derivatives by chemical reaction.

- *Highly sensitive.* Detection limits for many absorbing species range from 1 to 10 mg L^{-1}, but can often be extended to 0.1 or even 0.01 mg L^{-1}.
- *Highly accurate.* Relative errors in concentration are generally in the range from 1% to 5%, and these can be reduced to <1% in the hands of a highly skilled analyst.
- *Moderately to highly selective.* Quite often it is possible to find a wavelength at which only the analyte species absorbs. If an interfering compound absorbs at the same wavelength, it is possible to remove the interfering compound by simple chemical procedures (complexation, precipitation, *etc.*).
- *Easy and convenient to use.* Measurements can be performed rapidly, allowing for a large number of samples to be processed. Colorimeters using natural visible light or battery operated filter photometers are widely used as portable instruments for on-site field measurements.

There are two types of UV/visible spectrophotometers available: single-beam and double-beam instruments. A typical *single-beam* instrument is illustrated in Figure A7. Light from a lamp is directed onto the sample held in a cell, commonly called a *cuvette*, after passing through a zwavelength selector. The wavelength selector, also called a *monochromator*, can be a filter, a prism or a diffraction grating, and it serves to isolate the analytical wavelength that is to be used with the analyte in question. Some of the light is absorbed by the analyte in the

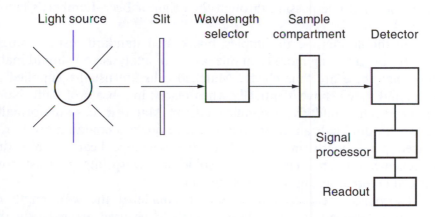

Figure A7 *Typical components of a single-beam UV/visible spectrophotometer*

sample and the intensity of the transmitted light is measured by a detector. The detector is usually a photomultiplier tube, converting light photons into electrons, which can be recorded as an electrical signal. Most instruments offer a choice of displays on the readout: "absorbance", "transmittance" or "% transmittance". The instrument is usually operated in the "absorbance" mode. Sample cells made of plastic or ordinary glass can be used for visible light, while more expensive quartz cells are required for the UV region. Cells with 1 cm path lengths are commonly used. Longer cells (*e.g.* 5 cm) can be used to increase the sensitivity. The instrument is zeroed by placing a water blank in the sample compartment. In *double-beam* instruments the light from the source is split into two parallel beams, one of which is directed onto the blank, while the other is directed onto the sample. The instrument is zeroed simultaneously while making the measurement, thus correcting for errors which may arise due to changes in source light intensity. Most modern, bench-top UV spectrophotometers are of double-beam design. *Photometers* are single-beam visible spectrophotometers that employ a filter as the wavelength selector. They are supplied with several filters which can be interchanged, depending on the colour of the absorbing solution. Photometers are inexpensive, rugged, portable and easy to use, as well as requiring little maintenance.

Beer–Lambert's law holds over a wide concentration range; however, deviations are possible when the coloured compound dissociates, associates, ionises or forms complexes. A calibration graph of absorbance against concentration prepared by analysing a series of standards should give a straight line passing through the origin if Beer–Lambert's law is obeyed.

You should analyse all samples, blanks and standards using a single cell. If you are using more than one cell in an analysis you should make sure that these are "matched". Matched cells are usually supplied in pairs and they behave identically when placed in a beam of light. Non-matched cells could give a significant error. Matched cells are normally used with double-beam instruments, where one cell containing water is placed permanently in the path of the reference beam. Never dry matched cells in an oven. Wipe any solution, fingerprints and dust from the cell before inserting in the instrument.

Absorbance measurements should be made at the wavelength of maximum absorption, λ_{max}. Wavelengths to be used are given in the experimental procedures. However, it is worthwhile checking that this is indeed the wavelength of maximum absorption. To do this, determine the entire spectrum of the absorbing species after full colour development. Either use a scanning double-beam instrument, or measure the absorbance at different wavelengths using a single-beam instrument and plot a graph of absorbance against wavelength. Determine λ_{max}. The spectrum of pararosaniline methyl sulfonic acid, the species resulting from the reaction of SO_2 with the colour-forming reagent specified in Section 3.2, is illustrated in Figure A8. The maximum is at 560 nm.

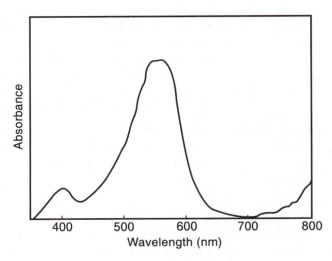

Figure A8 *Visible spectrum of pararosaniline methyl sulfonic acid*

Atomic Absorption Spectrometry (AAS)

Atomic absorption spectrometry is widely used for the determination of metals in environmental analysis. A schematic diagram of a typical atomic absorption spectrometer is illustrated in Figure A9. The source of radiation is a hollow cathode lamp, which contains a cathode constructed of the same metal as that being analysed. This emits the wavelengths characteristic of the metal, and a different lamp is required for each metal. Multi-element lamps are also available. The light from the lamp is directed through a flame and onto a monochromator, which selects the preferred analytical wavelength. The light from the monochromator is detected by a photomultiplier tube and converted to an electrical signal. The sample is aspirated in the flame where the water is evaporated and the metal-containing compounds are volatilised and dissociated into ground state atoms. The ground state atoms absorb the radiation from the hollow cathode lamp and are excited to higher energy levels. Some atoms are also thermally excited but their fraction is so small that it causes no errors in the analysis.

Temperatures of flames used in AAS are given in Table A8. An acetylene–air flame is suitable for the analysis of most metals. Higher temperatures of an acetylene–N_2O flame are required for some metals that form refractory oxides (*e.g.* Al, Ti).

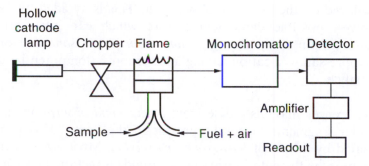

Figure A9 *Components of a single-beam flame atomic absorption spectrometer*

Table A8 *Flame temperatures in AAS*

	Temperature (K)	
Fuel	*Air*	*N_2O*
Acetylene, C_2H_2	2400	3200
Hydrogen, H_2	2300	2900
Propane, C_3H_8	2200	3000

AAS is prone to both spectral and chemical interferences. *Spectral interferences* arise from the overlap between the wavelength of the selected resonance line and the line emitted by some other metal present in the sample. There are very few cases of spectral interference in AAS as the absorption lines are very narrow, about 0.005 nm. In any case, a different line may be selected at which the interfering metal does not absorb. *Chemical interferences* may inhibit the atomisation process by one of two processes:

- *Stable compound formation* either from incomplete dissociation or from the formation of refractory compounds. An example of the former is the determination of calcium in the presence of phosphate. This may be solved by using an excess of *releasing agent*, such as lanthanum or strontium, which will preferentially combine with phosphate and release the calcium. Aluminium, vanadium and titanium may form stable refractory compounds (*e.g.* Al_2O_3) and these can be dissociated in higher temperature flames (*e.g.* acetylene–nitrous oxide).
- *Ionisation* of ground state atoms in the flame according to:

$$M \rightarrow M^+ + e^-$$

will reduce the extent of absorption. This is avoided by using the lowest possible flame temperature which can give satisfactory dissociation or by adding an *ionisation suppressant*. An ionisation suppressant is a cation having a lower ionisation potential than the analyte.

Interferences may also arise from *background absorption* caused by molecular fragments or smoke present in the flame. This problem is eliminated by means of *background correction*. Most AAS instruments are equipped with deuterium arc background correction, and you should use it if necessary. A good way to see if background correction is required is to analyse a set of calibration standards and samples, once without background correction and once with the deuterium arc lamp switched on. Plot the two calibration graphs and compare the results. If there is no significant difference you do not require to use background correction for that particular analysis.

Still other problems can be caused by different physical characteristics (density, viscosity, *etc.*) between the sample and the solvent used for preparing calibration standards, so-called *matrix effects*. These can be minimised by *matrix matching*, *i.e.* making sure that the standards and samples do not differ much in their bulk composition, and by using the same solvent for both samples and standards. The method of standard addition rather than conventional calibration, will also eliminate this problem.

Graphite furnace AAS (GFAAS) is now widely available in many laboratories. GFAAS involves electrothermal atomisation inside a graphite tube furnace instead of flame atomisation. Owing to the small volume of the tube (9 × 50 mm), small volumes (1–100 μL) of sample can be analysed and the technique is more sensitive than flame AAS. Detection limits for most metals are 100–1000 times lower with GFAAS than with flame AAS. If GFAAS is available in your laboratory, you may use it to analyse those samples which contain metal concentrations below the detection limit of flame AAS. Get the laboratory technician or demonstrator to show you how to use the instrument.

Atomic Emission Spectrometry (AES)

In AES, atoms are thermally excited in the flame and on returning to the ground state they emit radiation which can be detected. The AAS instrument may also be operated in the atomic emission mode simply by turning off the hollow cathode lamp. The sample is aspirated in the flame and the resonance wavelength selected on the monochromator in the same way as in flame AAS. The intensity of the emitted radiation is recorded. There is usually a switch on the instrument to indicate whether it is operating in the AAS or AES mode. Switch to the AES mode if using atomic emission. You may compare the sensitivity of AAS and AES for some metals by analysing calibration standards and samples using both techniques. Dedicated flame photometers capable of measuring Na, K, Mg and Ca are widely available in environmental laboratories. You may use a flame photometer for these analyses instead of AAS if an instrument is available in your laboratory.

Inductively coupled plasma AES (ICP-AES) is being increasingly used in environmental analysis and many laboratories have purchased, or are in the process of purchasing, ICP. This instrument uses an argon plasma, consisting mainly of argon ions and electrons, to atomise the sample. Temperatures as high as 10 000 K can be achieved with ICP. This leads to a much greater thermal excitation of metal atoms, as indicated by the relationship between the population of states and temperature discussed earlier on. If an ICP instrument is available in your laboratory, you may use it under direction of the laboratory technician or demonstrator.

ION CHROMATOGRAPHY (IC)

Ion chromatography (IC) is a specific type of chromatography involving separation of ions on an ion exchange resin. *Chromatography* is used to separate different chemical components in a mixture. As such, it is purely a separation method and it is incapable of analysing samples by itself. In order to determine components in a sample, chromatography has to be combined with a detector capable of responding to some physical or chemical property of the analytes. Modern chromatography instruments normally come equipped with one or more detectors.

During chromatographic separation, components are partitioned (*i.e.* distributed) between a *stationary phase* and a *mobile phase*. The mobile phase carries the components through the stationary phase and these are separated because each component interacts with the stationary phase to a different degree. Thus components which interact strongly with the stationary phase move slower than those components which interact weakly. There are many different types of chromatography, depending on the nature of the mobile and stationary phases (*e.g.* gas–liquid chromatography, liquid–liquid chromatography). *Elution* is the name given to the process of solute separation and the mobile phase is referred to as the *eluent*. The results of a chromatographic analysis are displayed on a *chromatogram*, which gives the detector response as a function of time.

In *ion exchange chromatography* the mobile phase is a liquid solution and the stationary phase is an ion exchange resin packed inside a column. A process of ion exchange takes place between eluting ions in solution and ions of the same charge on the surface of the ion exchange resin. Cationic exchange sites on the resin are commonly the sulfonic acid group (strongly acidic), $-SO_3^-H^+$, and the carboxylic acid group (weakly acidic), $-COO^-H^+$. Anionic exchange sites include the tertiary amine group (strongly basic), $-N(CH_3)_3^+OH^-$, and the primary amine group (weakly basic), $-NH_3^+OH^-$. Consider a solution of NaCl passing through an ion exchange resin. The cation exchange equilibrium can be described by the following equation:

$$Res-SO_3^-H^+ + Na^+ \text{ (solution)} \rightleftharpoons Res-SO_3^-Na^+ + H^+ \text{ (solution)}$$

while the anion exchange equilibrium can be described by:

$$Res-N(CH_3)_3^+OH^- + Cl^- \text{ (solution)}$$
$$\rightleftharpoons Res-N(CH_3)_3^+Cl^- + OH^- \text{ (solution)}$$

The *selectivity coefficient, K*, is defined as:

$$K = \frac{[Na^+]_r[H^+]_s}{[Na^+]_s[H^+]_r} \quad \text{for cation exchange, and}$$

$$K = \frac{[Cl^-]_r[OH^-]_s}{[Cl^-]_s[OH^-]_r} \quad \text{for anion exchange}$$

where subscripts r and s denote resin and solution phase, respectively. Values of selectivity coefficients provide a guide to the relative affinities of ions for particular resins. If a mixture of different cations, A, B, *etc.*, is passed through an ion exchange column, the ions may be separated if they have different affinities for the resin. For example, if ion B is held more strongly by the resin than ion A, ion A will emerge first from the column.

A schematic illustration of a typical ion chromatograph is shown in Figure A10. An electrolyte solution (*e.g.* $NaHCO_3$ for anion analysis) serves as the eluent. The eluent is pumped through a separator column containing the ion exchange resin and a suppressor column. The sample is introduced into the eluent upstream of the separator column. Analyte ions are separated on the column and elute into a cell where the conductivity is determined by a conductivity meter. The function of the suppressor column is to lower the background conductivity of the eluent so that the conductivity due to the eluting analyte ions may be determined. For example, in the case of $NaHCO_3$ eluent, the suppressor column contains a strong acid ion-exchange resin that converts sodium bicarbonate to dilute carbonic acid:

$$Res–SO_3^- H^+ + NaHCO_3 \text{ (solution)} \rightleftharpoons Res–SO_3^- Na^+ + H_2CO_3 \text{ (solution)}$$

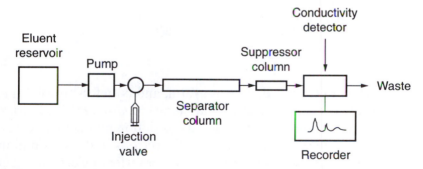

Figure A10 *Components of a typical ion chromatograph*

Single column ion chromatographs are also available. These are capable of measuring the small difference in conductivity between the eluted sample ions and the background electrolyte. Low-capacity ion-exchange resins that allow elution with dilute eluent are used. Furthermore, eluents with low conductivities are used.

An ion chromatogram shows the response of the conductivity detector as a function of time from the injection of the sample into the eluent stream. A typical chromatogram is shown in Figure 2.6. Each ion is identified by its retention time. The peak heights and the areas under the peaks can be used for quantitative analysis. A series of calibration standards is analysed and the peak height or peak area plotted against the ionic concentration. The concentration of the ion in the sample can then be determined from the calibration graph. There are several manual methods for calculating peak areas. For symmetric peaks, the area can be calculated as:

$$\text{Area} = \text{height} \times \text{width at half-height}$$

or

$$\text{Area} = \text{width at base} \times \text{height}/2$$

Modern chromatographs are equipped with integrators or on-line computers which print the peak heights and areas together with the chromatogram. Ion chromatography is widely used in analysis of environmental samples, especially rainwater and surface water samples.

ELECTROANALYTICAL METHODS

Electroanalytical methods are based on reduction/oxidation (redox) reactions taking place at two electrodes, reduction at one electrode and oxidation at the other. For example, consider reactions at Cu and Zn electrodes:

Reduction: $Cu^{2+}(aq) + 2e^- \rightarrow Cu(s)$

Oxidation: $Zn(s) \rightarrow Zn^{2+}(aq) + 2e^-$

where (aq) refers to dissolved, aqueous form, and (s) refers to solid (*i.e.* electrode) form. The above equations are termed *half-reactions*. During reduction a reactant gains electrons and its oxidation number decreases, whereas during oxidation a reactant loses electrons and its oxidation number increases. Redox reactions involve the transfer of electrons from one species to another, and two half-reactions combine to give a *redox couple*. The overall reaction, or redox couple, is derived by adding the two half-reactions. For the above case:

Overall reaction $Cu^{2+}(aq) + Zn(s) \rightarrow Cu(s) + Zn^{2+}(aq)$

In this reaction, Cu^{2+} acts as the oxidising agent, or *oxidant*, while Zn acts as a reducing agent, or *reductant*. The oxidising agent gains electrons and is itself reduced, while the reducing agent loses electrons and is oxidised. The electrode at which oxidation occurs is called an anode (Zn in the above example), while the electrode at which reduction occurs is called a cathode (Cu in the above example).

The electrode potential is defined as the potential difference between an electrode and its solution in a half-reaction. However, the potential of a single half-reaction cannot be measured and the potential is measured relative to a *standard hydrogen electrode* (SHE), which is assigned a potential of 0 V. The potential difference measured between a half-reaction and the SHE is called the *standard electrode potential, $E°$*. Standard electrode potentials can take both $+ve$ and $-ve$ values. Species with high $+ve$ values of $E°$ tend to be strong oxidising agents, while species with low $-ve$ values of $E°$ tend to be strong reducing agents. Standard electrode potentials refer to standard conditions of $1 \, mol \, L^{-1}$ concentration, 1 atm pressure and 25 °C. The electrode potential at non-standard conditions can be calculated from the standard electrode potential by means of the *Nernst* equation. For a half-reaction:

$$\text{Oxidant} + ne^- \rightleftharpoons \text{Reductant}$$

the Nernst equation is:

$$E = E° - \frac{RT}{nF} \ln \frac{[\text{reductant}]}{[\text{oxidant}]}$$

where R is the universal gas constant, T is the absolute temperature in K, n is the number of electrons transferred in the reaction, F is the Faraday constant and ln is the natural logarithm (*i.e.* $\log_e$). At 25 °C, the above equation reduces to:

$$E = E° - \frac{0.0591}{n} \log_{10} \frac{[\text{reducant}]}{[\text{oxidant}]}$$

Activities, rather than concentrations, should ideally be used in the above equation. For a general overall reaction:

$$aA + bB \rightleftharpoons cC + dD$$

the Nernst equation is:

$$E = E^\circ - \frac{0.0591}{n} \log_{10} \frac{[C]^c[D]^d}{[A]^a[B]^b}$$

where E° is the standard cell potential for the overall reaction, and it can be calculated from the standard electrode potentials of the half-reactions.

It is apparent from the above equations that the potential of electrodes and half-cells (*i.e.* half-reactions) depends on the activity, and hence concentration, of the species involved. This relationship serves as the basis of quantitative electroanalytical techniques.

Potentiometry

The two electrodes in the above example (Zn and Cu) can be connected by a wire and the two half-cells can be connected by a *salt bridge*, consisting of an electrolyte solution (*e.g.* KCl). Since electrons are produced at the Zn anode and consumed at the Cu cathode, an electrical current will flow through the wire. The salt bridge enables the half-cells to maintain electrical neutrality by allowing charged ions to move between the half-cells. Such cells, which produce an electrical current, are called *galvanic* cells.

Galvanic cells are used in potentiometry, an analytical technique that is based on the measurement of potential (or voltage) between two electrodes when they are immersed in a solution. *Ion selective electrodes* (ISE), of which the pH electrode is an example, involve the determination of analyte concentrations in solution by means of potentiometry. These so-called *indicator electrodes* are used together with a *reference electrode* to measure the potential. The *saturated calomel electrode* and the *silver–silver chloride electrode* are commonly used as reference electrodes in potentiometric analysis. Consider an ISE constructed of a metal M, which is immersed in a solution containing its own ions, M^{n+}, together with a reference electrode. An electrode potential is established, the value of which is given by the Nernst equation as:

$$E = E° - \frac{0.0591}{n} \log_{10} \frac{1}{[M^{n+}]}$$

The resultant cell potential ΔE is measured:

$$\Delta E_{cell} = E_{ISE} - E_{ref}$$

Since the potential of the reference electrode E_{ref} is known, the electrical potential E_{ISE} can be calculated:

$$E_{ISE} = \Delta E_{cell} + E_{ref}$$

Also, as the standard electrode potential for the metal M is known, the activity of M^{n+} can easily be calculated. For dilute solutions the activity is virtually identical to the concentration.

Measurement of pH is the most widely used application of potentiometry in analytical chemistry. The pH electrode, also called a glass electrode, consists of a thin membrane of special glass (63% SiO_2, 28% Li_2O, 4% BaO, 3% La_2O_3 and 2% Cs_2O) which is highly selective for H^+ ions. The internal element of the electrode consists of a Ag/AgCl electrode immersed in a pH 7 buffer saturated with AgCl. The glass electrode is used in conjunction with a reference electrode, usually the saturated calomel electrode. The electrode potential of the glass electrode is given by:

$$E = K + 0.0591 \log_{10}[H^+]$$

and since $pH = -\log_{10}[H^+]$, then:

$$E = K - 0.0591 \, pH$$

where K is a constant dependent on the composition of the glass and the individual character of each electrode. The value of K may vary slightly with time.

Combination pH electrodes, which contain the indicator electrode and a standard reference electrode (Ag/AgCl) combined in a single unit, are more practical to use. A combination pH electrode is illustrated in Figure A11.

Changes in temperature can affect the potential between the indicator

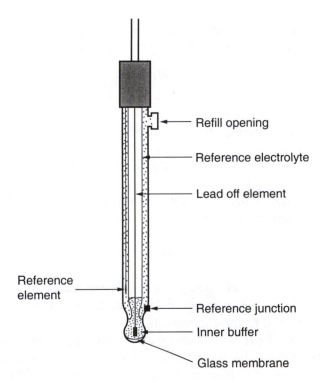

Figure A11 *Combination pH electrode*

and reference electrodes. Most pH meters come with a temperature probe which should be inserted into the sample when taking measurements. The meter then automatically compensates for the temperature. pH meters and electrodes should be calibrated with buffer solutions before use. Normally, a two-point calibration is used. This involves repeatedly inserting the electrode in pH 7 and pH 4 buffers and adjusting the reading until the meter reads the correct pH without need for further adjustment. You should follow the operating instructions that come with the pH meter.

Some basic operating hints which could improve your measurements are listed below:

- Calibrate the meter and electrode using a buffer solution with a pH close to that of the test sample, *i.e.* pH 7 and pH 4 buffers if you expect to be measuring pH values ⩽7, and pH 7 and pH 9 buffers if you expect to measure pH values ⩾7.
- Do not use buffers after their expiry date and do not pour used buffer back into bottles.

- After calibrating the electrode you may check the reading by measuring the pH of a dilute acid (*e.g.* HCl) or alkali (*e.g.* NaOH) solution of known pH.
- Always rinse electrodes with laboratory water and blot dry with tissue paper (do not wipe) before inserting in a buffer or sample solution.
- When not in use, store the electrode by immersing in water, buffer solution or electrode filling solution (KCl).
- Do not allow the electrode to dry out as this may cause permanent damage. Keep the electrode topped up with the correct filling solution.
- If the electrode response is sluggish or inaccurate, soak the tip of the electrode in 0.1 M HCl overnight.
- Record the pH to two decimal places only. Although some digital meters display three decimal places, the last decimal place is unrealistic since the accuracy of the pH measurement is no better than 0.01.

Glass electrodes are prone to overestimate the pH in highly acidic media of pH <0.5. This *acid error* is of little consequence in environmental samples, which are never that acidic. In basic solutions, pH electrodes respond to alkali metal ions (*e.g.* Na^+) as well as to H^+. The magnitude of this *alkaline error* increases with increasing pH at pH >9 and varies from electrode to electrode. For example, a Corning electrode may underestimate the pH by about 0.14 pH units at pH 10 and by about 0.8 pH units at pH 12 when immersed in a 0.1 mol L^{-1} Na^+ solution. Special pH electrodes are available for measuring high pH samples containing high Na^+ concentrations. Again, the alkaline error should cause no problems in most environmental samples, which generally have much lower concentrations of Na^+ and lower pH values.

The difference in ionic strength between calibration buffers and environmental samples can cause some problems. Buffer solutions tend to have higher ionic strength than many environmental samples. Several solutions to this problem have been proposed:

- Adding a neutral salt such as KCl to the sample to match the ionic strength of the buffers. This may, however, introduce impurities present in the reagent which may have a disproportionately large effect on the pH of dilute environmental samples, and should be discouraged.
- Using buffer solutions of low ionic strength.

● Using specialised pH electrodes filled with a low concentration of reference electrolyte.

The pH electrodes may respond slowly in dilute solutions with low ionic strength, such as rainwater and drinking water. Readings may sometimes take several minutes to stabilise. Some analysts recommend that samples should be stirred on a magnetic stirrer while taking pH measurements. This can lead to errors and should be discouraged. Others measure the pH in still solutions. The best practice is to stir the solution until it is well mixed and then allow it to come to rest before taking a pH measurement. Another problem with very dilute environmental samples with low ionic strength is the possibility of interferences which could affect the pH. For example, absorption of gases present in laboratory air (*e.g.* CO_2, NH_3) may affect the pH of some samples.

Ion selective electrodes are commercially available for a number of other ions (Br^-, Cl^-, F^-, I^-, NO_3^-, Na^+, NH_4^+, Ca^{2+}, Cu^{2+}, Pb^{2+}, S^{2-}, CN^-). These are constructed of metals, glass membranes, polymers or crystals, and are capable of measuring concentrations as low as 10^{-7}–10^{-6} mol L^{-1}. Gas sensing electrodes are also available for NH_3, SO_2, NO_2, H_2S, HCN, CO_2, HF and Cl_2. These electrodes consist of a tube containing the indicator and reference electrodes immersed in an electrolyte solution, isolated from the gas phase by a thin gas-permeable membrane through which the gaseous analyte can diffuse. When using indicator electrodes, it is common practice to add a supporting electrolyte, called TISAB (total ionic strength adjustment buffer), to both the samples and standards in order to provide a constant ionic strength. Quantitative analysis can be accomplished by means of calibration graphs or standard addition. The response of an ion selective electrode is a logarithmic function of activity (or concentration).

Potentiometric Titration

Potentiometric measurements can be used to monitor the progress of a titration and determine the end point more accurately than with visual, colour-change indicators. The pH electrode can be used to monitor acid–base titrations. A silver electrode can be used to monitor the concentration of Ag^+ ions during a precipitation titration involving the titration of halide ions (*e.g.* Cl^-) with silver nitrate ($AgNO_3$). Ion selective electrodes can also be used to monitor complexation and redox titrations. The experimental arrangement for performing a potentiometric titration is illustrated in Figure A12. An indicator and a reference electrode are

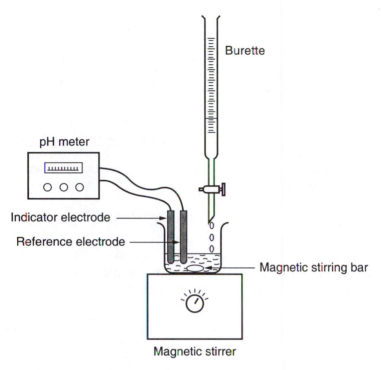

Figure A12 *Equipment required for potentiometric titration*

inserted in the sample solution, which is stirred on a magnetic stirrer throughout the titration. The electrodes are connected to a potentiometer or electronic voltmeter. The meter can be set to read potential or pH directly. Titrant is added from a burette, first in relatively large increments (1–2 mL) until the end point is approached and then in smaller increments (*e.g.* 0.1 mL). After each addition of titrant, sufficient time is allowed for the potential to stabilise. The amount of titrant added and the potential are recorded after each addition. As the end point is approached, the rate of change in potential increases.

A table of results is prepared. The first column should give the total volume of titrant added and the second column should give the corresponding potential in V or mV. From the data in these two columns the change in potential per unit volume of titrant ($\Delta E/\Delta V$) should be calculated and entered into a third column. The titration curve and the first-derivative curve should be plotted as shown in Figures A13 (a) and (b). The end point may be determined as the point of inflexion on the titration curve, or more accurately, as the maximum in the first-derivative curve. It is also possible to calculate and plot a second-derivative curve ($\Delta^2 E/\Delta V^2$). Automatic titrators are commercially avail-

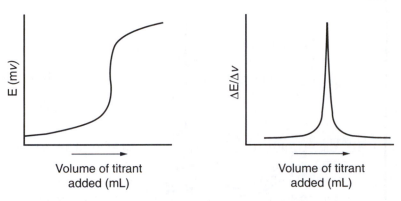

Figure A13 *Potentiometric titration curve and first derivative curve*

able and these can plot the titration curve while automatically adding titrant to the sample. These can also plot the first- and second-derivative curves.

Conductimetry

Conductimetry is the term used to refer to the measurement of conductivity. According to Ohm's Law:

$$I = E/R$$

where I is the current flowing in a conductor in amperes, E is the electromotive force in volts and R is the resistance of the conductor in ohms (Ω). The *conductance* (G) is defined as the reciprocal of the resistance ($1/R$) and it is measured in units of siemens (S) or reciprocal ohms (Ω^{-1}) also called "mho". The *resistivity* ρ, also sometimes called specific resistance, for a homogeneous material is defined as:

$$\rho = R \times a/l$$

where l is the length of the material and a is the cross-sectional area. The *conductivity* κ, also called *specific conductance*, is defined as the reciprocal of the *resistivity* ($1/\rho$) and it is measured in units of $\Omega^{-1}\,cm^{-1}$ or mho cm^{-1}. At any given temperature the conductivity of a solution depends on the ions present and their concentration. The *molar conductivity* (or molar conductance) Λ of an electrolyte is defined as:

$$\Lambda = 1000\kappa/C$$

where C is the concentration in mol L^{-1}. Units of Λ are Ω^{-1} cm^2 mol^{-1} or mho cm^2 mol^{-1}. The value of Λ extrapolated to zero concentration is called *molar conductivity at infinite dilution* Λ^∞. Values of Λ^∞ for different ions can be found in standard tables. The H^+ ion has a higher value of Λ^∞ than any other ion.

The conductivity is measured by inserting a cell containing two platinum electrodes into the test solution. These are connected to a conductivity meter and the result is usually displayed in μS cm^{-1} or μmho cm^{-1}. The instrument is calibrated using a standard KCl solution. The temperature can affect the conductivity and many instruments are equipped with a temperature probe and the meter automatically corrects conductivities to 25 °C. If temperature compensation is not available, the conductivity may be corrected using the following equation:

$$\kappa \text{ at } 25\,^\circ\text{C} = \frac{\kappa \text{ at } T}{1 + 0.0191(T - 25)}$$

where T is the temperature of the sample in °C. Some instruments display several ranges of conductivity. Make sure you use the appropriate conductivity range. When not in use, soak the electrode in water. Rinse the electrode with water and wipe gently between measurements.

LABORATORY NOTEBOOK

It is imperative that the student keeps all the details of his/her practical work in an orderly and well organised notebook. This will not only help the student when looking at raw data to perform calculations, referring back to past experiments and preparing reports, but will also assist the laboratory instructor in monitoring the student's progress. Furthermore, keeping a good notebook is excellent training for those who will find employment in an analytical laboratory after completion of their studies. In government and industrial laboratories the notebook may need to be inspected by management. The purpose of the notebook is that anyone should be able to follow what the analyst did and identify any problems that may have been encountered during the original analysis.

All observations, experimental details, measurements and calculations should be kept in a bound, hard-covered notebook, preferably A4 in size. The student should number, date and sign each page of the notebook. Any incorrect data should be deleted by drawing a single line through it, and not by erasing or removing it from the notebook. It is preferable to keep data in tabular form as much as possible.

The general format of the notebook will be the same as for one used in analytical chemistry, but in addition to analytical data the environmental analyst has to enter information relating to the collection of samples. The exact location of the sampling site, and the date and precise time of sampling, have to be recorded. The analyst should also note any other observations which may be of relevance, such as proximity of any pollution sources, meteorological observations in case of air sampling, *etc.*

The notebook should be organised according to a systematic format. A suggested format is given below:

- Date of experiment
- Name of experiment
- Objective
- Raw data
- Calculations
- Results
- Conclusions

Environmental Standards

Table A9 *WHO air quality guidelines*

Pollutant	Time-weighted average[a]	Averaging time
SO_2	500	10 min
	350	1 h
	100–150[b]	24 h
	40–60[b]	1 yr
CO	30	1 h
	10	8 h
NO_2	400	1 h
	150	24 h
O_3	150–200	1 h
	100–120	8 h
Black smoke	100–150	24 h
	40–60[b]	1 yr
Total suspended particulates	150–230[b]	24 h
	60–90[b]	1 yr
Thoracic particles (PM_{10})	70[b]	24 h
Pb	0.5–1	1 yr

[a] All concentrations in $\mu g\ m^{-3}$, except CO in $mg\ m^{-3}$.
[b] Guideline values for combined exposure to SO_2 and suspended particulate matter (they may not apply to situations where only one of the components is present).

Table A10 *US national ambient air quality standards (NAAQS)*

Pollutant	Averaging time	Primary standard[a]	Secondary standard[b]	Measurement method
SO_2	3 h	–	$1300\ \mu g\ m^{-3}$ (0.5 ppm)	West–Gaeke method or equivalent
	24 h	$365\ \mu g\ m^{-3}$ (0.14 ppm)		
	Annual average	$80\ \mu g\ m^{-3}$ (0.03 ppm)		
NO_2	Annual average	$100\ \mu g\ m^{-3}$ (0.05 ppm)	$100\ \mu g\ m^{-3}$ (0.05 ppm)	Colorimetry using Saltzman method or equivalent
CO	1 h	$40\ mg\ m^{-3}$ (35 ppm)	$40\ mg\ m^{-3}$ (35 ppm)	Non-dispersive infrared spectrometry
	8 h	$10\ mg\ m^{-3}$ (9 ppm)	$10\ mg\ m^{-3}$ (9 ppm)	
O_3	1 h	$235\ \mu g\ m^{-3}$ (0.12 ppm)	$235\ \mu g\ m^{-3}$ (0.12 ppm)	Chemiluminescence or equivalent
PM_{10} ($d \leqslant 10\ \mu m$)	24 h	$150\ \mu g\ m^{-3}$	$50\ \mu g\ m^{-3}$	Size-selective Hi-Vol sampler
	Annual average	$50\ \mu g\ m^{-3}$	$50\ \mu g\ m^{-3}$	
Pb	3 months	$1.5\ \mu g\ m^{-3}$	$1.5\ \mu g\ m^{-3}$	AAS

[a] To protect health.
[b] To protect public welfare.

Table A11 *Ambient air quality standards in some Asian countries*[a]

Pollutant	Averaging time	Malaysia	Thailand	Indonesia	Philippines	Singapore
SO_2	10 min	500	–	–	–	–
	1 h	350	–	900	850	–
	24 h	105	300	300	370	365
	Annual	–	100	60	–	80
NO_2	30 min	–	–	–	300	–
	1 h	320	320	–	–	–
	Annual	–	–	100	–	100
CO	1 h	35	50	30	35	40
	8 h	10	20	10	10	10
O_3	30 min	–	–	–	200	–
	1 h	200	200	160	–	235
	8 h	120	–	–	–	–
Suspended matter	24 h	260	330	230	180	260
	Annual	90	100	90	–	75
Pb	24 h	–	10	2	–	–
	3 months	1.5	–	–	20	1.5

[a] All concentrations in $\mu g\ m^{-3}$ except CO in $mg\ m^{-3}$. Indonesia figures are draft standards.

Table A12 *Guidelines and standards for drinking water quality*[a]

Parameter	WHO	EU	US EPA	Canada	UK	Denmark	Russia	Japan
Ag	–	0.01	0.1	0.05	0.01	0.01	–	–
Al	0.2	0.2 (0.05)	0.05–0.2	–	0.2	0.2	0.5	0.2
As	0.01	0.05	0.05	0.025	0.05	0.05	0.01	0.01
B	0.7	(0.1)	2.0	5.0	0.01	(100)	0.7	–
Cd	0.003	0.005	0.005	0.005	–	0.005	0.003	0.01
Cl^-	250	200 (25)	250	250	–	300	350	200
Cr	0.05	0.05	0.1	0.05	0.05	0.05	0.05	0.05
Cu	1–2	–	1.0	1.0	–	0.1	2.0	1.0
CN^-	0.07	0.05	0.2	0.2	0.05	0.05	0.07	0.01
F	1.5	1.5	2.0–4.0	1.5	1.5	1.5	<1.5	0.8
Fe	0.3	0.2 (0.05)	0.3	0.3	0.2	0.2 (0.05)	0.3	0.3
Hg	0.001	0.001	0.002	0.001	0.001	0.001	0.001	0.0005
K	–	12 (10)	–	12	10	–	–	–
Mn	0.1–0.5	0.05 (0.02)	0.05	0.05	0.05	0.05 (0.02)	0.5	0.01–0.05
Mo	0.07	–	–	–	–	–	–	0.07
Na	200	150 (20)	20	–	150	175 (20)	–	200
Ni	0.02	0.05	0.1	–	0.05	0.05	0.02	0.01
NH_4^+	1.5	0.5 (0.05)	–	–	0.5 (0.05)	2.0	–	–
NO_3^-	50 as NO_3^-	50 (25) as NO_3	10.0 as N	10.0 as N	50 as NO_3	50 as NO_3	45 as NO_3	10 as N
NO_2^-	3 as NO_2^-	0.1 as NO_2	1.0 as N	3.2 as N	0.1 as NO_2	0.1 as NO_2	3.0 as NO_2	10 as N
P	–	5 (0.4)	–	2.2	0.687	–	–	–
Pb	0.01	0.05	0.015	0.01	0.05	0.05	0.01	0.05
Sb	0.005	0.01	0.006	–	0.01	0.01	–	0.002
Se	0.01	0.01	0.05	0.01	0.01	0.01	0.01	0.01
SiO_2	–	–	–	–	–	–	–	–
SO_4^{2-}	250	250 (25)	250	500	250	250	500	–
Zn	3	(0.1)	5	5	–	0.1 5.0	1	1.0
TDS	1000	1500	500	500	1000	(1500)	1000	500
DO	–	–	–	–	–	–	4.0	–
BOD (as O_2)	–	–	–	–	–	–	3.0	–
pH	6.5–8.5	6.5–8.5	6.5–8.5	6.5–8.5	5.5–9.5	<8.5 (7–8)	6.0–9.0	5.8–8.6

[a] Units in $mg\ L^{-1}$ except pH in pH units. WHO = guideline value. US EPA = maximum contamination level. Canada = maximum acceptable concentration. EU, UK, Denmark, Japan = maximum admissible concentration. Values in brackets for EU, UK and Denmark = guide values where given. TDS = total dissolved solids, DO = dissolved oxygen, BOD = biochemical oxygen demand. Guidelines and standards for trace organic pollutants (benzo[*a*]pyrene, lindane, PCBs, *etc.*) have also been adopted but these are not listed here.

Table A13 *Proposed water quality standards in the UK (EQS)*[a]

| | Water (μg L^{-1}) Protection of aquatic life | | | |
| | *Freshwater* | | *Saltwater* | |
Element	*AA*[b]	*MAC*[c]	*AA*[b]	*MAC*[c]
Arsenic	50	–	25	–
Chromium	2–20[d]	–	5	–
Copper	0.5–12[d]	–	5	–
Lead (inorganic)	4–20[d]	–	10	–
Nickel	8–40[d]	–	15	–
Vanadium	20–60[d]	–	100	–
Zinc	8–50[d]	–	10	–

Source: Department of the Environment (1991); Baker (1994).
[a] Status: recommendations to the national government.
[b] AA = annual average.
[c] MAC = maximum allowable concentration.
[d] Depending on hardness (mg L^{-1} CaCO$_3$).

Table A14 *Environmental quality criteria in Canada: water quality criteria*

| | Water (μg L^{-1}) | | | | |
Element	*Freshwater aquatic life*	*Marine aquatic life*	*Livestock water*	*Irrigation water*	*Drinking water*
Arsenic	50	–	500	100	50
Barium	–	–	–	–	1000
Beryllium	–	–	100	100	–
Cadmium	0.01–0.06[b,c]	0.1[a,b]	80[b,c]	5[b,c]	5
Chromium	2	–	1000	100	50
Cobalt	–	–	1000	50	–
Copper	2–4[c]	–	500–5000[c]	200[c]	1000
Lead	1–7[c]	–	100[c]	200[c]	50
Mercury	0.1	–	3	–	1
Molybdenium	–	–	500	10	–
Nickel	25–150[c]	–	1000[c]	200[c]	–
Selenium	1	–	50	20	10
Vanadium	–	–	100	100	–
Zinc	30	–	50 000	1000–5000	5000

Source: CCME (1994).
[a] Status: recommendations to sub-national authorities.
[b] Draft.
[c] Depending on hardness of water (mg L^{-1} CaCO$_3$).

Table A15 *Environmental quality criteria in the European Union (EU): water quality criteria per function*

Element	Freshwater for fish (µg L⁻¹) Guideline[a] Salmonids	Guideline[a] Cyprinids	Mandatory[b] Salmonids	Mandatory[b] Cyprinids	As source of drinking water (µg L⁻¹) A1[c] Guideline	A1[c] Manda-tory	A2[d] Guideline	A2[d] Manda-tory	A3[e] Guideline	A3[e] Manda-tory	Consumption (µg L⁻¹) Guideline	Maximum admissible concentration
Antimony	—	—	—	—	—	—	—	—		—	—	10
Arsenic	—	—	—	—	10	50	—	50	50	100	—	50
Barium	—	—	—	—	—	10	—	1000	—	1000	—	—
Cadmium	—	—	—	—	1	5	1	5	1	5	—	5
Chromium (total)	—	—	—	—	—	50	—	50	—	50	—	50
Copper	5–112	5–112	—	—	20	50	50	—	1000	—	100	—
Lead	—	—	—	—	—	50	—	50	—	50	—	50
Mercury	—	—	—	—	0.5	1	0.5	1	0.5	1	—	1
Nickel	—	—	—	—	—	—	—	—	—	50	—	50
Selenium	—	—	—	—	—	10	—	10	—	10	—	10
Zinc	—	—	30–5000	300–2000	500	3000	1000	5000	1000	5000	100	—

Source: CEC DG XI (1992), Directives 75/440/EEC, 76/160/EEC and 78/659/EEC.
[a] Guidelines: recommendations to the Member States; legal in the sense the member states may not set criteria which are less stringent.
[b] Mandatory; legal.
[c] A1 = Category A1: simple physical treatment and desinfection, *e.g.* rapid filtration and desinfection.
[d] A2 = Category A2: normal physical treatment, chemical treatment and desinfection, *e.g.* prechlorination, coagulation, flocculation, decantation, filtration, desinfection (final chlorination).
[e] A3 = Category A3: intensive physical and chemical treatment, extended treatment and desinfection, *e.g.* chlorination to break-point, coagulation, flocculation, decantation, filtration, adsorption (activated carbon), desinfection (ozone, final chlorination).

Table A16 *Environmental quality criteria in the EU: water quality criteria per type of water*

Element	Quality objectives[a] (μg L^{-1})	Type of water
Cadmium	5	Inland water and estuary waters affected by discharge
Cadmium	2.5	Territorial and internal coastal waters affected by discharge
Mercury	1	Inland surface waters affected by discharge
Mercury	0.5	Estuary waters affected by discharge
Mercury	0.3	Territorial and internal coastal waters

Source: CEC DG XI (1992), Directives 84/491/EEC, 86/280/EEC, 83/513/EEC, 84/156/EEC, 82/176/EEC and 85/613/EEC.
[a] Status: legal.

Table A17 *Drinking, surface and irrigation quality criteria* (mg L^{-1}) *for heavy metals*

			Irrigation water		
			Continuous use[b]		Short-term use[b]
Heavy metal	Drinking water criteria[a]	Surface water FWPCA[b,d]	FWPCA any soil	NAS coarse-textured soil	FWPCA fine-textured soil
Ag	–	0.05	–	–	–
As(III)	0.190	0.05	1.0	0.1	10.0
B	–	1.0	0.75	0.75	2.0
Ba	2.0	1.0	–	–	–
Cd	0.005	0.01	0.005	0.01	0.05
Co	–	–	0.2	0.05	10.0
Cr(IV)	0.1	0.05	5.0	0.1	20.0
Cu	1.3	1.0	0.2	0.2	5.0
Pb	0.005	0.05	5.0	5.0	20.0
Mn	–	0.05	2.0	0.2	20.0
Mo	–	–	0.005	0.01	0.05
Ni	–	–	0.5	0.2	2.0
Se	0.050	0.01	0.05	0.02	0.05
Sb	0.05[c]	–	–	–	–
Zn	–	5.0	5.0	2.0	10.0

After Adriano (1986); Bandman *et al.* (1989); and US EPA (1991).
[a] US EPA (1991).
[b] Adriano (1986).
[c] Bandman *et al.* (1989).
[d] FWPCA, Federal Water Pollution Control Association.

Table A18 *Environmental quality criteria in the UK: soil quality criteria*[a]

| Element | Soil (mg kg^{-1}) Threshold | | |
	Domestic gardens, allotments, play areas	Landscapes, buildings, hardcovers	Any uses where plants are grown
Arsenic	10	40	–
Cadmium	3	15	–
Chromium	600	1000	–
Copper	–	130	
Lead	500	2000	–
Mercury	1	20	–
Nickel	–	–	70
Selenium	3	6	–
Zinc	–	–	300

Source: Visser (1993).
[a] Status: recommendations to the national government.

Table A19 *Environmental quality criteria in the EU: soil and sludge quality criteria, maximum admissible concentration*

Element	Soil Limit value (mg kg^{-1} d.m.)[a,b]	Sludge for agricultural use Limit value (mg kg^{-1} d.m.)[a]
Cadmium	1–3	20–40
Copper	50–140	1000–1750
Lead	50–300	750–1200
Mercury	1–1.5	16–25
Nickel	30–75	300–400
Zinc	150–300	2500–4000

Source: Visser (1993); ECE DG XI (1992), Directive 86/278/EEC.
[a] Status: recommendations to the Member States; legal in the sense that Member States may not set criteria which are less stringent.
[b] d.m., dry matter.

Table A20 *Maximum admissible concentrations of toxic metals in soil on which sewage sludge is applied*

Country	Concentration (mg kg^{-1})						
	Cd	Cr	Cu	Hg	Ni	Pb	Zn
USA	20	1500	750	8	210	150	1400
Sweden	0.5	30	40	0.5	1.5	40	100
Norway	1	100	50	1	30	50	150
Finland	0.5	200	100	0.2	60	60	150
Denmark	0.5	30	40	0.5	15	40	100
UK	3	400	135	1	75	300	300
France	2	150	100	1	50	100	300
Germany	1.5	100	60	1	50	100	200
EU	1–3	100–150	50–140	1–1.5	30–75	50–300	150–300

Source: Adapted from B. J. Alloway (1995).

Table A21 *Environmental quality criteria in Canada: interim environmental quality criteria for contaminated sites*[a]

Element	Water (µg L^{-1}) Assessment criteria	Soil (mg kg^{-1})			
		Assessment criteria	Remediation criteria		
			Agriculture	Residential/ parkland	Commercial/ industrial
Antimony	–	20	20	20	40
Arsenic	5	5	20	30	50
Barium	50	200	750	500	2000
Beryllium	–	4	4	4	8
Cadmium	1	0.5	3	5	20
Chromium (total)	15	20	750	250	800
Cobalt	10	10	40	50	300
Copper	25	30	150	100	500
Lead	10	25	375	500	1000
Mercury	0.1	0.1	0.8	2	10
Molybdenum	5	2	5	10	40
Nickel	10	20	150	100	500
Selenium	1	1	2	3	10
Thallium	–	0.5	1	–	–
Tin	10	5	5	50	300
Vanadium	–	25	200	200	–
Zinc	50	60	600	500	1500

Source: CCME (1991).
[a] Status: recommendations to sub-national authorities.

Table A22 *Environmental quality criteria in Canada: interim sediment quality assessment values*[a]

| | Sediment (μg kg^{-1} d.m.) | | | |
| | Freshwater aquatic life | | Marine aquatic life | |
Element	TEL[b]	PEL[c]	TEL[b]	PEL[c]
Arsenic	5900	17 000	7240	41 600
Cadmium	596	3530	676	4210
Chromium	37 300	90 000	52 300	160 000
Copper	35 700	196 600	18 700	108 000
Lead	35 000	91 300	30 200	112 000
Mercury	174	486	130	700
Nickel	18 000	35 900	15 900	42 800
Zinc	123 000	314 800	124 000	271 000

Source: Environment Canada (1994).
[a] Status: draft and interim recommendations to the national goverment.
[b] Threshold effect level.
[c] Probable effect level.

Table A23 *The maximum permissible content of nitrate in vegetables and fruits in Russia*

Type of crop	Maximum permissible nitrate content (mg NO$_3^-$ kg^{-1} wet weight)
White cabbage	500
Carrot	250
Beetroot	1400
Cucumbers	150
Tomatoes	150
Lettuce	2000
Potatoes	250
Water melons	60
Apples	60
Spinach	2100

Table A24 *Guideline values for Cd, Pb and Hg in foodstuffs recommended by the German Federal Health Agency in* mg kg^{-1} *fresh weight, or*[a] mg L^{-1}

Foodstuff	Cadmium	Lead	Mercury
Wheat	0.1	0.3	0.03
Rye	0.1	0.4	0.03
Rice	0.1	0.4	0.03
Potatoes	0.1	0.25	0.02
Green vegetables (except kale, herbs, and spinach)	0.1	0.8	0.05
Kale	–	2.0	–
Pot herbs	–	2.0	–
Spinach	0.5	–	–
Sprout vegetables	0.1	0.5	0.05
Fruit vegetables	0.1	0.25	0.05
Root vegetables	0.1	–	–
Celery	0.2	–	–
Pomaceous fruits	0.05	0.5	0.03
Fruits with stones	0.05	0.5	0.03
Berries, small fruits	0.05	0.5	0.03
Citrus fruits	0.05	0.5	0.03
Other fruits	0.05	0.5	0.03
Hard/shelled fruits	0.05	0.5	0.03
Milk[a]	0.0025	0.03	0.01
Condensed milk	0.05	0.3	0.01
Cheese	0.05	–	0.01
Cheese (except hard cheese)	–	0.25	–
Hard cheese	–	0.5	–
Eggs	0.05	0.25	0.03
Beef	0.1	0.25	0.03
Veal	0.1	0.25	0.03
Pork	0.1	0.25	0.03
Ground meat	0.1	0.25	0.03
Chicken meat	0.1	0.25	0.03
Bovine liver	0.5	0.8	0.03
Calf liver	0.5	0.8	0.1
Pork liver	0.5	0.8	0.1
Bovine kidney	0.8	0.8	0.1
Calf kidney	0.8	0.8	0.1
Pork kidney	0.8	0.8	0.1
Meat products	0.25	0.25	0.05
Sausages	0.25	0.25	0.05
Fish	0.5	0.5	1.0
Fish products	0.5	0.5	1.0
Canned fish	1.0	1.0	1.0
Refreshing drinks[a]	0.05	0.2	0.01
Wine[a]	0.01	0.3	0.01
Beer[a]	0.03	0.2	0.01

Source: E. Merian (1991).

REFERENCES

D. C. Adriano, "Trace Elements in the Terrestrial Environment", Springer, New York, 1986, pp. 1–42.

B. J. Alloway, "Heavy Metals in Soils", Blackie, London, 1995.

A. L. Bandman, B. A. Filov and A. A. Potechin (eds.), "Dangerous Chemical Compounds, vol. I. Inorganic Compounds of Elements of V–VIII Groups", Chemistry Publishing House, Leningrad, 1989 (in Russian).

M. G. C. Baker, D. Williams and C. S. Murgatroyd, 1994, "Evaluation of Priority Substances in the Aquatic Environment", Final Report to the Department of the Environment, Report no. DoE 3663/1.

CCME, "Interim Canadian Environmental Quality Criteria for Contaminated Sites", Report CCME EPC-CS3, 1991.

CCME, "Canadian Water Quality Guidelines", 1994.

CEC DG XI, "European Community Environment Legislation", vol. 2, Air, vol. 7, Water, Council Directives 75/440/EEC, 76/160/EEC, 78/659/EEC, 82/176/EEC, 82/884/EEC, 83/513/EEC, 84/156/EEC, 84/491/EEC, 85/613/EEC, 86/280/EEC, 1992.

Department of the Environment, "National Environmental Quality Standards for Dangerous Substances in Water: Joint Consultation Paper", November 1991.

E. Merian (ed.), "Metals and their Compounds in the Environment – Occurrence, Analysis and Biological Relevance", VCH, Weinheim, 1991, p. 704.

Environment Canada, "Interim Sediment Quality Assessment Values", Manuscript report, September 1994.

US EPA, "1991 Water Quality Criteria Summary", Poster of the Office of Science and Technology, May 1, 1991.

W. J. F. Visser, "Contaminated Land Policies in Some Industrialized Countries", TCB report R02, 1993.

Recommended Groundwater and Leachate Monitoring for Landfill Sites

Table A25 *Recommended monitoring[a] of groundwater for landfills receiving the specified wastes*

Municipal solid waste	Residue from incinerators burning municipal waste	Fly ash and furnace ash	Foundry waste	Paper mill sludge
Alkalinity	Alkalinity	Alkalinity	Alkalinity	Alkalinity
Cl	B	B	COD	Cl
COD	Cd	COD	Conductivity	COD
Conductivity	Cl	Conductivity	F	Conductivity
Hardness	COD	Hardness	Hardness	Hardness
pH	Conductivity	pH	Na	NH_3-N
	Hardness	SO_4^{2-}	pH	NO_3^--N
	pH		SO_4^{2-}	NO_2^--N
	Pb			pH
				SO_4^{2-}
VOC[b]	VOC	VOC	VOC	VOC

[a] All wells should be monitored every 6 months.
[b] VOC, volatile organic compounds.

Table A26 *Recommended chemical monitoring[a] of landfill leachate for landfills receiving the specified wastes*

Municipal solid waste + residue from incinerators burning municipal solid waste	Fly ash and furnace ash	Foundry waste	Paper mill sludge
Alkalinity	Alkalinity	Alkalinity	Alkalinity
BOD	B	BOD	BOD
Cd	BOD	Cd	Cd
Cl	Cd	Cl	Cl
COD	Cl	COD	COD
Conductivity	COD	Conductivity	Conductivity
Fe	Conductivity	F	Fe
Hardness	Fe	Fe	Hardness
Hg	Hardness	Hardness	Hg
Mn	Hg	Hg	Mn
Na	Mn	Mn	Na
NH_3-N	pH	Na	NH_3-N
N total	Pb	pH	N total
Pb	Se	Pb	Pb
pH	SO_4^{2-}	SO_4^{2-}	pH
SO_4^{2-}	TSS	TSS	SO_4^{2-}
TSS	VOC	VOC	TSS
VOC			VOC

[a] Monitoring should be carried out every 6 months.
[b] VOC, volatile organic compounds.

APPENDIX V

Statistical Tables

Table A27 *Critical values of the Student's t-distribution, t, and the correlation coefficient, r (P = 0.05)*

Degrees of freedom[a]	t	r
1	12.71	0.997
2	4.303	0.950
3	3.182	0.878
4	2.776	0.811
5	2.571	0.754
6	2.447	0.707
7	2.365	0.666
8	2.306	0.632
9	2.262	0.602
10	2.228	0.576
11	2.201	0.553
12	2.179	0.532
13	2.160	0.514
14	2.145	0.497
15	2.131	0.482
16	2.120	0.468
17	2.110	0.456
18	2.101	0.444
19	2.093	0.433
20	2.086	0.423

[a] Degrees of freedom $= n - 1$ for t, and $n - 2$ for r.

Table A28 *Critical values of the rejection quotient,* Q *(P = 0.05)*

Sample no., n	Q critical
4	0.831
5	0.717
6	0.621
7	0.570
8	0.524
9	0.492
10	0.464

Subject Index